Genomics Protocols

METHODS IN MOLECULAR BIOLOGY™

John Walker, SERIES EDITOR

Genomics Protocols

Second Edition

Edited by

Mike Starkey
Animal Health Trust,
Lanwades Park, Newmarket,
Suffolk, United Kingdom

Ramnath Elaswarapu
Oxford Gene Technology Ltd.,
Oxford, United Kingdom

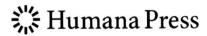 Humana Press

Editors
Mike Starkey
Animal Health Trust
Lanwades Park, Newmarket
Suffolk, United Kingdom

Ramnath Elaswarapu
Oxford Gene Technology Ltd.
Oxford, United Kingdom

Series Editor
John M. Walker
University of Hertfordshire
Hatfield, Herts
United Kingdom

ISBN: 978-1-58829-871-3   e-ISBN: 978-1-59745-188-8

Library of Congress Control Number: 2007938729

Printed on acid-free paper

9 8 7 6 5 4 3 2 1

springer.com

Preface

The genomics research world has moved on since the first volume of *Genomics Protocols* in the Methods in Molecular Biology series was published in 2001. This progression is reflected in the themes of the chapters that adorn the current volume. It is true that there is the same apparently eclectic mixture of subjects as in the former volume. However, undeniably, the emphasis has switched from gene identification to functional genomics and the characterization of genes and gene products.

Although essentially a volume of "wet lab" techniques, in silico approaches also are represented through three chapters addressing the detection of genome sequence variation and the prediction of gene function, respectively.

We make no apologies for including texts that may appear in volumes devoted to other of the "omics." Rather, we feel that the majority of these relative newcomers are derivatives of genomics, and from the perspective of being able to see "the bigger picture," a great value is to be gained from considering these research disciplines collectively.

For some of the specialist techniques, it is difficult to avoid the use of a specific/particular commercial platform or piece of equipment. For these procedures, the authors have focused on the generic issues (e.g., data analysis), which are all too often given inadequate coverage in user manuals. While not wishing to advocate particular systems, by including these topics we acknowledge the importance, and possibly the uniqueness, of the technologies involved.

Through the choice of some of the topics, we tried to remember that not everyone is engaged in investigation of organisms for which an annotated sequenced genome or a high-density genetic map is available. Similarly, lest we forget that not all researchers are affiliated to well-equipped laboratories, Chapter 5 describes a relatively inexpensive and low-tech approach to genetic mapping using microsatellites.

To our minds, the jewel in the crown of the Methods in Molecular Biology series remains the "Notes" section of each chapter. In spite of the complexity of some of the subject matter, the authors have endeavored to ensure that as little technical detail as possible is left to either the reader's imagination or to the extensivity of their organization's library.

The first chapter addresses an issue that all too often is a prerequisite for comprehensive genome analysis: high fidelity whole genome amplification. Chapters 2–4 focus on a spectrum of procedures that are a prelude to genetic linkage analysis (Chapter 5).

The molecular profiling of genomic DNA in a variety of guises is encompassed in Chapters 6–9. Chapters 10–12 are devoted to transcriptional profiling, and focusing on small samples and the use of formalin-fixed archival tissue, deal with two of the thornier issues confronting researchers. The profiling of microRNAs (Chapter 12) is an exciting area of interest in the study of the regulation of gene expression in disease.

The line between proteomics and functional genomics can be a little fuzzy, but Chapters 12–17 describe alternative strategies for protein profiling, while Chapters 18–26 arguably constitute approaches for gene characterization and the identification of protein interaction networks. The volume concludes with a detailed tutorial (Chapter 27) about an exciting genomics technology useful in the elucidation of gene function and with huge potential therapeutic applications.

So there you have it. All that remains is for us to thank a group of redoubtable authors, the series editor, John Walker, and all those involved at Humana Press. We truly hope that you find the book informative and a valuable addition to your laboratory bookshelf.

Mike Starkey
Ramnath Elaswarapu

Contents

Contributors

Lutgarde Arckens
Laboratory of Neuroplasticity and Neuroproteomics, Katholieke Universiteit Leuven, Leuven, Belgium

Larissa Belov
Medsaic Pty Ltd, National Innovation Centre, Eveleigh, Australia

Marina Bibikova
Illumina Inc., San Diego, CA

Helen Butler
Wellcome Trust Centre for Human Genetics, University of Oxford, Oxford, United Kingdom

Lina Cekaite
Department of Tumor Biology, Institute for Cancer Research, Rikshopitalet–Radiumhospitalet Medical Center, Montebello, Oslo, Norway

Lynda Chin
Center for Applied Cancer Science, Belfer Institute for Innovative Cancer Science, Dana-Farber Cancer Institute, Harvard Medical School, Boston, MA

Byung-Kwan Cho
Department of Bioengineering, University of California–San Diego, La Jolla, CA

Richard I. Christopherson
School of Molecular and Microbial Biosciences, University of Sydney, Sydney, Australia

John E. Collins
Wellcome Trust Sanger Institute, Cambridge, United Kingdom

Louise A. Dawson
Institute of Cellular and Molecular Biology, University of Leeds, Leeds, United Kingdom

Wen Ding
Institute for Biological Sciences, National Research Council, Ottawa, Ontario, Canada

Nicole Draper
Oxford Gene Technology Ltd, Yarnton, Oxford, United Kingdom

Per O. Ekstrøm
The Norwegian Radium Hospital, Oslo, Norway

Peter Ellmark
Department of Immunotechnology, Lund University, Lund, Sweden

S. Fabio Falsone
Institute of Pharmaceutical Sciences, University of Graz, Graz, Austria

Jian-Bing Fan
Illumina Inc., San Diego, CA

Bin Feng
Center for Applied Cancer Science, Belfer Institute for Innovative Cancer
Science, Dana-Farber Cancer Institute, Harvard Medical School, Boston, MA

Toni Gabaldón
Bioinformatics Department, CIPF, Valencia, Spain

Bernd Gesslbauer
Institute of Pharmaceutical Sciences, University of Graz, Graz, Austria

Opher Gileadi
Structural Genomics Consortium, Botnar Research Centre, University of Oxford,
Oxford, United Kingdom

Stephen D. Ginsberg
Center for Dementia Research, Nathan Kline Institute, and Departments of
Psychiatry and Physiology and Neuroscience, New York University School of
Medicine, Orangeburg, NY

Arsalan S. Haqqani
Institute for Biological Sciences, National Research Council, Ottawa, Ontario,
Canada

Runtao He
National Microbiology Laboratory, Public Health Agency of Canada Manitoba,
Manitoba, Canada

Lan Huang
Department of Physiology & Biophysics and Department of Developmental and
Cell Biology, University of California Irvine, Irvine, CA

Trupti Joshi
Digital Biology Laboratory, Computer Science Department and Christopher S.
Bond Life Sciences Center, University of Missouri–Columbia, Columbia, MO

Peter Kaiser
Department of Biological Chemistry, College of Medicine, University of
California Irvine, Irvine, CA

John F. Kelly
Institute for Biological Sciences, National Research Council, Ottawa, Ontario,
Canada

Eric M. Knight
Department of Bioengineering, University of California–San Diego,
La Jolla, CA

Andreas J. Kungl
Institute of Pharmaceutical Sciences, University of Graz, Graz, Austria

Sigrun Langbein
Department of Urology, Uro-Oncology, Academic Medic Centrum (AMC),
University of Amsterdam, Amsterdam, The Netherlands

Xuguang Li
Centre for Biologics Research, Biological and Genetic Therapies Directorate,
Health Canada, Ottawa, Canada

Guan Ning Lin
Digital Biology Laboratory, Computer Science Department,
Christopher S. Bond Life Sciences Center, University of Missouri–Columbia,
Columbia, MO

David Meierhofer
Department of Biological Chemistry, College of Medicine, University of
California Irvine, Irvine, CA

Cathryn Mellersh
Animal Health Trust, Suffolk, United Kingdom

Snehal Naik
Molecular Imaging Center, Mallinckrodt Institute of Radiology, and Department
of Molecular Biology and Pharmacology, Washington University School of
Medicine, St. Louis, MO

Bernhard Ø. Palsson
Department of Bioengineering, University of California–San Diego,
La Jolla, CA

Arturas Petronis
The Krembil Family Epigenetics Laboratory, Centre for Addiction and Mental
Health, Toronto, Ontario, Canada

David Piwnica-Worms
Molecular Imaging Center, Mallinckrodt Institute of Radiology, and Department
of Molecular Biology and Pharmacology, Washington University School of
Medicine, St. Louis, MO

Alexei Protopopov
Center for Applied Cancer Science, Belfer Institute for Innovative Cancer
Science, Dana-Farber Cancer Institute, Harvard Medical School, Boston, MA

Jiannis Ragoussis
Wellcome Trust Centre for Human Genetics, University of Oxford, Oxford,
United Kingdom

Janna Saarela
National Public Health Institute, Helsinki, Finland

Axel Schumacher
Epigenetics Lab, Department of Medicine II, Klinikum rechts der Isar,
Technical University Munich, Munich, Germany

Binesh Shrestha
Structural Genomics Consortium, Botnar Research Centre, University of Oxford,
Oxford, United Kingdom

Kaisa Silander
National Public Health Institute, Helsinki, Finland

Mouldy Sioud
Department of Immunology, Institute for Cancer Research, Rikshopitalet -
Radiumhospitalet Medical Center, Montebello, Oslo, Norway

Carol Smee
Structural Genomics Consortium, Botnar Research Centre, University of Oxford,
Oxford, United Kingdom

Zhao Song
Digital Biology Laboratory, Computer Science Department,
Christopher S. Bond Life Sciences Center, University of Missouri–Columbia,
Columbia, MO

Danica B. Stanimirovic
Institute for Biological Sciences, National Research Council, Ottawa, Ontario,
Canada

Minoru Takata
Shinshu University School of Medicine, Matsumoto, Japan

Badar A. Usmani
Institute of Cellular and Molecular Biology, University of Leeds, Leeds,
United Kingdom

Gert Van den Bergh
Laboratory of Neuroplasticity and Neuroproteomics, Katholieke Universiteit
Leuven, Leuven, Belgium

Victor Villalobos
Molecular Imaging Center, Mallinckrodt Institute of Radiology, and Department
of Molecular Biology and Pharmacology, Washington University School of
Medicine, St. Louis, MO

Xiaorong Wang
Department of Physiology and Biophysics and Department of Developmental and
Cell Biology, University of California Irvine, Irvine, CA

Jessica Wang-Rodriguez
University of California, San Diego, Department of Pathology and the Veterans
Administration San Diego Healthcare System, San Diego, CA

Andreas Weinhäusl
ARCS, Austrian Research Center, Seibersdorf, Austria

Adrian Woolfson
University of Cambridge School of Clinical Medicine, Addenbrooke's Hospital,
Cambridge, United Kingdom

Dong Xu
Digital Biology Laboratory, Computer Science Department,
Christopher S. Bond Life Sciences Center, University of Missouri–Columbia,
Columbia, MO

Joanne M. Yeakley
Illumina, Inc., San Diego, CA

Chao Zhang
Digital Biology Laboratory, Computer Science Department,
Christopher S. Bond Life Sciences Center, University of Missouri–Columbia,
Columbia, MO

Chapter 1
Whole Genome Amplification with Phi29 DNA Polymerase to Enable Genetic or Genomic Analysis of Samples of Low DNA Yield

Kaisa Silander and Janna Saarela

Abstract In many large genetic studies, the amount of available DNA can be one of the criteria for selecting samples for study. In the case of large population cohorts, selecting samples based on their DNA yield can lead to biased sample selection. In addition, many valuable clinical and research sample collections exist in which the amount of DNA is very small. Unbiased whole genome amplification (WGA) of such unique samples enables genomewide scale genetic studies that would have been impossible otherwise. Multiply primed rolling circle amplification (MPRCA) and multiple displacement amplification (MDA) methods are based on the same principle. The DNA amplification is non-PCR based and uses Φ29 DNA polymerase and random hexamer primers for unbiased whole genome amplification. MDA is used for linear DNA molecules, such as genomic DNA. This chapter reviews the various applications in which whole genome amplified DNA can be used, the types of commercial kits available, and the quality control steps necessary before using the DNA in the genetic studies.

Keywords multiply primed rolling circle amplification; multiple displacement amplification; WGA; DNA yield; genotyping

1 Introduction

1.1 Multiple Displacement Amplification and Other Methods for Whole Genome Amplification

A gold standard method for creating infinite DNA source is the immortalization of peripheral blood lymphocytes by transformation with Epstein-Barr virus. However, the method is laborious and expensive and thus not applicable to large-scale studies. Moreover, it does not allow amplification of the already existing genomic DNA sample collections. Several strategies have been utilized to amplify and preserve existing DNA samples. Most of the earlier methods have been based on cyclic amplification with Taq polymerase and utilized primers

From: Methods in Molecular Biology, Vol. 439: *Genomics Protocols: Second Edition*
Edited by Mike Starkey and Ramnath Elaswarapu © Humana Press Inc., Totowa, NJ

directed to repeated sequence elements, such as human Alu1-repeats [1], or random (PEP [2]) or partly degenerate (DOP [3]) primers. Both the originally described primer extension preamplification (PEP) and degenerate oligonucleotide primed polymerase chain reaction (DOP-PCR) techniques have been further developed and refined to increase the fidelity and the length of the amplification products [4, 5]. Another subgroup of whole-genome amplification methods relies on first generating fragments of the genomic DNA by restriction enzyme cleavage [6–8], random shearing of genomic DNA [9] or nick translation [10], followed by ligation of adaptors, which allow amplification using universal PCR primers. However, all these methods are limited by features of the Taq polymerase: typical amplification fragment length of <3 kb and the error rate of 3×10^{-5} (9×10^{-6} in combination with Pфo polymerase, reviewed in 11). These methods also suffer from incomplete coverage and uneven amplification of the genomic loci; 10^{2}–10^{4}- and 10^{3}–10^{6}-fold amplification biases have been described using PEP and DOP-PCR methods, respectively.

The multiple displacement amplification (MDA) method is based on isothermal amplification using the strand displacement activity of the Φ29 DNA polymerase [12].

MDA provides several improvements to the previously described methods. High processivity and fidelity of the Φ29 DNA polymerase allow highly uniform amplification across the genome: Dean et al. [12] observed less than a threefold amplification bias among the eight chromosomal loci monitored. However, their results indicated that repetitive sequences were underrepresented in the MDA product. The average amplification was 10,000-fold with the average product length of >10 kb. The MDA-amplified genomic DNA was shown to be suitable for several common genetic analyses, including single nucleotide polymorphisms (SNP) genotyping, chromosome painting, Southern blotting, restriction length polymorphism analysis, subcloning, and DNA sequencing [11]. In addition to high quality genomic DNA, the amplification can be carried out directly from biological samples, like whole blood or tissue culture cells [12, 13], plasma or serum [14], and laser-capture microdissected tissue [15]. Further, MDA amplification of low-copy-number and highly degraded DNA samples for forensic testing has been evaluated [16]. In addition to human samples, MDA method has also been utilized for canine [17], insect [18], and microbial samples [19]. Figure 1.1 shows the principle of WGA with Φ29 DNA polymerase and random hexamer primers.

1.2 Applications in Which Whole Genome Amplified DNA Have Been Used

Table 1.1 summarizes some of the subsequent studies evaluating the usability of MDA products in different genetic analyses. MDA-amplified genomic DNA has been successfully utilized in several downstream analyses, including sequencing, mutation detection, SNP and STR genotyping, Southern blot analysis, copy

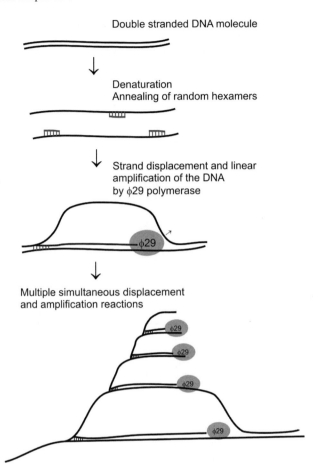

Double stranded DNA molecule

Denaturation
Annealing of random hexamers

Strand displacement and linear
amplification of the DNA
by ϕ29 polymerase

Multiple simultaneous displacement
and amplification reactions

Fig. 1.1 Principle of whole genome amplification with Φ29 DNA polymerase and random hexamer primers

number analysis (CGH, LOH), and forensic profiling. Unfortunately, many of these studies utilized only few amplified/unamplified sample pairs in the analysis, thus the results may not be fully generalizable. Dickson et al. [20] obtained genotypes of 15 MDA-amplified high-quality genomic DNA samples for a panel of 768 STR markers and observed genotyping call and accuracy rates to be only marginally lower than in the paired unamplified genomic DNAs (call rates: 95.0% versus 96.5%; concordance: 97.8% versus 99.2%). Pooling of three replicate MDA products improved both the call rate (96.2%) and genotyping accuracy (concordance 98.3%) slightly. A set of 34 widely distributed STR markers was found to cause a significant part of the failures/discrepancies. Pask et al. [21] described the genotyping data of 86 genomic DNA/MDA-DNA pairs produced by three different SNP genotyping platforms (TaqMan assays: 95 SNPs, Invader assays: 13 SNPs and a BeadArray multiplex: 345 SNPs). The genotype call rates

Table 1.1 Applications of WGA-DNA for various DNA analysis methods

Method	Assay*	References
Sequencing	Dye terminator method with capillary electrophoresis on the ABI 377, ABI 3700 or ABI 3730 DNA sequencer (Applied Biosystems)	23, 27, 31–34
Genotyping (STR)	AmpFLSTR Profiler Plus (Applied Biosystems)	13
	AMpFISTR Identifiler (Applied Biosystems)	28
	Weber screening sets 13 and 52 (Genotyping service at Marshfield Clinic)	20
	ABI Prism Linkage Mapping Set v2.5 (Applied Biosystems)	15, 31
Genotyping (SNP)	TaqMan Assay (Applied Biosystems)	13, 15, 21, 28, 31, 32, 35
	Pyrosequencing (Pyrosequencing AB)	31
	MassARRAY (4 SNPs)	34
	Allele-specific primer extension on microarray (8 SNPs)	34
	DNA ligase-mediated single nucleotide discrimination and signal amplification by rolling circle amplification (RCA) (10 SNPs)	12
	Minisequencing on microarray (45 SNPs)	22
	BeadArray (384 SNPs) (Illumina)	21
	Illumina linkage panel version 1 (2,320 SNPs) (Illumina)	36
	Human Mapping 10 K Array Xba131 (11,560 SNPs) (Affymetrix)	23, 25
	Human Mapping 100 K Set (116,204 SNPs on two arrays) (Affymetrix)	24, 26
Amplified fragment length polymorphism (AFLP)	AFLP® Core Reagent Kit (Invitrogen)	19
Comparative genomic hybridization (CGH)		12, 37
Forensic profiling	GammaSTR multiplex kit (Promega), AmpFLSTR Profiler Plus (Applied Biosystems)	16, 38
Loss of heterozygosity (LOH) analysis	Human Mapping 10 K Array and 100 K Set (Affymetrix), TaqMan Assay and Linkage Mapping set (Applied Biosystems)	15, 23, 26
Mutation analysis (PCR)		33, 39
Preimplantation genetic diagnosis (PGD)		40
Southern blot		12, 33

*Number of SNPs genotyped in a single multiplex is shown for the multiplexed SNP genotyping methods

were comparable among different SNP platforms and only marginally lower in MDA-DNA samples compared to genomic DNA. However, more of the MDA DNA samples (10.5%) than genomic DNA samples (5.7%) were excluded because they did not perform consistently on the BeadArray platform. The concordance rate between genomic DNA and MDA-DNA was very high in all the platforms (98.8–99.9%), although the clustering of the data points for the MDA samples was less distinct in the majority of the TaqMan assays. Lovmar et al. [22] amplified 15 genomic DNA samples of CEPH individuals and genotyped 45 SNP using the multiplexed, four-color fluorescent minisequencing technique in a microarray format. The genotyping success, signal-to-noise ratio, and power of genotype discrimination were found to be comparable between the unamplified and MDA-amplified samples. Paez et al. [23] evaluated the genome coverage and fidelity of the MDA reaction in 14 cell line-derived samples using the Affymetrix 10K SNP platform. They also monitored for differences between samples subjected and not subjected to denaturation treatment with alkali prior to amplification. The samples treated with a denaturing step prior to MDA showed call rates comparable to unamplified samples (92.06% versus 92.93%, respectively), while the nondenatured samples had substantially lower call rates (88.92%). Paez et al. [23] identified six regions (1q24, 4q35, 6p25, 7q36, 10q26, and 18p11), containing maximum of 5.64 Mb genomic sequences, which showed significantly reduced signal intensity and consistent no-call designations in MDA samples. They also assessed the SNP call concordance in ten pairs of replicate samples and found a concordance of 99.85% among paired replicates, 99.59% when comparing nondenatured MDA samples to unamplified samples and 99.80% when alkaline-denatured MDA samples were compared to unamplified samples. Paez et al. [23] further evaluated the fidelity of the MDA method by sequencing 20 MDA amplified and paired unamplified DNA samples. Other, more exploratory studies also demonstrated the usability of MDA-amplified DNA in the highly multiplexed Affymetrix GeneChip SNP platforms [24–26]. Further, Mai et al. [27] evaluated the performance of 23 MDA amplified clinical samples for sequencing and observed a 100% concordance for detection of point mutations in the alpha1, alpha2, and beta globin genes.

Increased amount of starting genomic DNA in the WGA reaction improved MDA-DNA genotyping performance in a study by Bergen and coworkers [28], in which 27 genomic DNA samples extracted from lymphoblastoid cell lines were amplified and genotyped for 15 STR and 49 SNP markers. A small set of MDA amplified samples ($n = 4$) analyzed using the Affymetrix 10K SNP array suggested that pooling three MDA reaction products with low amount of input genomic DNA (6.7 ng) provides marginally lower call and accuracy rates than using a larger amount of genomic DNA (20 ng) in the original MDA amplification [25]. Therefore, based on the recent publications, the MDA method appears to be a reliable method for faithful and balanced amplification of a genome, especially when a large enough amount of high-quality genomic DNA or biological sample material is available.

1.3 Why Low Yield DNA Samples Should Be Studied

In many large genetic studies, the amount of available DNA can be one of the criteria for selecting samples for study. In cases of epidemiological population cohorts, selecting samples based on their DNA yield can lead to biased sample selection. For example, association between higher DNA yield and several phenotypes, including daily smoking and HDL (high density lipoprotein)-cholesterol levels recently has been shown [29]. This suggests that exclusion of DNA samples with very low DNA yield may lead to biased results [29]. In addition, in many valuable clinical and research sample collections, the amount of DNA is very small, for example, if only small tissue samples were originally obtained or the source of DNA is from archived samples. Unbiased whole genome amplification of such unique samples enables genomewide scale genetic studies that would not have been possible otherwise.

2 Materials

2.1 DNA

Low-yield DNA samples are those in which the amount of DNA is not enough for the requirements of the genetic study planned to be performed. For example, DNA isolated from buccal swabs or small tissue samples. Even DNA isolated from EDTA-blood samples can occasionally be of low yield (<10 μg DNA obtained from ≥5 ml blood). Depending on the source of the sample, the length of storage, isolation methods, and the like, such DNA samples also can be fragmented. For whole genome amplification, the DNA can be isolated with any commercially available methods or with standard laboratory protocol, such as a phenol-chloroform extraction. However, it is important that the DNA solution does not contain any substances that may inhibit the DNA synthesis reaction by Φ29 DNA polymerase. Further, most WGA kits also have protocols for direct WGA from nonpurified cell lysate of various sources.

To obtain good-quality MDA DNA, it is important that the amount of genomic DNA used for a WGA reaction is at least 10 ng. Some downstream analysis methods may require even larger amounts of genomic DNA input into the WGA reaction. Using lower amounts of DNA can lead to randomly biased amplification of the genome, which may eventually lead to errors in genotyping. However, kits now are available in which, according to manufacturer, the amount of genomic DNA required can be as low as 1 ng (e.g. illustra GenomiPhi DNA Amplification Kit of GE Health Care Life Sciences).

2.2 Kits

Various WGA kits are available that use Φ29 DNA polymerase. See Table 1.2 for a list of some of the currently available kits. Kit selection may be based on the amount of the existing genomic DNA, the amount of the desired MDA DNA, the degradation state of the DNA samples, the source of the DNA sample, and whether the WGA process must be automated. In addition, it is possible to buy only the Φ29 DNA polymerase, with or without reaction buffer, and optimize a

Table 1.2 Φ29-based WGA kits currently available

Kit	Produced by	DNA sample origin	Amount of template genomic DNA	Target/specificity	WGA product yield	WGA product size	Comment
Low-yield DNA sample							
Illustra GenomiPhi™ DNA Amplification Kit*	GE Health Care Life Sciences	Purified DNA or nonpurified cell lysate.	Minimum 1 ng		4–7 µg	Not specified	Samples from formalin-fixed tissue and templates stored under undesirable conditions for long periods produce poor amplification. Process can be automated so that all liquid handling steps are carried out by a robotic system.
Good-quality DNA, various sources, resulting in standard WGA-DNA yield							
Repli-g Screening Kit**	Qiagen	Variety of clinical samples, including purified DNA, whole blood, and tissue culture cells	>10 ng (suspended in TE), or 0.5 µL whole blood or cell material	Can be done in a 96-well plate format, enabling automation of the WGA process	Approximately 5 µg	>10 kb (range 2–100 kb)	Smaller amounts of DNA (1–10 ng) can be used if the DNA is of high quality.
Repli-g Mini kit***	Qiagen	Variety of clinical samples, including purified DNA, whole blood, and tissue culture cells	>10 ng (suspended in TE), or 0.5 µL whole blood or cell material		Approximately 10 µg	>10 kb (range 2–100 kb)	Protocols for the amplification of DNA from dried blood cards, buccal cells, tissue, serum, plasma, and laser microdissected cells are available.

(continued)

Table 1.2 (continued)

Kit	Produced by	DNA sample origin	Amount of template genomic DNA	Target/specificity	WGA product yield	WGA product size	Comment
Illustra GenomiPhi V2 DNA Amplification Kit*	GE Health Care Life Sciences	Purified DNA or nonpurified cell lysate (blood cells, buccal swab, mouth wash, also blotted on paper)	1–10 ng	A mini-scale genomic DNA preparation. Lowers reaction time and reduces nonspecific amplification	4–7 µg	>10 kb	No DNA amplification in no-template controls.
Good-quality DNA resulting in high-yield WGA-DNA							
Repli-g Midi kit***	Qiagen	Variety of clinical samples, including purified DNA, whole blood, and tissue culture cells	>10 ng (suspended in TE), or 0.5–1.0 µL whole blood or cell material	For high WGA-DNA yield	Approximately 40 µg	>10 kb (range 2-100 kb)	Protocols for the amplification of DNA from dried blood cards, buccal cells, tissue, serum, plasma, and laser microdissected cells are available.
Illustra GenomiPhi HY DNA Amplification Kit*	GE Health Care Life Sciences	High-quality DNA, isolated or in cell lysate	≥10 ng	For high WGA-DNA yield	Approximately 45 µg	>10 kb	Process can be automated. Use of degraded DNA, such as from formalin-fixed paraffin-embedded samples, may result in amplification bias

* http://www1.gelifesciences.com/APTRIX/upp01077.nsf/Content/genome_illustra_redirect–phi29_dna_polymerase_whole_genome
** http://www1.qiagen.com/Products/GenomicDnaStabilizationPurification/replig/REPLIgsScreeningKit.aspx?ShowInfo=1
*** http://www1.qiagen.com/Products/GenomicDnaStabilizationPurification/replig/RepliGMiniMidiKits.aspx

protocol specific to your applications. Each company also offers a vast range of supportive material on their Web pages. This chapter presents results obtained using two kits: the GenomiPhi kit (GE Healthcare) and the Repli-g Mini kit (Qiagen).

2.3   DNA Purification Columns

Many downstream applications using WGA DNA can be hindered if the WGA DNA mixture contains unused hexamer primers and unincorporated dNTPs. The hexamer primers also might be a source of contamination in the laboratory if opened in pre-PCR space. Therefore, we recommend that the WGA samples be purified using, for example, spin column chromatography. In most WGA kits, the protocol contains an optional column purification step with suggestions for the commercial column to use.

2.4   Additional Laboratory Reagents and Solutions

1. Standard PCR reagents (Taq polymerase, dNTPs, PCR reaction buffer, MLS)
2. Filter tips of various volumes.
3. 1X TE buffer.
4. 70% EtOH solution in a squirt bottle.
5. 0.5-mL Thermo-Tubes, thin-walled with flat caps (MLS).
6. Thermal cyclers (MLS).
7. Water bath.
8. NanoDrop. (NanoDrop Technologies).
9. 2100 Bioanalyser (Agilent technologies).
10. DNA 12000 LabChip kit (Agilent technologies).
11. Fluorometer (MLS).
12. Quant-iT PicoGreen dsDNA Reagent (Invitrogen).
13. Agarose.
14. Ethidium bromide or SYBR safe (Invitrogen).
15. SRY-F primer 5′-ATAAGTATCGACCTCGTCGGAA-3′
16. SRY-R primer 5′-CACTTCGCTGCAGAGTACCGA-3′
17. HTR2C-F primer 5′-GTGGTTTCAGATCGCAGTAA-3′
18. HTR2C-R primer 5′-ATATCCATCACGTAGATGAGAA-3′

2.5 Laboratory Space Requirements and Laboratory Equipment

WGA work should be done *only* in a designated space in the laboratory to avoid contamination. It is recommended that this working space would be marked with "WGA" or "MDA" labels. In addition, designate a set of pipettes, a vortex, a centrifuge, tube racks, and ice box for WGA work, all properly labeled. Use only filter tips in WGA work. Cover the laboratory table with a new paper cloth. Throw the used paper cloth and wipe the table with 70% EtOH once the work is finished.

3 Methods

3.1 DNA Sample Preparation

Measure the DNA concentration with a reliable system, preferably using the PicoGreen quantitation method. At least 10 ng of genomic DNA should be used for each WGA reaction, unless you use a specific protocol suitable for smaller amounts of DNA. Dilute the DNA to a concentration of 10 ng/μL, either in dH$_2$O or in 1X TE buffer. If the DNA has a concentration lower than 10 ng/μL, pipet an amount of DNA corresponding to 10 ng into a 0.5-mL tube or a well in a 96-well plate, then dry down the DNA (for example, at 50 °C). For some WGA kits, it is also possible to use a tissue lysate for DNA amplification—consult the WGA protocol you are using regarding this option. For low-yield DNA samples, it is recommended to do at least two independent WGA reactions from each sample, that is, a total of 20 ng of DNA is required. Label the WGA-reaction tubes clearly or prepare a map of the DNA samples in your 96-well plate. If the concentration of your DNA sample is very low, consider using a specific kit for low amounts of DNA (see Table 1.2).

3.2 Whole Genome Amplification

Select the kit based on your needs, as stated earlier (see Sect. 2.2), and follow the protocol as detailed in the kit or use a modification suitable for your applications. The incubation steps are most easily done using a 96-well PCR machine, but it is also possible to use water bath for the incubation steps, and an ice box for the cooling steps. Perform all pipetting steps in the WGA-designated working space using WGA-designated pipettes and equipment. Replace gloves often and clean the table with EtOH once you have finished working.

Figure 1.2 describes the basic steps of whole genome amplification with Φ29 DNA polymerase and random hexamer primers.

Fig. 1.2 Steps in the WGA reaction

Step 1:
Double-stranded genomic DNA is denatured
by heat or alkali treatment

Step 2:
Annealing of a mix of random
hexamer primers to the DNA

Step 3:
Addition of Φ29 DNA polymerase
and DNA synthesis (+30°C, 16h)

Step 4:
Deactivation of Φ29 DNA polymerase
(+65°C, 3-5 min)

Step 5:
Clean-up of WGA-DNA sample from
unincorporated dNTPs and unused
hexamer primers (optional)

Step 6:
Quality control checks of WGA-DNA
(optional)

3.3 Cleanup of WGA Products

This step is optional but recommended, to avoid problems during downstream applications and contamination of the laboratory. Spin columns can be used for a fast cleanup of the WGA products from unused hexamer primers and unincorporated dNTPs. Follow the instructions and recommendations of the manufacturer.

3.4 Recommended Quality Checks

Several quality checks are recommended to ensure that your WGA DNA amount is appropriate, the quality of the amplified DNA is good, and the amplification of the genome is not biased.

3.4.1   DNA Concentration Measurement

Use PicoGreen or any other reliable method you have in your laboratory to measure the amount of double-stranded WGA DNA obtained. Depending on the kit you are using and the expected yield, dilute your sample 5–100 times for DNA concentration measurement. In our experience, the WGA DNA yield using GE Health Care's GenomePhi kit is approximately 4–5 μg. Other kits may yield more or less WGA-DNA. In our experience, the same kit usually gives similar yields for all samples, no matter what the amount of genomic DNA was used as starting material. In fact, water control samples result in similar amount of WGA-DNA as samples in which genomic DNA was used as template. Therefore, once you know the approximate yield from the kit you are using, this step can be skipped.

3.4.2   Evaluation of Size of WGA Product

The size distribution of the WGA products can be studied, for example, with Agilent 2100 Bioanalyzer, DNA 12000 LabChip kit. Here, 1 μL of WGA-DNA diluted to 50 ng/μL can be run according to the manufacturer's instructions. An example of size distribution analysis for WGA products using two different kits, GenomiPhi kit and Repli-g Mini kit, is shown in Fig. 1.3, for two low yield DNA samples, a CEPH (Centre d'Etude du Polymorphisme Humain) control DNA sample and a water control sample. WGA DNA from both kits was cleaned using a column as specified by the manufacturer, prior to running the sample on a bioanalyzer. For GenomiPhi, the size distribution was between 4 and 6 kb; while for Repli-G, the average fragment size was somewhat larger but there was a wider range of product sizes. However, it is important to note that, also in water control samples which have no DNA template, there is synthesis of junk DNA with equal amount and size distribution as for true DNA samples. The junk DNA produced from water control samples does not give any results in downstream applications (see Sect. 3.4.3.1).

3.4.3   PCR Tests

3.4.3.1   Sex-PCR

The amount of WGA DNA obtained and the sizes of WGA fragments provide only information whether an amplification reaction has happened but do not inform on the quality of DNA. We therefore suggest that the quality of the amplified DNA, especially for low-yield DNA samples, be tested using sex-specific PCR, verifying that the result corresponds with the sample information. In this test (for human DNA), two fragments were amplified, a 92 bp fragment from SRY gene on the Y chromosome and a 158 bp fragment from the H2RTC gene on the X chromosome. The PCR products then were separated on an EtBr- or SYBR safe stained agarose gel and visualized under ultraviolet light. For this test, SRY-F/R and HTR2C-F/R primers were used in the PCR.

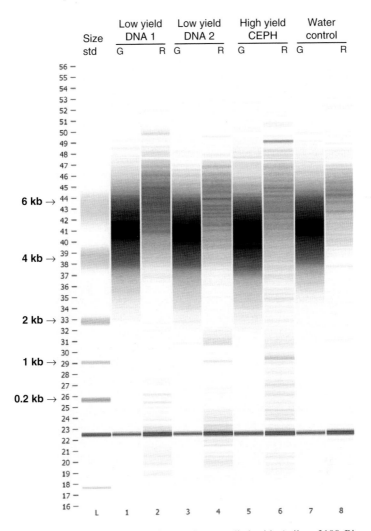

Fig. 1.3 The size distribution of the WGA products studied with Agilent 2100 Bioanalyzer. WGA-DNA was produced from two low-yield DNA samples, a CEPH control DNA sample and a water control sample, using two different kits, GE Health Care's GenomiPhi kit (G) and Qiagen's Repli-g Mini kit (R). The WGA DNA from both kits was cleaned using a column as specified by the manufacturer, prior to running the sample on a bioanalyzer

The PCR was done in a volume of 15 μL, and the reaction mix contained 1X BFR buffer, 200 μM of dNTPs, 1.5 mM of MgCl$_2$, 0.8 U of AmpliTaq Gold (Applied Biosystems), 200 nM of each PCR primer, and 20 ng of (WGA) DNA.

In Fig. 1.4, lanes 2–3 show the results of sex PCR for two WGA samples, sample WGA1 is a male and sample WGA2 is a female. The WGA-water control sample (lane 4) does not yield any PCR fragment, although the amount of WGA-DNA obtained is similar to that obtained from the genomic DNA sample.

PCR program	Temperature	Time
	95 °C	11 min
5 cycles of	95 °C	30 s
	65 (–1.0/cycle)	1 min
15 cycles of	95 °C	30 s
	60 °C (–0.5/cycle)	30 s
	72 °C	30 s
14 cycles of	95 °C	30 s
	53 °C	30 s
	70 °C	30 s
Final steps	72 °C	6 min
	10 °C	Forever

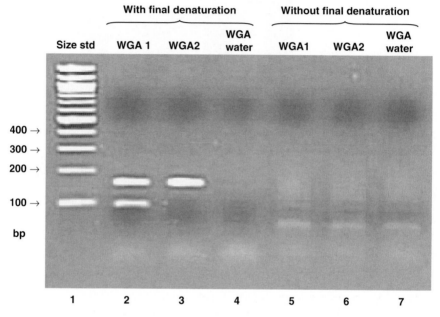

Fig. 1.4 Sex-specific PCR test for two WGA DNA samples produced using as a template two low-yield genomic DNA samples and a water control. In this test, two fragments are amplified, a 92 bp fragment from SRY gene on the Y chromosome and a 158 bp fragment from the H2RTC gene on the X chromosome. The PCR products were separated on an EtBr-stained agarose gel and visualized under ultraviolet light. Lanes 5–7: the WGA protocol lacked the last denaturation step to inactivate the Φ29 enzyme; lanes 2–4: the same samples were used following the final denaturation step. The WGA-water control sample does not yield any PCR fragment, although the amount of WGA DNA obtained was similar to that obtained from the genomic DNA sample

3.4.3.2 Quantitative PCR

When the genomic DNA sample is of very low concentration or somewhat degraded, the whole genome amplification may result in biased amplification. It therefore is important to assess whether the WGA DNA is representative of the genomic DNA and there is no bias in amplification. For this purpose, PCR-based

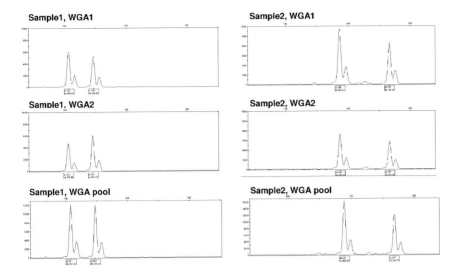

Fig. 1.5 Genotyping results for one microsatellite, AC00818_1 (in the PDE4D gene [30]) for two different WGA samples from low-yield genomic DNA, each sample amplified twice (WGA1 and WGA2) using the GenomiPhi kit. The WGA pool sample was created by combining equal amounts of WGA DNA from the two WGA replicates of the same sample. One of the PCR primers was fluorescently labeled, and the PCR fragments were separated on a 3730 DNA Analyzer. The samples were analyzed using GeneMapper software (Applied Biosystems). For sample 1 the allelic imbalance is visible between the two WGA repeats, while the other sample shows balanced amplification of the two WGA repeats. Pooling the two WGA reactions from each sample (see Sect. 3.5) provided a more balanced genotyping result for sample 1

genotyping assays can be used. For example, several highly polymorphic microsatellite markers can be genotyped to determine that the allele ratio is correct and there are no extra alleles. Alternatively, quantitative real-time PCR assay can be used to determine that allele ratios are identical.

Figure 1.4 shows genotyping results for one microsatellite, AC00818_1 (in the PDE4D gene [30]) for two different WGA samples, each sample amplified twice using the GenomiPhi kit. For one sample, an allelic imbalance is visible between the two WGA repeats, while the other sample shows balanced amplification of the two WGA repeats. Pooling the two WGA reactions from each sample (see Sect. 3.5) provides a more balanced genotyping result for sample 1.

3.5   Pooling of WGA Products

Performing two or more WGA reactions for each sample and pooling equal amounts of WGA DNA from these two reactions can help in case of an unbalanced WGA reaction and especially is recommended for samples in which the amount of genomic DNA was very low, <10 ng. Because the bias usually is random, most often the amplification bias in one reaction is different from that in the other reaction and pooling the two reactions results in accurate genotyping without allele dropouts. An example of biased WGA reactions from the same genomic DNA sample, which is improved by pooling, is illustrated in Fig. 1.5.

3.6   Storage of WGA Products

Make a working dilution of your WGA DNA, either in dH$_2$O or in 1X TE buffer. Store the rest of the WGA DNA at −20 °C. Avoid, as much as possible, freeze-thaw cycles, which may damage your DNA. In our experience, the WGA DNA can be reliably used even after 18 months of storage.

4   Notes

1. Water controls usually yield the same amount of WGA DNA as genomic DNA samples, and the size distribution of fragments is similar. The only way to verify that there is no contamination is with a downstream test such as PCR of a specific locus. One commercial kit (GenomiPhi V2 of GE Health Care Life Sciences) promises that no DNA is synthesized in water control samples.
2. If you have very low amounts of DNA, a biased amplification and allele dropout in genotyping may occur. Contamination also is a bigger problem with low amounts of DNA. It is recommended that at least 10 ng of DNA is used as starting material. With a lesser amount of genomic DNA, the genotyping results may be unreliable. It is important to vigorously check the quality of WGA DNA obtained from very low DNA amounts before using it in your research project (see Sect 1.3.4.3). In our experience, if the WGA and pooling were successful, genotyping results obtained from WGA DNA may be more reliable than genotyping results obtained using very low quantities of the original genomic DNA.
3. The inactivation step at the end of WGA reaction is very important, to degrade the exonuclease activity of the Φ29 enzyme. If the Φ29 enzyme is not inactivated properly, unspecific PCR products (smear) may result. Figure 4, lanes 5 and 6, show two WGA DNA samples in which the last inactivation step was not done. The sex-PCR test failed for both samples. Following the inactivation step for these two samples (lanes 2 and 3), the sex-PCR test was successful.

References

1. Nelson DL, Ledbetter SA, Corbo L, Victoria MF, Ramirez-Solis R, Webster TD, Ledbetter DH, Caskey CT (1989) Alu polymerase chain reaction: A method for rapid isolation of human-specific sequences from complex DNA sources. Proc Natl Acad Sci USA 86:6686–6690
2. Zhang L, Cui X, Schmitt K, Hubert R, Navidi W, Arnheim N (1992) Whole genome amplification from a single cell: Implications for genetic analysis. Proc Natl Acad Sci USA 89:5847–5851

3. Telenius H, Carter NP, Bebb CE, Nordenskjold M, Ponder BA, Tunnacliffe A (1992) Degenerate oligonucleotide-primed PCR: General amplification of target DNA by a single degenerate primer. Genomics 13:718–725

4. Dietmaier W, Hartmann A, Wallinger S, Heinmoller E, Kerner T, Endl E, Jauch KW, Hofstadter F, and Ruschoff J (1999) Multiple mutation analyses in single tumor cells with improved whole genome amplification. Am J Pathol 154:83–95

5. Kittler R, Stoneking M, Kayser M (2002) A whole genome amplification method to generate long fragments from low quantities of genomic DNA. Anal Biochem 300:237–244

6. Klein CA, Schmidt-Kittler O, Schardt JA, Pantel K, Speicher MR, Riethmuller G (1999) Comparative genomic hybridization, loss of heterozygosity, and DNA sequence analysis of single cells. Proc Natl Acad Sci USA 96:4494–4499

7. Lucito R, Nakimura M, West JA, Han Y, Chin K, Jensen K, McCombie R, Gray JW, Wigler M (1998) Genetic analysis using genomic representations. Proc Natl Acad Sci USA 95:4487–4492

8. Vos P, Hogers R, Bleeker M, Reijans M, van de Lee T, Hornes M, Frijters A, Pot J, Peleman J, Kuiper M et al. (1995) AFLP: A new technique for DNA fingerprinting. Nucleic Acids Res 23:4407–4414

9. Tanabe C, Aoyagi K, Sakiyama T, Kohno T, Yanagitani N, Akimoto S, Sakamoto M, Sakamoto H, Yokota J, Ohki M et al (2003) Evaluation of a whole-genome amplification method based on adaptor-ligation PCR of randomly sheared genomic DNA. Genes Chromosomes Cancer 38:168–176

10. Langmore JP (2002) Rubicon Genomics, Inc. Pharmacogenomics 3:557–560

11. Lovmar L, Syvanen AC (2006) Multiple displacement amplification to create a long-lasting source of DNA for genetic studies. Hum Mutat 27:603–614

12. Dean FB, Hosono S, Fang L, Wu X, Faruqi AF, Bray-Ward P, Sun Z, Zong Q, Du Y, Du J et al (2002) Comprehensive human genome amplification using multiple displacement amplification. Proc Natl Acad Sci USA 99:5261–5266

13. Hosono S, Faruqi AF, Dean FB, Du Y, Sun Z, Wu X, Du J, Kingsmore SF, Egholm M, Lasken RS (2003) Unbiased whole-genome amplification directly from clinical samples. Genome Res 13:954–964

14. Lu Y, Gioia-Patricola L, Gomez JV, Plummer M, Franceschi S, Kato I, Canzian F (2005) Use of whole genome amplification to rescue DNA from plasma samples. Biotechniques 39:511–515

15. Rook MS, Delach SM, Deyneko G, Worlock A, Wolfe JL (2004) Whole genome amplification of DNA from laser capture-microdissected tissue for high-throughput single nucleotide polymorphism and short tandem repeat genotyping. Am J Pathol 164:23–33

16. Ballantyne KN, van Oorschot RA, Mitchell RJ (2007) Comparison of two whole genome amplification methods for STR genotyping of LCN and degraded DNA samples. Forensic Sci Int 166:35–41

17. Short AD, Kennedy LJ, Forman O, Barnes A, Fretwell N, Wiggall R, Thomson W, and Ollier WE (2005) Canine DNA subjected to whole genome amplification is suitable for a wide range of molecular applications. J Hered 96:829–835

18. Gorrochotegui-Escalante N, Black WCT (2003) Amplifying whole insect genomes with multiple displacement amplification. Insect Mol Biol 12:195–200

19. Gadkar V, Rillig MC (2005) Suitability of genomic DNA synthesized by strand displacement amplification (SDA) for AFLP analysis: Genotyping single spores of arbuscular mycorrhizal (AM) fungi. J Microbiol Methods 63:157–164

20. Dickson PA, Montgomery GW, Henders A, Campbell MJ, Martin NG, James MR (2005) Evaluation of multiple displacement amplification in a 5 cM STR genome-wide scan. Nucleic Acids Res. 33:e119

21. Pask R, Rance HE, Barratt BJ, Nutland S, Smyth DJ, Sebastian M, Twells RC, Smith A, Lam AC, Smink LJ et al (2004) Investigating the utility of combining phi29 whole genome amplification and highly multiplexed single nucleotide polymorphism BeadArray genotyping. BMC Biotechnol 4:15

22. Lovmar L, Fredriksson M, Liljedahl U, Sigurdsson, S, Syvanen AC (2003) Quantitative evaluation by minisequencing and microarrays reveals accurate multiplexed SNP genotyping of whole genome amplified DNA. Nucleic Acids Res 31:e129

23. Paez JG, Lin M, Beroukhim R, Lee JC, Zhao X, Richter DJ, Gabriel S, Herman P, Sasaki H, Altshuler D et al (2004) Genome coverage and sequence fidelity of phi29 polymerase-based multiple strand displacement whole genome amplification. Nucleic Acids Res 32:e71

24. Montgomery GW, Campbell MJ, Dickson P, Herbert S, Siemering K, Ewen-White KR, Visscher PM, Martin NG (2005) Estimation of the rate of SNP genotyping errors from DNA extracted from different tissues. Twin Res Hum Genet 8:346–352

25. Tzvetkov MV, Becker C, Kulle B, Nurnberg P, Brockmoller J, Wojnowski L (2005) Genome-wide single-nucleotide polymorphism arrays demonstrate high fidelity of multiple displacement-based whole-genome amplification. Electrophoresis 26:710–715

26. Zhou X, Temam S, Chen Z, Ye H, Mao L, Wong DT (2005) Allelic imbalance analysis of oral tongue squamous cell carcinoma by high-density single nucleotide polymorphism arrays using whole-genome amplified DNA. Hum Genet 118:504–507

27. Mai M, Hoyer JD, McClure RF (2004) Use of multiple displacement amplification to amplify genomic DNA before sequencing of the alpha and beta haemoglobin genes. J Clin Pathol 57:637–640

28. Bergen AW, Qi Y, Haque KA, Welch RA, Chanock SJ (2005) Effects of DNA mass on multiple displacement whole genome amplification and genotyping performance. BMC Biotechnol 5:24

29. Alanne M, Salomaa V, Saarela J, Peltonen L, Perola M (2004) DNA extraction yield is associated with several phenotypic characteristics: Results from two large population surveys. J. Thromb Haemost 2:2069–2071

30. Gretarsdottir S, Thorleifsson G, Reynisdottir ST, Manolescu A, Jonsdottir S, Jonsdottir T, Gudmundsdottir T, Bjarnadottir SM, Einarsson OB, Gudjonsdottir HM et al (2003) The gene encoding phosphodiesterase 4D confers risk of ischemic stroke. Nat. Genet. 35:131–138

31. Holbrook JF, Stabley D, Sol-Church K (2005) Exploring whole genome amplification as a DNA recovery tool for molecular genetic studies. J Biomol Tech 16:125–133

32. Jiang Z, Zhang X, Deka R, Jin L (2005) Genome amplification of single sperm using multiple displacement amplification. Nucleic Acids Res 33:e91

33. Luthra R, Medeiros LJ (2004) Isothermal multiple displacement amplification: A highly reliable approach for generating unlimited high molecular weight genomic DNA from clinical specimens. J Mol Diagn. 6:236–242

34. Silander K, Komulainen K, Ellonen P, Jussila M, Alanne M, Levander M, Tainola P, Kuulasmaa K, Salomaa V, Perola M et al (2005) Evaluating whole genome amplification via multiply-primed rolling circle amplification for SNP genotyping of samples with low DNA yield. Twin Res Hum Genet 8:368–375

35. Dean FB, Nelson JR, Giesler TL, Lasken RS (2001) Rapid amplification of plasmid and phage DNA using Phi 29 DNA polymerase and multiply-primed rolling circle amplification. Genome Res 11:1095–1099

36. Barker DL, Hansen MS, Faruqi AF, Giannola D, Irsula OR, Lasken RS, Latterich M, Makarov V, Oliphant A, Pinter JH et al (2004) Two methods of whole-genome amplification enable accurate genotyping across a 2320-SNP linkage panel. Genome Res 14:901–907

37. Ng G, Roberts I, Coleman N (2005) Evaluation of three methods of whole-genome amplification for subsequent metaphase comparative genomic hybridization. Diagn Mol Pathol 14:203–212

38. Sorensen KJ, Turteltaub K, Vrankovich G, Williams J, Christian AT (2004) Whole-genome amplification of DNA from residual cells left by incidental contact. Anal. Biochem. 324:312–314

39. Thompson MD, Bowen, RA, Wong BY, Antal J, Liu Z, Yu H, Siminovitch K, Kreiger N, Rohan TE, Cole DE (2005) Whole genome amplification of buccal cell DNA: Genotyping concordance before and after multiple displacement amplification. Clin Chem Lab Med 43:157–162

40. Handyside AH, Robinson MD, Simpson RJ, Omar MB, Shaw MA, Grudzinskas JG, Rutherford A (2004) Isothermal whole genome amplification from single and small numbers of cells: A new era for preimplantation genetic diagnosis of inherited disease. Mol Hum Reprod 10:767–772

Chapter 2
Scanning for DNA Variants by Denaturant Capillary Electrophoresis

Per O. Ekstrøm

Abstract Analysis and detection of DNA variation is important in any field of biology. Hence, numerous methods have been developed to analyze DNA. A simple yet effective way of analyzing DNA is by denaturant capillary electrophoresis (DCE). The method is in theory applicable to 95% of the human genome. The method involves three steps; fragment design, PCR amplification and allele separation. The allele separation can in principle be performed with any DNA capillary sequencing instrument.

Keywords DGGE (denaturant gradient gel electrophoresis), CDCE (constant denaturant capillary electrophoresis), CTCE (cycling temperature capillary electrophoresis), PCR (polymerase chain reaction), dsDNA - double strand DNA, ssDNA - single strand DNA

1 Introduction

Analysis of DNA variation, either somatic or inherited mutation, is important in all fields of biology. As all life forms, from virus to humans, are products of their genetic code, any minor alteration in this code may induce severe or life-threatening changes in the organism. Alternatively, adaptation and survival advantage in a given environment may ensue. Thus, analysis of DNA variation has become a major field of research in biology. In 1953, Watson and Crick postulated the structure of DNA [1], which may be regarded as the beginning of the genomic era. This was followed by the description of PCR in 1971 by Kleppe et al. [2], demonstration of Sanger et al. sequencing reactions with termination nucleotides in 1977 [3], and the separation of base substitutions by denaturant gradient gel electrophoresis (DGGE) in 1983 [4, 5]. The familiar PCR amplification, using thermostable polymerases, was first demonstrated in 1985 [6–8]. This discovery accelerated analysis of DNA and development of methods able to analyse DNA. One type of method widely used is the melting gel technique, which has been reviewed extensively by Bjørheim and Ekstrøm

From: Methods in Molecular Biology, Vol. 439: *Genomics Protocols: Second Edition*
Edited by Mike Starkey and Ramnath Elaswarapu © Humana Press Inc., Totowa, NJ

[9, 10]. The method is based on a theoretical calculation of DNA melting by statistical mechanics calculation [11] and separation of variant DNA during electrophoresis, where the migration of less stable fragments (i.e., heteroduplexes) are retarded as compared to the wild-type sequence. The method, in theory, is applicable to about 95% of the human genome, where the melting temperature is below the stability of the artificial high melting domain (*see* **Note 1**), known as the GC-clamp [12–14] (*see* **Note 2**). The method, which initially was developed in "slab" gel form, has been transferred successfully to six different commercial DNA capillary sequencing instruments. DCE requires three steps, design of the appropriate fragment (software simulation), PCR amplification, and allele separation by electrophoresis.

Under the conditions for which DNA fragments are appropriately designed for DCE, PCR amplifiable, and given appropriate denaturing conditions (temperature and chemical), heteroduplexes are separated from the wild-type fragments. Therefore, DCE is an excellent method for scanning DNA for unknown variants.

2   Materials

2.1   *Statistical Mechanics Calculation of DNA Melting Probabilities*

Online Internet services, http://www.biophys.uni-duesseldorf.de/local/POLAND/poland.html (*see* **Note 3**).

2.2   *PCR*

1. 10 × buffer IV (ABgene, Epsom, UK) (*see* **Note 4**).
2. 25 mM MgCl$_2$.
3. 4 × 2.5 mM dNTPs.
4. Taq polymerase (ABgene) (*see* **Note 5**).
5. Primers.
6. DNA template.

2.3   *Allele separation*

Commercial capillary DNA sequencing instrument (*see* **Note 6**).

3 Methods

Scanning DNA for variants by DCE requires three steps: fragment design, PCR amplification of the target sequence, and allele separation by capillary electrophoresis. Evaluation of the DNA sequence of interest by use of statistical mechanics calculates the melting properties of the DNA and provides information as to whether the fragments can be analyzed by DCE and where the high melting temperature tail, the GC-clamp (*see* **Note 2**), should be attached. Furthermore, the simulation of allele melting temperatures yields information on their expected separation by DCE. The purpose of PCR is to produce enough copies of the target sequence, incorporating the label and attaching the high temperature melting domain. DCE for allele separation alleles was first performed with in-lab assembled instruments, where optical parts like mirror, lenses, laser, and detectors were mounted onto an optical bench [15–19]. Thus, all handling, gel replacement, sample loading, and recording of electrophoresis were done manually. However, the manual handling and technical skill of the users resulted in detection of mutants fractions down to 10^{-6} [20, 21]. This detection limit is superior to most methods. With the introduction of automated capillary DNA sequencing instruments, more samples could be processed without intervention from the operator [22–35], while upholding a limit of detection of 10^{-3} [22, 25, 26].

3.1 Target Selection

1. Find DNA sequence of interest in a public database (*see* **Note 7**).
2. Select primers for target sequence. The length of the target sequence should be in the range of 80–210 bp (*see* **Note 8**). Primers can be selected using the following website: http://frodo.wi.mit.edu/primer3/input.htm.
3. Cut and paste the sequence to be amplified into a DNA melting simulation bioinformatics tool (*see* **Note 3**). Observe the theoretical melting profile of the target sequence; simulate GC-clamp (*see* **Note 2**) and known variants (*see* **Note 9**). Figure 2.1 demonstrates the 50% melting probability of the Kras gene exon 1, simulated by four computer programs. Please note the difference in melting temperatures; this is a result of algorithm approximations and differences between salt concentration and dissociation constant settings in each program.

3.2 PCR

1. Optimize the PCR reaction, by use of concentration combinations of Mg^{2+}, primers, and dNTPs (*see* **Note 10**). This should be performed in combination with different annealing temperatures.

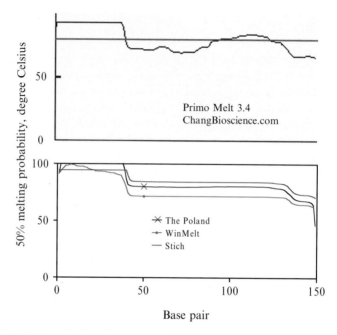

Fig. 2.1 Statistical mechanics calculation of DNA melting, Kras exon 1, obtained with four programs

2. Amplify samples by PCR, using the conditions identified by optimization. Table 2.1 provides an example of the composition of a "standard" PCR reaction mix (*see* **Note 11**). Perform PCR as follows: 95 °C, 5 min; 35 cycles of 95 °C, 30 s, annealing temperature, 30 s, 72 °C, 60 s (the annealing temperature varies between target sequences and is dependent on the primer design).
3. Create heteroduplexes by denaturing the PCR product at 95 °C for 2 min, followed by reannealing at 65 °C for 30 min (*see* **Note 12**) (this step can be programmed at the end of the amplification thermocycling program).

Table 2.1 20-µL PCR mix, volume, and concentrations of reagents

	Concentration	Volume (µL)	Desired concentration
H_2O		12.2	
10 × buffer	10 ×	2	1 ×
$MgCl_2$	25 mM	2	2.5 mM
Primer 1	5 µM	1.2	0.3 µM
Primer 2 (6-fam-gc)	5 µM	0.6	0.15 µM
DNTP (2.5 mM of each)	10 mM	0.8	400 µM
DNA	50 ng/µL	1	2.5 ng/µL
Taq DNA polymerase	5 U/µL	0.2	0.05 U/µL

3.3 Denaturant Capillary Electrophoresis

1. Set the running temperature of the capillary electrophoresis instrument according to the theoretical melting temperature of the low melting domain of the target sequence (*see* **Note 13**). For instruments with inappropriate temperature control, temperature cycling (cycling temperature capillary electrophoresis) can be used [29, 36].
2. Prepare the instrument for electrophoresis by replacing the polymer using a syringe pump or by applying high-pressure nitrogen (instrument dependent).
3. Inject the PCR product by applying a high voltage for a short period of time, such as 10 kV for 20 s (*see* **Notes 14 and 15**).
4. Perform electrophoresis and record the fluorescent signal.
5. Inspect the electropherogram as raw data in the sequencing analysis software supplied with the instrument. Figures 2.2 and 2.3 are two examples of allele separation and repeatability.
6. Using the internal standard (*see* **Note 15**), call the genotypes. Figure 2.4 demonstrates genotyping by DCE.

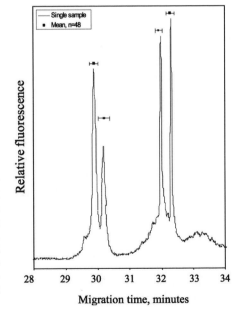

Fig. 2.2 Allele separation and repeatability of a one base pair deletion in the PPRA gamma gene (OMIM #601487). DCE was performed with temperature cycling between 50 °C and 47 °C, repeated 20 times [29, 36]. The average migration time of 48 runs (black square) is shown with 1 standard deviation for each duplex. Migration times were normalized to the sum of the four-migration times of each peak maximum [36]

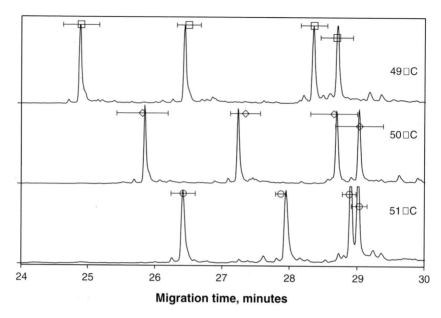

Migration time, minutes

Fig. 2.3 Separation of homoduplexes and heteroduplexes (NCBI ref SNP ID: rs10887), using the CEQ 8000 instrument (Beckman Coulter) at different temperature settings. The CEQ8000 instrument hardware was used without modification, and standard CEQ capillary arrays (33 cm), polymer, and separation buffer were used. The sample injection was standardized to 2.5 kV for 15 s while the separation was performed under the standard voltage setting of 4,000 V. Replicates ($n = 32$) of the samples were analyzed at three temperature settings. The average migration time of 32 runs is shown with 1 standard deviation for each peak (homoduplex and heteroduplex). Please note the baseline separation between the homoduplexes at all temperatures

4 Notes

1. Figure 2.5 depicts the distribution of 50% melting probabilities in 58440 bp from chromosome 1. Here, 95% of the melting temperature is below 77 °C, which is below the stability of the GC-clamp. Please note that the resolution on the x-axis is 0.1 °C.

2. Standard GC-clamp with 6-Fam label:
   5′-6-FAM-CGCCCGCCGCGCCCCGCGCCCGTCCCGCCGCCCCGC CCG-3′
   If this sequence is used as it is on the "forward" or "reverse" primer, a standard 20-mer primer with a different label can be used to make the fragment-specific internal standard (*see* **Note 15**). For large-scale studies featuring many fragments, primer cost reduction can be achieved by in-PCR attachment of a GC-clamp and labeling. This is illustrated in Fig. 2.6. The purpose of the GC-clamp is to prevent strand dissociation during electrophoresis. When the low melting

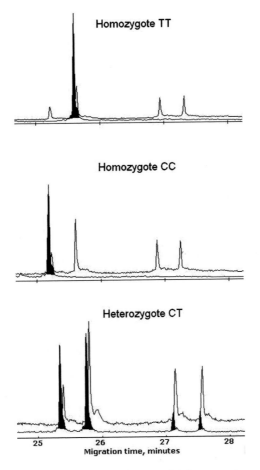

Fig. 2.4 Genotyping according to an internal standard. Three samples are genotyped (solid peaks) and scored against an internal standard (open peaks) for the SNP with NCBI reference number rs375384. The four peaks in the internal standard are the two homoduplexes and the two heteroduplexes (mismatch of Watson and Crick strands) formed during PCR and heteroduplex formation, respectively. However, when using DCE for genotyping, the user should be aware that any new or additional events in the target sequence many alter the peaks positions in the electropherogram

domain is subjected to the appropriate denaturing condition (chemical and temperature), this part of the fragment will be in a state of equilibrium between dsDNA and ssDNA. Because of the branching, the DNA fragment will be retarded during electrophoresis. Therefore, fragments with minor differences (substitution, deletion, or insertion) generally have a different average velocity through the polymer from the consensus sequence, which results in the observed separation of the alleles.

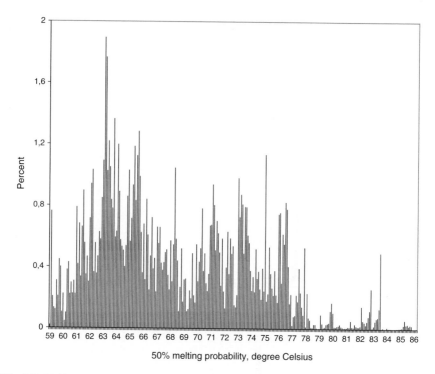

Fig. 2.5 Melting probabilities of 54440 bp from chromosome 1

In PCR labeling and GC extension, Kras exon 1 amplification and allele separation by CTCE

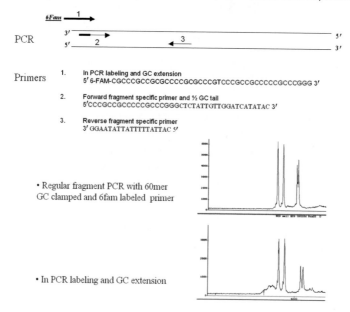

Fig. 2.6 In-PCR labeling and GC extension

3. Calculation of the melting properties of the DNA of interest can be performed by various software packages or online Internet programs. Information about commercial software or online bioinformatics tools can be found via the following Internet links:

http://bioweb.pasteur.fr/seqanal/interfaces/dan.html.

http://www.changbioscience.com/primo/primomel.html.

http://stitchprofiles.uio.no/ctemp_prof.htm.

http://bioinformatics.org/meltsim/wiki/Main/HomePage.

http://www.biophys.uni-duesseldorf.de/local/POLAND/poland.html.

http://www.medprobe.com/uk/sequencher-demo.html.

We have been successfully using WinMelt (Medprobe, Oslo, Norway) to simulate changes in thermodynamics and used the theoretical melting temperature as a guide for the temperature setting during electrophoresis. We found a linear relationship between stability changes and separation achieved by DCE [36].

4. The buffer and $MgCl_2$ are prepared in the laboratory to keep the PCR cost as low as possible. As a $10 \times$ PCR buffer, we regularly use sterile-filtered 750 mM Tris-HCl (pH 8.8 at 25 °C), 200 mM $(NH_4)_2SO_4$, with 0.1% (v/v) Tween 20 added to the buffer after sterile filtration. This buffer is the same as ABgene´s buffer IV (ABgene, Epsom, UK). $MgCl_2$ is added into the reaction mixture from a separate vial, because this facilitates PCR optimization with respect to Mg^{2+} concentration.

5. The selection of enzyme should be based on the detection limit needed for the study. Using Taq polymerase gives an error rate in the range of 1×10^{-4} to 1×10^{-5} misincorporations per bp per cycle, while PFU (Stratagene, La Jolla, CA) has been shown the produce tenfold fewer errors [37]. Therefore, studying low mutant fractions requires the use of PFU polymerase. However, we successfully combined the two enzymes (ratio 10:1, Taq:PFU), preserving the processivity of TAQ and the fidelity of PFU.

6. We tested six capillary DNA sequencing instruments, from different manufacturers, and found that alleles can be sufficiently separated by DCE with ABI 310 and ABI 3100 (Applied Biosystems, Foster City, CA), CEQ 8000 Genetic Analysis (Beckman Coulter, Fullerton, CA), MegaBACE 1000 and MegaBACE 4000 (GE Healthcare Bio-Sciences Corp., Piscataway, NJ), and SCE 2410 (Spectrumedix, State College, PA) [22–27, 29, 30, 32–35, 38, 39]. Standard capillaries and sieving matrix for each instrument were used, making this method readily available to any user with access to such instruments.

7. An example of a bioinformatics tool that can be used to access a nucleic acid sequence is found at http://snpper.chip.org/. This "database" presents gene search, dbSNP data, sequence search options, and annotated sequences (with intron, exon, coding, and SNP) in a very straightforward format. For illustration purposes, annotated "epidermal growth factor receptor isoforma"canbefoundat the following link: http://snpper.chip.org/bio/show-sequence/?TYPE=U&GENE=17905.

8. Primers can be selected using the following website: http://frodo.wi.mit.edu/primer3/input.htm

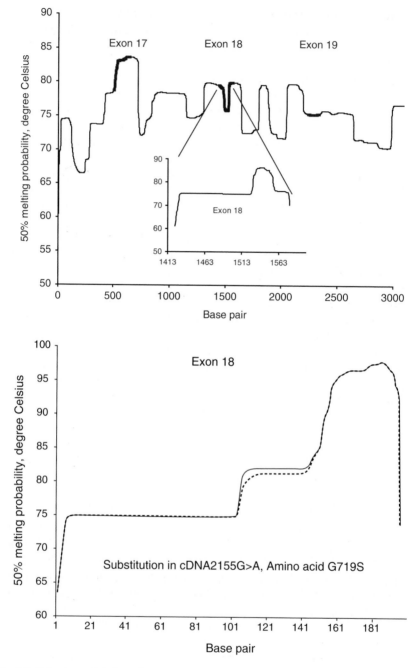

Fig. 2.7 Selection and design of target sequence and stability difference between alleles as calculated by statistical mechanics

9. Figure 2.7 represents a point-by-point illustration of the process of fragment design simulation and primer selection, leading to a fragment suitable for DCE analysis. Assume we want to analyze the EGFR gene (OMIM #131550) for mutations that render lung cancer patients susceptible to Iressa treatment. One of the mutational hotspots is sited in exon 18. In Fig. 2.7, the 50% melting probability (in °C) is plotted for the DNA sequence of exon 17 to 19. The bold line in the Fig. 2.7(a) graph indicates the exons. The enlargement of exon 18 contains 17 bp of priming sites in the intron on each side. From this melting profile, decision on the placement of the GC-clamp is made. As a rule of thumb, always place the GC-clamp on the side of the fragment with the "highest" average melting temperature. This prevents "hidden Death Valleys" which can result in pore peak resolution (31). Although the stability of the GC-clamp is high, the length of the low melting domain is limited. Experience has shown that the optimal low melting domain lengths are between 80 and 210 bp. Figure 2.7(b) illustrates a simulation of two different alleles in the target sequence. Note: The high temperature melting domain from the base pair 150 to 180 is the effect of the CG-clamp. By simply changing a base pair (SNP or mutation hot spot), a slightly different melting profile is revealed. This melting difference is sufficient to separate homoduplex alleles.

10. While optimizing the chemical conditions of the PCR is important for most assays involving amplification of DNA, the number of thermocycles also can affect the specificity of the amplification. The number of final copies (with a maximum of 10^{11} copies/μl) is a function of the starting copy number (No), amplification efficiency (eff) and the number of cycles (C). The following formula can be used to calculate the final copy number after PCR:

$$N = No \, (eff)^c$$

Because one of the primers is labeled with 6-fam, we generally use twice the amount of the unlabeled primer than the labeled primer. This is to prevent over-amplification of the labeled strand compared to the unlabeled (see **Note 16**). Table 1 is an example of a standard 20-μL PCR mix.

11. Troubleshooting a PCR reaction is beyond the scope of this chapter. The objective for primer design is to design primers specific for the target sequence, which anneal to the correct sites and facilitate amplification of the desired sequence. Once this is achieved, the labeled PCR products are ready for allele separation by DCE. However, a good laboratory practice is always to optimize each fragment with regard to Mg^{2+} concentration and annealing temperature.

12. Although many minor alterations (base substitution, deletion, and insertion) in DNA results in thermodynamic changes, some changes do not alter the thermodynamics of the target sequence [40]. These changes might be a G:C → C:G or a A:T → T:A transition, depending on the neighboring base sequence. To be able to detect those changes, a simple heteroduplex formation step (post-PCR) can be performed. Simply by denaturing the dsDNA and annealing the Watson and Crick strands slowly, highly unstable (as compared to matched base pair)

mismatches are formed (Fig. 2.7). A regularly used heteroduplex formation program of 95 °C for 2 min, followed by reannealing at 65 °C for 30 min, is sufficient for heteroduplex formation. Please note that this should be performed only on undiluted PCR products. Figure 2.8 depicts possible strand combination as a result of the denaturation and reannealing.

13. The denaturing temperature has to be adjusted for chemical denaturant in the polymer. Therefore, if the melting domain of the target sequence is theoretically melting at 75 °C and the polymer contains 7 M urea, the denaturing temperature is set to 54 °C [41] (75 °C – 21 °C, correction for chemical denaturant, ′3 °C/M urea). Sequencing polymer regularly contains 6 or 7 M urea. Consequently, the denaturing conditions are corrected accordingly. For sieving matrixes [42] formulated in the laboratory, chemical denaturant can be omitted. However, the temperature limitation or restriction on the various DNA capillary sequencing instruments dictates the use of chemical denaturants. For new fragments, a serial analysis at different temperatures for a heterozygote sample reveals the best separation temperature for the instrument used.

14. Crude or diluted PCR product in water (down to 1:50) is injected electrochemically, following the standard injection procedure recommended by the instrument manufacturer. Figure 2.9 demonstrates the relative small differences in injection efficiency introduced when PCR products are diluted in water. The PCR product was injected by applying 10 kV for 20 s and alleles were separated by CTCE in a MegaBACE 1000. The peaks are two homoduplexes and two heteroduplexes, respectively. Please note the nontemplate addition of adenosine, which is the stutter peak to the right of each duplex. We observed degradation of DNA diluted in water stored in 4 °C. Therefore, the diluted product should be stored

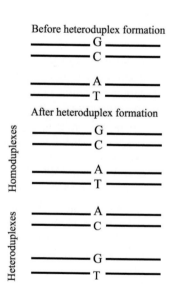

Fig. 2.8 Strand combination as a result of heteroduplex formation

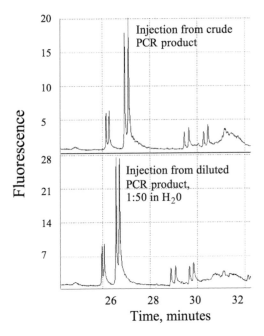

Fig. 2.9 Effect on electrochemical injects by diluting PCR product in H_2O

at −20 °C or below when not in use. The dilution of PCR products in "commercial loading solutions" or formamide is not recommended, as this may denature the dsDNA.

15. DCE has been used for the genotyping of DNA variants by separating alleles and comparing the pattern with an internal standard. The internal standard is made from a heterozygote sample by reamplification with a 20-mer primer specific for the GC-clamp (*see* **Note 2**). This primer has a different fluorophore attached to its 5′ end from the primer used to amplify the samples. As the GC-clamp is standardized, the same 20-mer labeling technique can be used for different fragments [27, 28, 30, 33]. Amplification of the internal standard with differential labeling is performed on a 1:1,000, or 1:10,000, diluted heterozygote PCR product in H_2O. Due to the high template copy number, fewer thermocycles are needed (less than 25). Primers:
Forward: 5′-Alternative label-CGCCCGCCGCGCCCCGCGCC3′
Reverse: Unlabeled primer
In Fig. 2.10, the bold lines represent the GC-clamp and the short lines with arrows indicate the forward and reverse primers. The alternative label is instrument dependent but could be, for example, carboxy-X-rhodamine (ROX).

16. We noted that labeled ssDNA molecules have a "tendency" to migrate into the heteroduplex region during DCE. Hence, using unbalanced primer concentrations (see Table 2.1) reduces the risk of amplifying the labeled ssDNA but

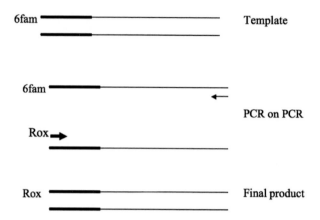

Fig. 2.10 Label "switching" used to construct an internal standard

allows the amplification of unlabeled ssDNA, which has no effect on the analysis.

17. The electrophoresis cost per sample (using standard conditions, as recommended by the instrument manufacturer) is about $1 (MegaBACE 1000). However, we optimized the DCE method using the MegaBACE 1000 with respect to cost per analysis. By formulating the sieving matrix and running buffer in the laboratory, we got more runs out of the capillary arrays and thereby reduced the polymer cost. Hence, the cost to us of running the electrophoresis with the MegaBACE 1000 is about 5 cents per sample.

References

1. Watson J, Crick F (1953) Molecular structure of nucleic acids—a structure for deoxyribose nucleic acid. Nature 171:737–738
2. Kleppe K, Ohtsuka E, Kleppe R, Molineux I, Khorana H (1971) Studies on polynucleotides. 96. Repair replication of short synthetic DNAs as catalyzed by DNA polymerases. J Mol Biol 56:341
3. Sanger F, Nicklen S, Coulson AR (1977) DNA sequencing with chain-terminating inhibitors. Proc Natl Acad Sci USA 74:5463–5467.
4. Fischer S, Lumelsky N, Lerman L (1983) Separation of DNA fragments differing by single base substitution—application to beta-degrees-thalassemia identification. DNA—J Molec Cell Bio, 2:171
5. Fischer SG, Lerman LS (1983) DNA fragments differing by single base-pair substitutions are separated in denaturing gradient gels: Correspondence with melting theory. Proc Natl Acad Sci USA 80:1579–1583
6. Saiki RK, Scharf S, Faloona F, Mullis KB, Horn GT, Erlich HA, Arnheim N (1985) Enzymatic amplification of beta-globin genomic sequences and restriction site analysis for diagnosis of sickle cell anemia. Science 230:1350–1354

7. Mullis K, Faloona F, Scharf S, Saiki R, Horn G, Erlich H (1986) Specific enzymatic amplification of DNA in vitro: The polymerase chain reaction. Cold Spring Harb Symp Quant Biol 51, Part 1:263–273

8. Kwok S, Mack DH, Mullis KB, Poiesz B, Ehrlich G., Blair D, Friedman-Kien A, Sninsky JJ (1987) Identification of human immunodeficiency virus sequences by using in vitro enzymatic amplification and oligomer cleavage detection. J Virol, 61:1690–1694

9. Bjørheim J, Gaudernack, G, Ekstrøm PO (2002) Melting gel techniques in single nucleotide polymorphism and mutation detection: From theory to automation. J Separation Science, 25:637–647

10. Bjørheim J, Ekstrøm PO (2005) Review of denaturant capillary electrophoresis in DNA variation analysis. Electrophoresis 26:2520–2530

11. Poland D (1974) Recursion relation generation of probability profiles for specific-sequence macromolecules with long-range correlations. Biopolymers 13:1859–1871

12. Myers RM, Fischer SG, Lerman LS, Maniatis T (1985) Nearly all single base substitutions in DNA fragments joined to a GC-clamp can be detected by denaturing gradient gel electrophoresis. Nucleic Acids Res 13:3131–3145

13. Myers RM, Fischer SG, Maniatis T, Lerman LS (1985) Modification of the melting properties of duplex DNA by attachment of a GC-rich DNA sequence as determined by denaturing gradient gel electrophoresis. Nucleic Acids Res 13:3111–3129

14. Sheffield VC, Cox DR, Lerman LS, Myers RM (1989) Attachment of a 40-base-pair G + C-rich sequence (GC-clamp) to genomic DNA fragments by the polymerase chain reaction results in improved detection of single-base changes. Proc Natl Acad Sci USA 86:232–236

15. Khrapko K, Hanekamp, JS, Thilly WG, Belenkii A, Foret F, Karger BL (1994) Constant denaturant capillary electrophoresis (CDCE): A high resolution approach to mutational analysis. Nucleic Acids Res 22:364–369

16. Kumar R, Hanekamp JS, Louhelainen J, Burvall K, Onfelt A, Hemminki K, Thilly WG (1995) Separation of transforming amino acid-substituting mutations in codons 12, 13 and 61 the N-ras gene by constant denaturant capillary electrophoresis (CDCE). Carcinogenesis 16:2667–2673

17. Bjørheim J, Lystad S, Lindblom A, Kressner U, Westring S, Wahlberg S, Lindmark G, Gaudernack G, Ekstrøm PO, Roe J et al. (1998) Mutation analyses of KRAS exon 1 comparing three different techniques: Temporal temperature gradient electrophoresis, constant denaturant capillary electrophoresis and allele specific polymerase chain reaction. Mutation Research—Fundamental and Molecular Mechanisms of Mutagenesis 403:103–112

18. Ekstrøm PO, Børresen-Dayle AL, Qvist H, Giercksky KE, Thilly WG (1999) Detection of low-frequency mutations in exon 8 of the TP53 gene by constant denaturant capillary electrophoresis (CDCE). Biotechniques 27–12

19. Ekstrøm PO, Wasserkort R, Minarik M, Foret F, Thilly WG (2000) Two-point fluorescence detection and automated fraction collection applied to constant denaturant capillary electrophoresis. Biotechniques 29:582

20. Li-Sucholeiki XC, Khrapko K, Andre PC, Marcelino LA, Karger BL, Thilly WG (1999) Applications of constant denaturant capillary electrophoresis/high-fidelity polymerase chain reaction to human genetic analysis. Electrophoresis 20:1224–1232

21. Li-Sucholeik, XC, Thilly WG (2000) A sensitive scanning technology for low frequency nuclear point mutations in human genomic DNA. Nucleic Acids Res 28:e44

22. Bjørheim J, Ekstrøm PO, Fossberg E, Børresen-Dale AL, Gaudernack G (2001) Automated constant denaturant capillary electrophoresis applied for detection of KRAS exon 1 mutations. Biotechniques 30:972–975

23. Minarik M, Bjørheim, J, Ekstrøm PO, Dains KM (2001) High-throughput mutation detection and screening using MegaBACE (TM) capillary array instrument for genetic analysis. Amer J Human Genetics 69:469–469

24. Bjørheim J, Gaudernack G, Ekstrøm PO (2001) Mutation analysis of TP53 exons 5-8 by automated constant denaturant capillary electrophoresis. Tumor Biology 22:323–327

25. Kristense, AT, Bjorheim J, Minarik M, Giercksky KE, Ekstrom PO (2002) Detection of mutations in exon 8 of TP53 by temperature gradient 96-capillary array electrophoresis. Biotechniques 33:650–653
26. Bjørheim J, Minarik M, Gaudernack G, Ekstrøm PO (2002) Mutation detection in KRAS exon 1 by constant denaturant capillary electrophoresis in 96 parallel capillaries. Analytical Biochemistry 304:200–205
27. Ekstrøm PO, Bjørheim J, Gaudernack G, Giercksky KE (2002) Population screening of single-nucleotide polymorphisms exemplified by analysis of 8000 alleles. J Biomolecular Screening 7:501–506
28. Bjørheim J, Abrahamsen TW, Kristensen AT, Gaudernack G, Ekstrøm PO (2003) Approach to analysis of single nucleotide polymorphisms by automated constant denaturant capillary electrophoresis. Mutation Research—Fundamental and Molecular Mechanisms of Mutagenesis 526:75–83
29. Minarik M, Minarikova L, Bjorheim J, Ekstrom PO (2003) Cycling gradient capillary electrophoresis: A low-cost tool for high-throughput analysis of genetic variations. Electrophoresis 24:1716–1722
30. Kristensen AT, Bjorheim J, Wiig J, Giercksky KE, Ekstrom PO (2004) DNA variants in the ATM gene are not associated with sporadic rectal cancer in a Norwegian population-based study. Int J Colorectal Dise, 19:49–54
31. Hinselwood DC, Abrahamsen TW, Ekstrøm PO (2005) BRAF mutation detection and identification by cycling temperature capillary electrophoresis. Electrophoresis 26:2553–2561
32. Hinselwood DC, Warren DJ, Ekstrøm PO (2005) High-throughput gender determination using automated denaturant gel capillary electrophoresis. Electrophoresis 26:2562–2566
33. Lorentzen AR, Celius EG, Ekstrøm PO, Wiencke K, Lie BA, Myhr KM, Ling V, Thorsby E, Vartdal F, Spurkland A et al (2005) Lack of association with the CD28/CTLA4/ICOS gene region among Norwegian multiple sclerosis patients. J Neuroimmunol, 166:197–201
34. Lind H, Zienolddiny S., Ekstrøm PO, Skaug V, Haugen A (2006) Association of a functional polymorphism in the promoter of the MDM2 gene with risk of nonsmall cell lung cancer. Int J Cancer 119:718–721
35. Harbo HF, Ekstrøm PO, Lorentzen AR, Sundvold-Gjerstad V, Celius EG, Sawcer S, Spurkland A (2006) Coding region polymorphisms in T cell signal transduction genes. Prevalence and association to development of multiple sclerosis. J Neuroimmunol 177:40–45
36. Bjørheim J, Minarik M, Gaudernack G, Ekstrøm PO (2003) Evaluation of denaturing conditions in analysis of DNA variants applied to multi-capillary electrophoresis instruments. J Separation Science, 26:1163–1168
37. Bracho MA, Moya A, Barrio E (1998) Contribution of Taq polymerase-induced errors to the estimation of RNA virus diversity. J Gen Virol 79, Part 12:2921–2928
38. Li-Sucholeik, XC, Tomita-Mitchell A, Arnold K, Glassner BJ, Thompson T, Murthy JV, Berk L, Lange C, Leong-Morgenthaler, PM, MacDougall D et al (2005) Detection and frequency estimation of rare variants in pools of genomic DNA from large populations using mutational spectrometry. Mutation Research—Fundamental and Molecular Mechanisms of Mutagenesis 570:267–280
39. Li Q, Deka C, Glassner B, Arnold K, Li-Sucholeiki X, Tomita-Mitchell A, Thilly W, Karger B (2005) Design of an automated multicapillary instrument with fraction collection for DNA mutation discovery by constant denaturant capillary electrophoresis (CDCE). J Separation Science 28:1375–1389
40. Bjørheim J, Gaudernack G, Giercksky KE, Ekstrøm PO (2003) Direct identification of all oncogenic mutants in KRAS exon 1 by cycling temperature capillary electrophoresis. Electrophoresis 24:63–69
41. Fodde R, Losekoot M (1994) Mutation detection by denaturing gradient gel electrophoresis (DGGE). Hum Mutat 3:83–94
42. Ekstrøm PO, Bjørheim J (2006) Evaluation of sieving matrices used to separate alleles by cycling temperature capillary electrophoresis. Electrophoresis 27:1878–1885

Chapter 3
Identification of SNPs, or Mutations
in Sequence Chromatograms

Nicole Draper

Abstract With the completion of the human genome sequencing project in 2001, the identification of novel markers is rapidly gaining importance. It is increasingly recognized that SNPs (single nucleotide polymorphisms) are good markers for disease susceptibility. SNPs are DNA sequence variations that occur when a single nucleotide in the genome sequence is altered in at least 1% of the population. SNPs may have no effect on cell function, but scientists believe that they could predispose people to disease or influence their response to a drug.

This chapter describes the method of using fluorescent based sequencing to detect SNPs and mutations. Sequencing provides information on the type and location of the SNPs with high accuracy. Researchers will need to provide information on the area of the genome they wish to sequence to design primers to PCR amplify the specific region.

Keywords Sequencing, mutations, polymorphisms (SNPs), introns, electropherogram

1   Introduction

Since the completion of the human genome sequencing project in 2001 [1], more and more scientists are sequencing samples to identify the mutations responsible for disease. Most common diseases are complex genetic traits, with multiple genetic and environmental components contributing to susceptibility. It has been proposed that common genetic variants, including single nucleotide polymorphisms (SNPs), influence susceptibility to common diseases. This proposal has begun to be tested in numerous studies of association between genetic variation at these common DNA polymorphisms and variation in disease susceptibility [2].

A polymorphism is a variation in DNA sequence that has an allele frequency of at least 1% in the population. Approximately 1 in 1,000 base pairs of the human genome differs between any two chromosomal homologues [3]. Studies

From: Methods in Molecular Biology, Vol. 439: *Genomics Protocols: Second Edition*
Edited by Mike Starkey and Ramnath Elaswarapu © Humana Press Inc., Totowa, NJ

comparing individuals in a population have shown that, on average, polymorphisms in occur every 200–300 hundred base pairs [4]. There are several types
of polymorphisms, the most common, SNPs, and repeat polymorphisms (microsatellites), insertions or deletions, which range extensively in size. It is estimated
that 93% of genes contain a SNP [5]. A variation (or SNP) is termed a *mutation*
when it leads to a change in amino acid sequence and consequently an alteration
to the protein and its function. High-throughput technology is now used to
genotype SNPs and genetic variants [6, 7], although it is still not practical
to genotype all SNPs in the genome for association with disease phenotypes.
Therefore, SNPs and mutations are selected on their ability to affect the function of a protein or its expression, as those polymorphisms with molecular
consequences are more likely to be involved in influencing disease risk.
However, many SNPs have no direct effect on cell function, but it is believed
that they could predispose people to diseases or influence their response to a
drug. The 5′ flanking regions upstream of genes are known to harbor transcriptional control elements and therefore are routinely screened for sequence
variants that may help define a clinical phenotype [8]. Recently, attention has
focused on the effects of polymorphisms in noncoding regions, such as 3′ and
5′ UTR (untranslated region), regulatory regions, and intronic regions, rather
than the traditional promoter and coding region mutations, as these regions also
can contribute to the control of gene expression [9]. The role of UTRs and
introns in gene regulation may have been underestimated [10].

Scientists believe SNP maps will help them identify the multiple genes associated with such complex diseases such as cancer, diabetes, vascular disease, and
some forms of mental illness. These associations are difficult to establish with conventional gene-hunting methods because a single altered gene may make only a
small contribution to the disease. By comparing patterns and frequencies of SNPs
in patients and controls, researchers hope to identify which SNPs are associated
with which diseases [11–13].

A number of methods are available for SNP or mutation identification, including
DNA sequencing [4, 14]; enzymatic digestion combined with gel-electrophoresis
(SNP-RFLP, restriction fragment length polymorphism) [15–17]; pyrosequencing
[18]; ARMS-PCR (amplification refractory mutation system) [16, 19], and confor-
mational analysis (WAVE) [20, 21]; all of which require an initial target sequence
PCR amplification. Another, more recent SNP detection method is by hybridization
to DNA or oligonucleotide microarrays. This method does not require prior PCR
amplification, as total genomic DNA is used directly in labeling and hybridization.
Additionally, there are several high-throughput methodologies for SNP detection,
including ABI TaqMan assays and mass spectrometry. The current methods have
been comprehensively reviewed by Wang et al. [22].

This chapter describes the laboratory and computational methods for identifying
SNPs and mutations employing fluorescent-based sequencing protocol using ABI's
3700 DNA analyzer. DNA sequencing is the preferred method for SNP and mutation detection, as it provides information on the type and location of the variants at
very high accuracy, whereas other methods require sequencing as an additional step

to confirm results due to higher false positive rates. However, sequencing can be a costly method and is less adaptable to multiplex and automation than other methods.

2 Materials

2.1 Preparation of Sequencing Template

1. The DNA template can be either purified plasmid DNA or PCR product. Total plasmid DNA is isolated from bacterial recombinant clones, whereas the PCR technique is used for amplifying the required target sequences using specific flanking primers and polymerase.
2. Qiagen Plasmid Mini Kit (Qiagen 12123).
3. LB broth medium: 1% tryptone, 0.5% yeast extract, 0.5% NaCl (Sigma Aldrich, L3022-1 kg).
4. Nuclease-free water (Sigma Aldrich W4502).
5. PCR primers for amplifying a specific region of the genome for SNP analysis, (e.g. vector primers T7 and SP6):
6. T7- 5′ AATACGACTCACTATAG 3′.
7. SP6- 5′ ATTTAGGTGACACTATAG 3′.
8. PCR buffer and reagents (Promega PCR Core System 1, M7660).
9. Gel electrophoresis equipment.
10. 1X TBE: 54 g Tris-base, 27.5 g boric acid, 20 mL of 0.5 M EDTA, pH 8.0. Make up the vol to 1 L with water (Sigma Aldrich, 93290-1 L).
11. Electrophoresis-grade agarose (BioRad, 161-3101 [125 g]).
12. Ethidium bromide (Sigma Aldrich, E8751-5G).
13. Agilent bioanalyser and DNA chips (Agilent, 5067-1506).
14. Gel Extraction Kit (Qiagen, 28704).
15. QIAquick PCR Purification Kit (Qiagen, 28104).
16. Incubating shaker.
17. Thermal cycler for PCR amplification.
18. 15-mL Falcon tubes (Fisher, CFT-420-079D).

2.2 Setting up Sequencing Reactions

19. PCR primers for sequencing reaction (e.g. T7 Universal primer): T7- 5′ TAA TAC GAC TCA CTA TAG GG 3′
20. ABI Prism Big Dye Terminator Cycle Sequencing Kit (ABI, 4337455).
21. SeqSaver Sequencing Premix Dilution Buffer (Sigma Aldrich, S3938) or BigDye Terminator v3.1 5X Sequencing Buffer (ABI, 4336699).

2.3   Purification of Sequencing Reactions

22. 0.5 M EDTA (Sigma Aldrich, E-7889).
23. 70% and 95% ethanol.
24. Vacuum centrifuge.
25. Hi Di Formamide (ABI, 4311320).

2.4   Sequencing of DNA

26. ABI 3700 DNA analyzer.

2.5   Analysis of Sequence Data

27. Computer, minimum requirement is Windows 98.
28. Chromas or similar sequencing software. Chromas is available for free download (noncommercial use).
29. Internet access.

3   Methods

3.1   Sequencing Template Preparation

The following description covers both PCR and plasmid sequencing methods and analysis of data for the detection of SNPs or mutations.

3.1.1   Plasmid Preparation

Several methods of DNA preparation and purification are available, all of which work reasonably well. However, the most commonly used kit, with a high rate of success, is the Qiagen kit for plasmid purification. The following protocol is suitable when small quantities of DNA is required:

30. PCR product preparation. Pick a colony and transfer into 5 mL of LB-broth medium (*see* **Note 1**).
31. Grow overnight at 37 °C with shaking at 220–250 rpm.
32. Centrifuge the overnight culture in 15-mL Falcon tube at 600 g for 10 min at room temperature.

33. Discard the medium and resuspend the pellet in (250 µL) Cell Resuspension Solution, which is provided in the Qiagen kit.
34. Follow the Qiagen Plasmid Miniprep Kit protocol, in the booklet provided.
35. Elute the DNA using 100 µL of nuclease-free water.

3.1.2 PCR Product Preparation

Before sequencing PCR products, it is important to run them on an agarose gel to ascertain the success of the PCR reaction.

36. Amplify the DNA samples using preoptimized PCR primers and PCR cycling conditions to amplify a region of the genome for SNP analysis.
37. Run samples on an agarose gel or Agilent Bioanalyser DNA chip to determine whether the PCR amplification has generated a single band.
38. Prepare 1% TBE agarose gel with ethidium bromide (1 µL of 1 mg/mL EthBr in 100 mL gel). Load 5 µL of the PCR product on the gel for each sample. Alternatively, load 1 µL of each PCR product into the wells of a DNA 7500 chip and run on the Agilent Bioanalyser following the protocol provided with the instrument.
39. If a single band is present, purify the PCR product using Qiagen PCR Purification columns, following the protocol provided. If more than one band is seen, load the remaining PCR sample onto a 1% TBE agarose gel, containing ethidium bromide. Purify the correct band by cutting out of the gel with a scalpel blade and use a Qiagen Gel Extraction Kit to isolate DNA from agarose slices (*see* **Note 2**).

3.1.3 Quantities of Template DNA Required for Cycle Sequencing Reactions

The following table indicates the quantities of starting DNA for sequencing reactions as recommended by Applied Biosystems (ABI), using its Big Dye Terminator sequencing reaction kit.

Template	Quantity
PCR product size:	
100–200 bp	1–3 ng
200–500 bp	3–10 ng
500–1,000 bp	5–20 ng
1,000–2,000 bp	10–40 ng
>2,000 bp	40–100 ng
Plasmid (single stranded)	50–100 ng
Plasmid (double stranded)	200–500 ng
Cosmid or BAC DNA	0.5–1.0 µg
Bacterial genomic DNA	2–3 µg

These template quantities are recommended for all types of primers. The amount of PCR product to use will be dependant on the length and purity of the products (*see* **Note 3**).

3.2 Setting up the PCR Reaction Using the Big Dye Cycle Sequencing Kit

This section describes how to prepare sequencing reactions with templates such as plasmids and PCR products. The first step in the sequencing of PCR products or plasmids is to incorporate fluorescent dNTPs. The automated sequencer detects fluorescence from four dyes that are used to identify A, G, C, and T extension reactions. Each dye emits light at a different wavelength when excited by laser energy, thus all four dyes and reactions can be detected and distinguished in a single gel lane. All the sequencing reactions described here are carried out using dideoxy terminator chemistry. Several versions of ABI sequencing kits are available, and all have easy to follow protocol manuals. The reaction described here uses the ABI Prism Big Dye Terminator v3.1 cycle sequencing kit (Applied Biosystems, UK). Dye terminator chemistry uses four 3′-dye labeled dideoxynucleotide triphosphates (dye terminators) each labeled with a different dye.

The exact base sequence of the primers used for the sequencing of a specific gene region are not relevant, but all sequencing primers should be designed with an approximate GC content of 50% and an annealing temperature of 50 °C. Sequencing reactions are carried out using a single primer (*see* **Note 4**). The reagents needed for sequencing PCR products are provided together as a "ready reaction" mix supplied by ABI, except for the primers and milliQ water, as shown in the following table. The dNTP mix in the ready reaction contains dITP in place of dGTP to minimize band compressions, and dUTP instead of dTTP, as dUTP improves the incorporation of the T terminators and results in a better T pattern.

1. For each sample add the following reagents in order:

Reagent	Quantity
Terminator Ready Reaction Mix	4.0 μL
SeqSaver dilution buffer*	4.0 μL
Template	See the table in Sect. 3.1.3
Primer (forward/reverse)	3.2 pmol
Sterile water	Make up to 20 μL

*(*See* **Note 5**.)

2. Keep the reaction mix on ice.
3. Vortex and briefly centrifuge.
4. Place the reaction tubes in a PCR machine.
5. Typical run parameters using a Thermal Cycler follow (*see* **Note 6**):

Lid preheat, 110 °C.

Temperature gradient to 96 °C at 1 °C/s.
Hold at 96 °C for 10 s.
Temperature gradient to 50 °C at 1 °C/s.
Hold at 50 °C for 5 s.
Temperature gradient to 60 °C at 1 °C/s.
Hold at 60 °C for 4 min.
Repeat cycle loop (25 cycles).
Store at 8 °C (*see* **Note 7**).

3.3   Purification of Sequencing Reactions

After the extracted samples have undergone the sequencing PCR to incorporate fluorescent dNTPs, the PCR product must be purified to remove any remaining dye terminators, as these interfere with the data analysis. Several methods are available for postreaction cleanup (*see* **Note 8**). However, a widely used small-scale method is described in detail next:

1. Pipette entire contents of PCR reaction into a 1.5 mL microfuge tube.
2. Add 2 μL of 250 mM EDTA.
3. Add 64 μL of nondenatured 95% ethanol.
4. Close the tubes and vortex briefly.
5. Leave at room temperature for 15 min (*see* **Note 9**).
6. Place the tubes in a microcentrifuge and mark their orientations. Centrifuge the tubes for 20 min at 20,000 g.
7. Carefully aspirate the supernatants with a separate tip for each sample and discard the tip. Pellets may or may not be visible (*see* **Note 10**).
8. Add 250 μL of 70% ethanol and vortex briefly.
9. Centrifuge the tubes for 20 min at 20,000 g.
10. Carefully aspirate the supernatants.
11. Repeat steps 8–10.
12. Place the tubes in the microcentrifuge in the same orientation as step 6 and spin for 10 min at 20,000 g.
13. Aspirate the supernatant carefully as in step 7.
14. Dry the samples in a vacuum centrifuge for 10–15 min or to dryness (*see* **Notes 11 and 12**).

3.4   Sequencing on ABI DNA Analyzer

The samples are now ready to load on an ABI DNA analyzer and can be resuspended to the desired volume for the sequencer. The detailed protocol for setting up and running the ABI 3700 DNA analyzer is beyond the scope of this chapter. Full instruction and training on the DNA analyzer would be provided by ABI on installation of the analyzer [23]. In many cases, DNA analyzers are maintained and run by core facilities and specialized technical staff; therefore, the individual researcher need not have full knowledge of the machine parameters and programs.

3.5   Analysis of Sequencing Results

Once the sequencing run is complete, the data are assimilated by computer using the software collection supplied with the DNA sequencer. The latest software upgrade is the ABI Sequencing Analysis Software v5.2 with KB Basecaller Software v1.2. The basic feature of the software is that it interprets the four fluorescent dyes (base calling), and produces the raw data nucleotide sequence as a display of peaks: an electropherogram. An electropherogram is a graph measuring the intensity of the wavelengths corresponding to the four fluorescent colors at every nucleotide position. The intensity is registered as a colored peak on the electropherogram. The highest peak at every specific nucleotide position thus is allocated as the nucleotide corresponding to the color of that peak. The colors allocated for each nucleotide are

Red = T (thymine).
Blue = C (cytosine).
Green = A (adenine).
Black = G (guanine).

The sequence can be read from the electropherogram as the definitive nucleotide order. If the software were unable to interpret a particular fluorescent peak a letter *N* is assigned for that base position. This could be caused by two bases being present in one position (heterozygosity) or weak peak height.

In addition to the chromatogram trace, automated DNA sequencers generate a text file of sequence data by calling the bases associated with each peak (base calling). It is important to manually check the electropherogram against the predicted base, as errors are reasonably common. Commonly, errors occur near the beginning and the end of any sequencing run.

3.5.1   Analysis Programs

A number of sequence analysis programs are available. The most popular free program that displays the electropherogram is Chromas Lite [24]. The program opens chromatogram files from Applied Biosystems and Amersham MegaBace DNA sequencers, in addition to opening SCF format chromatogram files created by ALF, Li-Cor, Visible Genetics OpenGene, Beckman CEQ 2000XL and CEQ 8000, and other sequencers (*see* **Note 13**).

3.5.2   Chromas Chromatogram Analysis

The following protocol is recommended for analyzing the sequence data using Chromas.

1. Load Chromas.
2. Open the sequencing file (.ab1). Click on File, Open.
3. Scan the chromatogram for sequence quality, that is, noise, check base calling and resolution by scrolling through the sequence.
4. Check the signal levels.
5. Search for Ns. Click on Edit, Search next N.
6. Search for heterozygotes and mutations by scanning the electropherogram for double peaks or anomalies.
7. Compare the sequence to a known sequence (from database or previous sequencing) and search for sequence differences.
8. Save sequence text file as FASTA format.
9. The saved text file can be used in a number of sequence alignment programs to identify novel variants. Often, a sequence generated from a clinical diseased sample is compared against the sequence of a normal sample. Any sequence differences then are investigated to see if they play a role in the etiology of the disease.

3.5.3   BLAST Analysis

The Basic Local Alignment Search Tool (BLAST) finds regions of local similarity between sequences. The program compares nucleotide or protein sequences to sequence databases and calculates the statistical significance of matches [25].

1. Load the search engine (http://www.ncbi.nlm.nih.gov/BLAST/).
2. Select Nucleotide-nucleotide BLAST (blastn).
3. Enter the sequence (cut and paste the FASTA sequence into search box).
4. Select the genome of interest to search sequence against (i.e., Human).
5. To search, press the BLAST button.
6. Press BLAST again on Format page. Wait for alignment results.
7. Search the alignment result for highlighted differences, that is, Ns or changed bases. These are polymorphism or mutations in the sequence.

BLAST analyses highlight regions in the newly generated sequence that do not match with the known sequence for that region in the GenBank database [26]. These variations in the sequence can be termed an *SNP* if it is a single base change, or they may be other types of variants such as microsatellites, insertions, or deletions. A single base change may be classified as a common polymorphism or SNP, however; if it changes an amino acid, it would be termed a *mutation*, as it may lead to a change in protein function. Figure 3.1 shows that a difference of one base

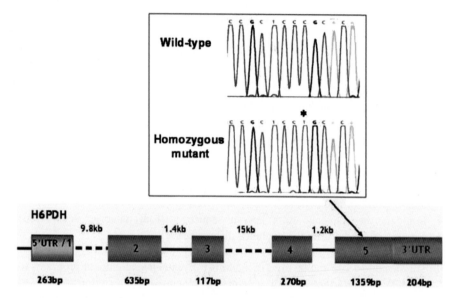

Fig. 3.1 Diagrammatic representation of H6PDH gene showing position of mutation. Chromatogram showing single C-T base pair mutation in exon 5 of H6PDH, which causes the change R453Q: (top) wild type; (bottom)- mutant; *indicates base change (*see* **Note 25**)

between sequences caused a homozygous mutation in exon 5 of the gene H6PDH, resulting in an amino acid change (R to Q), identified by BLAST analysis [27].

Individual BLAST analyses are not practical for analyzing multiple sequence results. For multiple sequence analyses, a number of programs are available whereby the known sequence can be uploaded and batches of unknown sequences can be aligned and compared to the known sequence, for example, DNAstar [28].

3.6   Interpretation of Sequencing Chromatograms

3.6.1   Overview

This section explains how to examine and interpret a DNA sequencing chromatogram. First, when examining an electropherogram, it is best to get an overview of how clean the sequence is. Second, the sequence should be checked for miscalls and truncated when errors become frequent.

The sequence chromatograph should have well-defined, evenly spaced peaks, each with only one color. Figure 3.2(A) shows an example of a good sequence chromatograph. The height if each peak may vary up to threefold, and baseline (noise) peaks may be present, but with a good template and primer, they should be minimal. However, when it becomes difficult to call peaks and allocate the correct

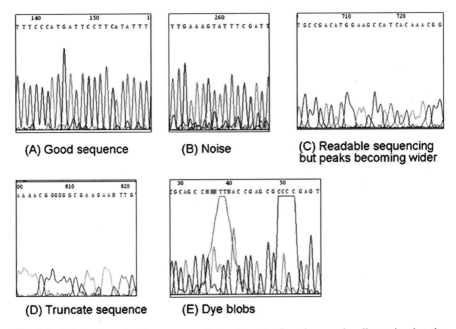

Fig. 3.2 (A) An example of an average chromatograph: there is some baseline noise, but the "real" peaks still are easy to call. (B) This example has a bit too much noise. Note the multicolored peaks and the oddly spaced interstitial peaks. (C) Readable sequence but the sequence is becoming unreliable. (D) Peaks are unreliable, the sequence should be truncated. (E) Artifact, dye blobs

base, the noise is too high. Figure 3.2(B) shows an example of a noisy chromatograph. The signal strength can be checked by viewing the signal intensity data for the chromatograph, which usually is printed on the top left hand of the trace, for example, Signal G:131 A:140 T:98 C:78. To view these numbers on the computer, open the trace file within a viewing program such as Chromas [24]. The numbers are "relative fluorescence units." Ideally, the signal intensity should be between 500 and 2,000 for best sensitivity and low noise (although this can vary significantly from instrument to instrument). If they are much less than 100, then it can be assumed that the noisy data are due to its weak signal.

Noisy data can be identified by the presence of multiple peaks and numerous Ns within the sequence. Noise can be caused by several common problems, including

1. Low DNA concentrations (*see* **Note 14**).
2. Presence of inhibitory contaminants (*see* **Note 15**).
3. Inefficient primer binding (e.g. low Tm) (*see* **Note 16**).

However, sometimes there is a complete lack of sequence data. This can be due to following reasons:

1. Lack of priming site in the sequence (*see* **Note 17**).
2. Low amounts or absence of DNA or primer in tube (*see* **Note 18**).
3. Inhibitory contaminants (*see* **Note 19**).

3.6.2   Miscalled Nucleotides

Sequences should be checked for any obvious errors in the base calling. Search the electropherogram for small peaks, N calls, and any misspaced peaks or nucleotides. Nucleotides that have been erroneously inserted into a sequence often appear to be oddly spaced relative to their neighboring bases, usually too close. Another common problem arises when the base caller attempts to interpret a gap in the sequence as a real nucleotide and calls a background peak from the baseline.

3.6.3   Heterozygous (Double) Peaks

Heterozygosity is displayed as a double peak at a single position, that is, two peaks overlaying the same position. This is common when sequencing a PCR product derived from diploid genomic DNA, where polymorphic positions show both nucleotides simultaneously. Usually the two peaks show 50% intensity of a single neighboring peak (homozygous) (*see* **Note 20**). Figure 3.3 shows an example of a PCR amplicon from genomic DNA, with a clear heterozygous single-nucleotide polymorphism. In this case, one allele carries a C, while the other has a T. However, a heterozgote can be missed by the base caller as the text sequence may show only a C. If all your other sequences also had a C at that location, the heterozgous SNP would be missed (*see* **Notes 21 and 22**).

The presence of multiple peaks within the sequence can be caused by numerous other factors that are not due to sequence polymorphisms. To help determine the

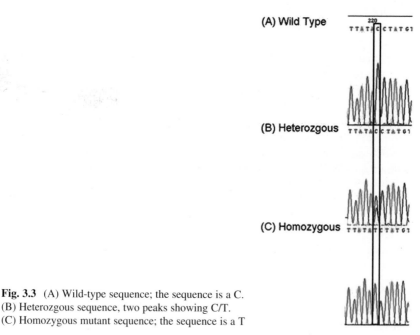

Fig. 3.3 (A) Wild-type sequence; the sequence is a C.
(B) Heterozgous sequence, two peaks showing C/T.
(C) Homozygous mutant sequence; the sequence is a T

cause, it can be useful to look at two aspects, where the multiple peaks begin and the overall signal strength of the sample (*see* **Note 23**).

3.6.4   Loss of Resolution Later in the Gel

Chromatograms stop giving accurate data toward the end of the trace, because as the gel progresses, it loses resolution. This is normal as peaks broaden and shift, making it difficult for the base caller program. The sequencer will continue attempting to "read" the data, but errors become more and more frequent. Generally, up to approximately 800 nucleotides, the sequence is quite reliable. The peaks will be broader and less well-resolved, but there still will be evident separation between them and usually few miscalls. Figure 3.2(C) shows data from later regions in a normal chromatogram. Approximately 900–1,000 nt is the very limit of resolution on an ABI 3700 DNA analyzer. It is possible to get only a general sense of the sequence at this point, as only a few base calls can be considered reliable. As the sequence progresses, the errors just described become more and more frequent. Where this occurs depends primarily on the quality of the template. When the error rate is too high, the remaining data should be ignored for analysis (*see* **Note 24**). Figure 3.2(D) shows an example of sequencing data where the chromatograph is truncated.

3.6.5   Truncated Sequences

Truncated sequences are characterized as abrupt or gradual and can be caused due to any of the following reasons:

1. Secondary structure (*see* **Note 25**).
2. Too much DNA (*see* **Note 26**).
3. Salts (*see* **Note 27**).

3.6.6   Homopolymeric Regions

Homopolymeric regions contain long stretches of a single nucleotide and can be difficult to sequence through accurately. Sequence data up to and including the polynucleotide region may be fine, but the last base of the poly region and all peaks following it may show a wavelike, stuttering pattern of double peaks that cannot be interpreted (*see* **Note 28**).

3.6.7   Sequencing Artifacts

The capillary-based system appears to be more susceptible to sequencing artifacts than the original slab-based gel sequencers, mostly due to the increased sensitivity of the instrument. There is a link between template preparation and the degree to which artifacts can be a problem. Dye blobs are a common sequencing artifact, caused by unincorporated dye terminator molecules that passed through the cleanup columns and remain in solution with the purified DNA loaded onto the sequencers. They are seen most often with samples that have low signal strength. In general, dye blobs appear as broad, undefined peaks with the true DNA peaks underneath and tend to occur relatively early in the data. An example of a dye blob is shown in Fig. 3.2(E).

4   Notes

1. Always start using a freshly grown culture (inoculate the previous day on an LB agar plate containing the appropriate antibiotic).
2. Elute DNA in water. Do not use a Tris-EDTA buffer, as EDTA inhibits the polymerase.
3. Cycle-sequencing reactions are made up in a final volume of 20 μL. This volume allows for up to 8 μL of DNA template and 4 μL of primer (0.8 pmol/μL). If the sample DNA is not concentrated enough and you need to add more than 8 μL of DNA template, then compensate for the additional volume by using a more concentrated solution of primer.
4. For each DNA sample, it is advisable to prepare a forward and reverse reaction covering the same region. This enables each sequence to be checked by comparing the forward- and reverse-complementary sequences to ensure 100% complementarity.
5. The Terminator reaction mix supplied with the Cycle Sequencing kit can be diluted to increase the number of samples that can be labeled. Dilution can provide for more than 200+ reactions from a 100 reaction kit (typical dilution: 4 μL of Reaction Mix + 4 μL of dilution buffer = 8 μL).
6. Due to variation in thermal cyclers it is recommended to test a couple of samples using the new program before preparing a large number of samples.
7. After cycling is completed, samples can be stored at −20 °C to prevent excitement of the fluorescent dyes.
8. To purify the sequencing samples, a number of high-throughput 96-well kits are available, such as UltraClear Sequencing Reaction Clean-Up Kit (Sigma, UC9601) or Centri-Sep 96-well plates (ABI, 4367819).
9. Precipitation times of <15 min result in a loss of very short extension products. Precipitation times >24 h increase the precipitation of unincorporated dye terminators.
10. The supernatant must be removed completely, as unincorporated dye terminators are dissolved in them.
11. Do not overdry.

12. After drying, it is recommended that 10 µL of Hi Di Formamide is added to each sample and left for 30 min to allow the DNA to resuspend before loading onto the 3700 DNA Analyzer.

13. A number of other sequence analysis programs are available that allow chromatograph visualization, such as DNAstar [28]. This has a sequence analysis program, called *Lasergene*, which is a comprehensive software for DNA and protein sequence analysis, contig assembly, and sequence project management. BioEdit (29) is another biological sequence alignment editor written for Windows 95/98/NT with many convenient features, making alignment, manipulation, and viewing of sequences relatively quick and easy.

14. Check the stock concentrations, calculations and dilutions.

15. The sample must be repurified.

16. In the cycle sequencing reaction, the primer and template annealing step occurs at 50 °C. Therefore, increasing the primer Tm by adding additional bases to the 5′ or 3′ end to raise the Tm to be within the range of 52–58 °C provides better sequence quality.

17. Check the plasmid maps and sequences to make sure the priming site is present in the vector.

18. Check the stock concentrations and dilutions.

19. The sequencing reaction is very sensitive to the presence of certain contaminants, some of which completely inhibit the sequencing enzyme. The sample may need to be repurified to sufficiently remove the inhibitory components. The following table is a list of potential inhibitors and the amounts that are tolerable (30):

Contaminant	Amount tolerated in sequencing reaction
RNA	1 µg
PEG	0.3%
NaOAc	5–10 mM
Ethanol	1.25%
Phenol	0%
CsCl	5 mM
EDTA	0.25 mM

20. Note that the basecaller may mark the base position as an N, or it may call only the larger of the two peaks.

21. To detect all of the polymorphic positions in a large SNP-detection project, a computer program should be used to scan the chromatograms; for example, the phred/phrap/polyphred/consed suite of programs.

22. To obtain good sequencing results, it is important to download and examine the chromatogram. By using the text data alone, many anomalies maybe missed and the data may be invalid.

   (a) In multiple priming sites involving vectors, the primer may have a secondary hybridization site that may be identical or closely related, with different nucleotide sequences following each site, giving superimposed bands within the sequence. In this case, a new primer design is required.

(b) In mixed plasmid prep, a plasmid prep that is contaminated by more than one product, such as two vectors, generally shows an early section of clean sequence data (common vector multiple cloning site sequence) followed by double peaks. Check that only a single colony is picked.

(c) A frame shift mutationcan occur when one or more bases are inserted or deleted into the template DNA and if multiple products are present in the sample. There is clean sequence up to the point of the mutation, followed by double peaks caused by the shift in the nucleotide sequence.

23. For SNP detection, only high-quality sequence data can be used.

24. GC-rich DNA is predisposed to secondary structure formation, causing hairpins that can restrict the passage of the sequencing polymerase and thus be very difficult to sequence through reliably. A secondary structure may appear as a sharp termination of signal with no sequence data after, or if the loop has been relaxed slightly, there may be some weaker peaks following that are still quite accurate. Adding a DNA denaturant such as DMSO to the sequencing reaction can help melt the duplex formation and allow the polymerase to pass through. Changing the cycle sequencing parameters to include a higher denaturation temperature (98 °C vs. 96 °C) and elimination of the 50 °C annealing step sometimes is useful.

25. Overloading of DNA exhibits early top-heavy peaks followed by rapidly weakening peak height and strength, leading to premature termination of signal.

26. Excessive amounts of salts give rise to premature termination with strong signal followed by progressively weakening signal. Salts have an inhibitory effect on the sequencing Taq polymerase, which can lead to an overabundance of short fragments. If salts are potentially a problem, an ethanol precipitation could be performed.

27. In repetitive regions, the nucleotide composition as well as the size of a repetitive region can play a large role in the success of sequencing through the area. In general, G-C and G-T (often seen in bisulfite-treated DNA) repeats are the most troublesome. Usually, it is possible to sequence partially through the repetitive region but the signal then begins to fade and eventually becomes unreadable. This may be due to premature dNTP depletion, secondary structure formation, or enzyme slippage. When sequencing cloned DNA with a homopolymer region, several options can be tried. In the BigDye Terminator sequencing chemistry, dTTP has been replaced with dUTP, which lowers the melting temperature of the DNA. An alternative sequencing chemistry can be used: dRhodamine chemistry, where dTTP is still in the reaction and generally gives better results through polyA regions. The addition of DMSO or alterations in cycle sequencing parameters can help. If the repeat region is not excessively large, sequencing from the opposite strand to complete the region can be successful.

References

1. Venter JC, Adams MD, Myers EW, Li PW, Mural RJ, Sutton GG et al (2001) The sequence of the human genome. Science 291:1304–1351
2. Hirschhorn JN, Lohmueller K, Byrne E, Hirschhorn K (2002) A comprehensive review of genetic association studies. Genet Med 4:45–61
3. Tabor HK, Risch NJ, Myers RM (2002) Opinion: Candidate-gene approaches for studying complex genetic traits: practical considerations. Nat Rev Genet 3:391–397
4. Kwok PY (2001) Methods for genotyping single nucleotide polymorphisms. Ann Rev Genomics Hum Genet 2:235–258
5. The International SNP Map Working Group (2001) A map of human genome sequence variation containing 1.42 million single nucleotide polymorphisms. Nature 409:928–933
6. Cargill M, Altshuler D, Ireland J, Sklar P, Ardlie K, Patil N et al (1999) Characterization of single-nucleotide polymorphisms in coding regions of human genes. Nat Genet 22:231–233
7. Halushka MK, Fan JB, Bentley K, Hsie L, Shen N, Weder A, Cooper R, Lipshutz R, Chakravarti A (1999) Patterns of single-nucleotide polymorphisms in candidate genes for blood-pressure homeostasis. Nat Genet 22:239–247
8. Shamsher MK, Chuzhanova NA, Friedman B, Scopes DA, Alhaq A, Millar DS, Cooper DN, Berg, LP (2000) Identification of an intronic regulatory element in the human protein C (PROC) gene. Hum Genet 107:458–465
9. Drysdale CM, McGraw DW, Stack CB., Stephens JC, Judson, RS, Nandabalan K, Arnold K, Ruano G, Liggett SB. (2000) Complex promoter and coding region beta 2-adrenergic receptor haplotypes alter receptor expression and predict in vivo responsiveness. Proc Natl Acad Sci USA 97:10483–10488
10. Cooper DN (2002) Introns, exons and evolution. Human Gene Evolution 3:107–138
11. Collins FS, Guyer MS, Chakravarti A (1997) Variations on a theme: Cataloguing human DNA sequence variation. Science 278:1580–1581
12. Lander ES (1996) The new genomics: Global views of biology. Science 274:536–539
13. Risch N, Merikangas K (1996) The future of genetic studies of complex human diseases. Science 273:1516–1517
14. Kwok PY, Carlson C, Yager TD, Ankener W, Nickerson DA (1994) Comparative analysis of human DNA variations by fluorescence-based sequencing of PCR products. Genomics 23:138–144
15. Powell BL, Haddad L, Bennett A, Gharani N, Sovio U, Groves CJ et al (2005) Analysis of multiple data sets reveals no association between the insulin gene variable number tandem repeat element and polycystic ovary syndrome or related traits. J Clin Endocrinol Metab 90:2988–2993
16. Draper N, Powell BL., Franks S, Conway GS, Stewart PM, McCarthy MI (2006) Variants implicated in cortisone reductase deficiency do not contribute to susceptibility to common forms of polycystic ovary syndrome. Clin Endocrinol (Oxf) 65:64–70
17. Speight G, Turic D, Austin J, Hoogendoorn B, Cardno AG, Jones L et al (2000) Comparative sequencing and association studies of aromatic L-amino acid decarboxylase in schizophrenia and bipolar disorder. Mol Psychiatry 5:327–331
18. Fakhrai-Rad H, Pourmand N, Ronaghi M (2002) Pyrosequencing: An accurate detection platform for single nucleotide polymorphisms. Hum Mutat 19:479–485
19. Ye S, Dhillon S, Ke X, Collins AR, Day IN (2001) An efficient procedure for genotyping single nucleotide polymorphisms. Nucleic Acids Res 29:e88-8
20. Kuklin A, Munson K, Gjerde D, Haefele R, Taylor P (1998) Detection of single-nucleotide polymorphisms with the WAVE DNA fragment analysis system. Genet Test 1:201–206
21. O'Donovan MC, Oefner PJ, Roberts SC, Austin J, Hoogendoorn B, Guy C, Speight G et al (1998) Blind analysis of denaturing high-performance liquid chromatography as a tool for mutation detection. Genomics 52:44–49

22. Wang L, Luhm R, Lei M (2007) SNP and mutation analysis. Adv Exp Med Biol 593:105–116
23. At www.appliedbiosystems.com
24. At http://www.technelysium.com.au/chromas_lite.html
25. At http://www.ncbi.nlm.nih.gov/BLAST
26. At http://www.ncbi.nlm.nih.gov/Genbank/index.html
27. Draper N, Walker EA, Bujalska IJ, Tomlinson JW, Chalder SM, Arlt W, Lavery GG et al (2003) Mutations in 11β-hydroxysteroid dehydrogenase type 1 and hexose-6-phosphäte dehydrogenase interact to cause cortisone reductase deficiency. Nat Genet 34:434–439
28. At http://www.dnastar.com
29. At http://www.mbio.ncsu.edu/BioEdit/bioedit.html
30. At http://www.roswellpark.org/Site/Research/Shared_Resources/Biopolymer_Resource/DNA_Sequencing/Sequencing_Basics

Chapter 4
BeadArray-Based Genotyping

Helen Butler and Jiannis Ragoussis

Abstract There is a demand for technologies that allow the interrogation of large numbers of SNP polymorphisms, both in whole-genome panels and in smaller custom designed sets, to attempt to elucidate the nature of complex disease through linkage and association studies. The Illumina BeadArray technology offers a flexible platform for such analyses through two assays: GoldenGate, as described in this chapter, and Infinium II. Both assays utilize the BeadArray technology, where targeted regions of DNA are immobilized on beads randomly arranged into arrays, and the SNPs visualized through fluorescent tags, which differentiate among alleles. The platform is scalable such that mid- to high-throughput projects can be designed and completed in either a manual or automated mode. Small amounts of starting material are required, and in the case of GoldenGate, some level of sample degradation can be tolerated. The assays are efficient and reliable giving high pass rates and concordance levels.

Keywords Genotyping, SNP (single nucleotide polymorphism), ASO (allele- specific oligonucleotide), LSO (locus-specific oligonucleotide)

1   Introduction

The completion of the human genome sequence in 2001 [1] and many other plant and animal genomes since then has paved the way to identify and record genetic variation. The most frequent forms of genetic variation are single nucleotide polymorphisms (SNPs). Several laboratories contributed to the identification of SNPs in human and other organisms (Celera, Baylor College of Medicine, Sanger Institute, The Whitehead Institute for Biomedical Research/MIT, and many others) at present resulting in about 25 million validated SNP entries for 15 organisms in db SNP (http://www.ncbi.nlm.nih.gov/projects/SNP/snp_summary.cgi). The SNPs have been used to investigate human variation in different populations within the framework of the human HapMap project

From: Methods in Molecular Biology, Vol. 439: *Genomics Protocols: Second Edition*
Edited by Mike Starkey and Ramnath Elaswarapu © Humana Press Inc., Totowa, NJ

(http://www.hapmap.org), which produced 3.5 million mapped SNPs [2]. In turn, these developments enabled the use of SNPs to study the genetics of complex disease [3]. To conduct such studies in a cost-efficient way, it was critical to develop reliable, high-throughput, low-cost genotyping assays [4]. Illumima used its proprietary BeadArray technology to develop the high-throughput genotyping assays GoldenGate [5] and Infinium [6, 7].

The GoldenGate assay is able to combine high levels of multiplexing (typically 384–1536 plex SNP assays) with 96 or 16 sample throughput capability. The 96 sample throughput is achieved by using the fiber optic Sentrix arrays, while the 16 sample assays are realized using a chip format. The method is based on allele-specific primer extension followed by oligonucleotide ligation and generic PCR for signal amplification. All reaction steps (Fig. 4.1) take place in a single tube and include biotinylation of the DNA and immobilization on streptavidin-coated particles, followed by annealing of three primers per SNP: one locus-specific (LSO) and two allele-specific oligonucleotides (ASO). The next step includes an extension reaction from the ASO toward the LSO followed by ligation. All oligo-nucleotides have sequences allowing a PCR step using generic primers as well as sequences complementary to the tags present in the beads. The successfully extended and ligated products are amplified by PCR with fluorescently labeled primers. The PCR products are denatured and hybridized to arrays of beads carrying sequences complementary to the locus-specific tags. A specifically developed 5-μm resolution scanner is used to detect the 30-μm coded fluorescent beads, and specially designed BeadStudio software is applied to produce the genotype calls. The technology offers excellent pass rates and compares well with alternative methods [8].

2   Materials

1. Hettich Rotanta 460 Bench Top Ambient Centrifuge (Jencons Scientific, Leighton Buzzard, UK), including 194–580, A5624, Swing-Out Rotor for 4 × 750 ml buckets.
2. 2×194–591, Microplate bucket for 194–580 motor, 2×194–593, carrier for 6×15 mm high microplates.
3. SpectraFluor Plus Picogreen Reader (Tecan, Mannerdorf, Switzerland).
4. 96-well microplates (black) 655076 Greiner Bio-One International AG, Kremsmuenster AUT).
5. Combi Thermosealer 240V AB-0384/240 (ABgene, Epsom, UK).
6. AB-0563/1000, 96-Well PCR Plate Carrier (ABgene).
7. AB-0724, 384-Well PCR Plate Carrier (ABgene).
8. LTS multichannel pipettes L8-200 and L8-20 and tips SR-L10F and SR-L200F (Rainin Instruments, Woburn, MA).
9. Thermo-Fast 96, Skirted Plate AB-0800 (ABgene).

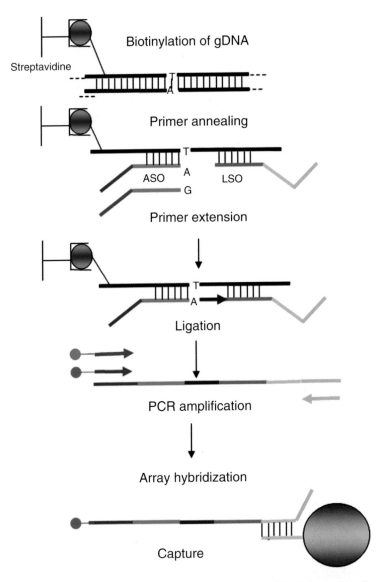

Fig. 4.1 The molecular biology steps in the GoldenGate assay. ASO allele specific oligonucleotide; LSO = locus specific oligonucleotide. In gray, blue/green, and yellow, generic PCR primers. Light blue sequence is the capture sequence (modified from [8])

10. Easy Peel Heat Sealing Foil AB-0745 (ABgene).
11. 384 well plates TF-0384 (ABgene).
12. Adhesive PCR film AB-0558 (ABgene).

13. Platesealers (Greiner Bio-One International AG).
14. Corning reservoirs CC901 (Appleton Woods, UK).
15. 384-well universal microplates (Labsystems Cliniplate Clear, ThermoElectron Corporation, Waltham, MA).
16. Multiscreen HTS 96 well filtration plate MSHVN4510 (Millipore, UK).
17. OmniTrays 242811 (Nunc, Scientific Laboratory Supplies, UK).
18. Tris-EDTA buffer (TE): 1 M Tris-HCl, pH 8.0, 0.1 M EDTA.
19. Titanium Taq polymerase (Takara Bio, Shiga 520–2193, Japan).
20. Quant-iT High-Sensitivity DNA Assay Kit (Invitrogen, Paisley, UK).

3   Methods

3.1   PicoGreen Assay for DNA Quantification

PicoGreen dsDNA quantitation reagent is an ultrasensitive fluorescent nucleic acid stain for quantifying double stranded DNA (dsDNA) in solution. It does not bind to single stranded DNA, RNA, proteins, or salts and, hence, is one of the most accurate and reliable ways to quantitate dsDNA. The linear detection range of the dye in the kit is between 25 pg/mL and 1,000 ng/mL of dsDNA. The quantitation assay is performed according to the manufacturer's instructions.

1. Prepare a standard curve (of DNA concentration versus a "raw fluorescence unit") using lambda DNA standard solutions (0, 5, 10, 20, 60, 100, 140, and 200 ng/mL).
2. Measure the fluorescence of duplicate samples of each DNA, and calculate the mean of the DNA concentration values derived from the standard curve.

3.2   Sample Preparation

1. Samples must be adjusted to 60–100 ng/μL and 5 μL of each provided. It is difficult to quantify DNA accurately and reproducibly, so we found that increasing the threshold ensures samples do not fall below the minimum quantity (250 ng) required for the assay. While resuspension in TE is recommended in the protocol, we often dilute using sterile water.
2. Samples used in the HapMap project are good candidates to use for positive controls. The genotype concordance can be checked for any HapMap SNP used with these samples by downloading the data from the HapMap website (http://www.hapmap.org).
3. Include replicates, either within a single plate to verify reproducibility or in multiple plates to check for reproducibility between batches of plates processed on different dates.

4. Position the positive control samples so that they can be used to orientate the plates and ensure there is no mix up during processing.

3.3   GoldenGate Assay (see Note 1)

The full protocol is available from Illumina (Document #11171817 Rev. B). We have not found it necessary to modify the protocol, but include some user notes to accompany the protocol. The method described here is a summarized overview of the main steps. It represents a two-and-a-half day process for one person processing 2 × 96 plates (with a variation shown for processing 16-sample BeadChips), assuming that the DNA samples have been previously quantified and adjusted to the correct concentration).

3.3.1   Day 1. 0900–1600—Pre-PCR (PCR Overnight)

1. Pipette 5-µL DNA samples (at 60–100 ng/µL) into a 96-well plate for processing.
2. Heat samples in the presence of 5 µL of MS1 reagent to incorporate biotin.
3. Precipitate using 5 µL of GS#-PS1 and 15 µL of 2-propanol, and resuspend each DNA in 10 µL of GS#-RS1 (see Note 2).
4. To each sample, add 10 µL of GS#-OPA reagent (containing the mixture of allele- specific and oligonucleotide-specific probes) and 30 µL of OB1 (containing the streptavidin-coated paramagnetic beads).
5. Hybridize at 70 °C, dropping to 30 °C, over 2 h.
6. Following hybridization, remove excess oligos by several cycles of washing (using buffers GS#-UB1, GS#-AM1), while using a magnetic plate (see Note 3) to retain the hybridized products attached to the streptavidin-coated beads.
7. Perform the extension and ligation reaction, by adding 37 µL of GS#-MEL to each sample and incubate at 45 °C for exactly 15 min.
8. Add each product to 30 µL of PCR mix, containing the universal primers (which binds to the tags on the probes) and uracil DNA glycosylase for contamination control (see Note 4).
9. Perform the PCR cycling overnight, holding the products at 15 °C following thermocycling (see Note 5).

3.3.2   Day 2. 1300–1700—Hybridization to Array (Overnight)

1. Incubate the PCR products at room temperature in the dark for 1 h with 20 µL of MPB, which contains beads to bind the PCR products.
2. Separate the PCR products from the PCR reaction mixture by filter plate centrifugation, and elute single stranded PCR products off the beads (and into 30 µL

of GS#-MH1 reagent) by the addition of 0.1 M NaOH to the filter plate containing the bead-immobilized products.

From this point onward, the protocol varies slightly according to whether the array format used is the 96-well Sentrix Array Matrix (SAM) (steps 3–5), or the 16-sample BeadChip (steps 6–9) (*see* **Note 6**).

3. For SAM, pipette the products into a 384-well Cliniplate. Fill additional wells of the Cliniplate with buffer GS#-UB2 to prevent dehydration during the hybridization process (*see* **Note 7**).
4. Condition the SAM using GS#-UB2 and 0.1 M NaOH, then fit it into the Cliniplate, such that the fiber bundles are submerged in the PCR mixtures (*see* **Note 8**).
5. Clamp the SAM to the Cliniplate using the SAM alignment fixture and place in the oven for hybridization at 60 °C for 30 min, followed by 16 h at 45 °C.
6. The advantage of using BeadChips is that the processing can be scaled up in batches of 16 samples, by reducing the volumes of GS#-UB2, GS#-MH1, and 0.1 M NaOH used accordingly. Filter the PCR products in the same way as when using a SAM (step 2).
7. Following elution of the PCR products into the GS#-INT reagent, place the samples into the oven, and incubate at 60 °C for 10 min.
8. Load the PCR products onto BeadChips fitted into a Hyb Cartridge.
9. Incubate the Hyb Cartridge at 60 °C for 30 min, followed by 16 h at 45 °C.
10. Add ethanol to the GS#-WC1 reagent, and leave it on a rocking platform to resuspend overnight, in preparation for use on day 3.

3.3.3   Day 3. 0900–1400—Scanning (1000–1400 Time on Scanner)

1. For a SAM, disassemble the hybridization plate from the SAM.
2. Prepare the SAM for imaging by washing twice with GS#-UB2 (1 min each) and once with IS1 solution (5 min), finally air dry for 20 min.
3. Load the SAM into the scanner for imaging (each SAM takes approximately 2 h to scan).
4. For BeadChips, prepare the BeadChips for imaging by washing twice with GS#-UB2 (5 min each) and once with GS#-WC1 solution (5 min), finally air dry for 5 min.
5. Lower a coverslip on top of $2 \times 18 \mu L$ of reagent GS#-AC1 (*see* **Note 9**), and after 5 min, load the BeadChip into the scanner for imaging (each BeadChip takes approximately 1 h to scan).

3.4   Analysis of Custom Assays

Following scanning of the arrays using the BeadScan software, raw data is obtained in the form of image files (IDAT) containing intensity data for the Cy3 and Cy5 fluorescence signals for each bead. Data are analyzed using Illumina's BeadStudio software (currently version 2.3.43). The following is a suggestion for the process to review and edit the data, finally printing a report containing the genotypes. This has most relevance when analyzing custom assays for which no cluster file is provided. Cluster files contain the positions of the call region (shaded area) for the three genotypes. These positions are specific for each assay and, once they have been determined for one set of samples, can be saved as a file, imported, and used for new sample sets. Illumina provides cluster files for some standard content panels (see Sect. 3.7 for more details on the use of cluster files).

3.4.1 Create a New Project

Load all samples into the project (use the sample sheet to include as much information about samples as possible, such as position of replicates, family relationships, genomic (gDNA) or whole genome amplified (WGA) DNA and sample group). Duplicate samples should be given unique names. This ensures that both samples are entered in the final report rather than just retaining a single genotype call for the duplicates per assay. This naming convention also facilitates tracking. The "sample group" field can be used to group the samples such as "control," "patient," and so forth; then the final reports can be generated for specific groups rather than the whole data set.

3.4.2 Run the Clustering Tool

This algorithm automatically clusters the samples into three groups where possible, two homozygous and one heterozygous group. The samples are then color coded in the SNP graph to denote the genotype assigned for each sample. The circles around each cluster indicate 2 standard deviations from the midpoint of the cluster and the shading delimits the area according to GenCall (GC, normally set to 0.25) within which the samples are called. Samples outside the shaded region are not given a genotype (GC value below 0.25). At this point, it is good to briefly view a random selection of the SNP graphs to check that the data quality is good; that is, the sample DNAs produce reasonably tight clusters of data.

3.4.3 Review the DNA Quality and Select Samples for Exclusion

Produce a plot of p10GC for the samples and use this to identify problematic DNAs. This can be done either via a DNA report opened in Microsoft Excel with the DNAs ranked according to p10GC score and plotted as a scatter plot (Fig. 4.2) or using the "line plot" tool within BeadStudio. The p10GC is the GC score for a SNP at the tenth percentile position when GC scores are ranked per sample.

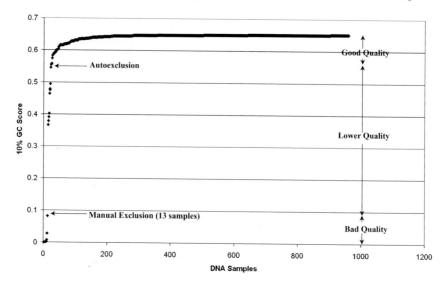

Fig. 4.2 A p10GC plot to asses DNA quality; used to determine samples to exclude from analysis. The plot is for a project containing 960 DNAs. The cutoff for the autoexclusion (22 samples) and manual exclusion (13 samples) is indicated by the arrow

The GC score is a confidence value assigned to the sample based on the distance of the sample from the midpoint of the cluster. It is a good indication of DNA performance in the assays with low scores indicating poor-quality DNA. Typically, the vast majority of samples are of high quality, have similar p10GC scores, and form a plateau of data points on the graph. In addition, a small number of samples will have considerably lower p10GC scores. As shown in Fig. 4.2, often a clear distinction is found between the very poorest-quality DNAs with the lowest p10GC scores and those of the main group. In this way, the distribution of p10GC scores can be used as a guide to aid the selection of samples for exclusion. This should be used alongside the "0/1" column of the "DNA Report," which indicates the samples to be excluded with a 0. The DNAs with very low p10GC scores (generally below 0.3) normally are excluded ("Bad DNA"). In the example that follows, this means excluding 13 samples from the project, containing 960 samples overall.

The samples that are of medium quality (i.e., that have p10GC scores significantly below the majority of samples) should be flagged in the SNP graph by using the "mark" sample function. This allows you to observe their positions when manually editing the cluster positions later in the analysis. We do not use the "autoexclude" function within the BeadStudio but rather employ the method just described, as the autoexclusion criteria appears to be more severe than necessary. In the example in Fig. 4.2, autoexclusion would have removed 22 samples out of 960. Often it is possible to retain samples that otherwise would be excluded (through the autoexclusion function) by basing the judgment on the p10GC plot and the 0/1 column.

Once a list of low-quality DNAs has been defined, these samples should be excluded using manual exclusion method, that is, the "exclude selected sample" feature from the "samples table." The assays then are reclustered without the low quality samples, so they have no bearing on the position of the groups. In addition, it may also be worth removing the medium-quality DNAs for the purposes of clustering then reincluding them after clustering. In this way, these samples will have no bearing on cluster position and can be genotyped after clustering.

3.4.4   Check the Integrity of the Data

It is possible to incorrectly orientate the plate during processing or hybridize the wrong array to a plate. While you can safeguard against this by careful execution of the laboratory work and implementing appropriate tracking procedures, you also can use sample information to confirm no mix-ups occurred. Check data integrity using gender check, inheritance, and replicates to see that the samples are located correctly in the plates for each Sentrix array. To check the gender, use the control panel for Gender and highlight the samples matching the U3, which are expected to be the females. These samples then are highlighted in the sample table, and this can be sorted and checked to verify there are no errors in gender assignment. To check the inheritance and replicate errors, produce a "Reproducibility and Heritability Report" under the analysis function. This report can be sorted by frequency to show samples with high Mendelian inheritance, or replicate error rate. If the discrepancies are biased toward one sample plate, this indicative of a plate mixup, whereas errors affecting a small number of samples within a plate suggests other causes that could be related to DNA quality or quantity.

3.4.5   Review and Edit Clusters

The most useful view of the clusters is the normalized polar SNP graph. It usually is necessary to examine only the nonnormalized polar, or the Cartesian plot, if the SNP needs more detailed editing. The main criteria used to assess the quality of a SNP assay is the GenTrain score (this is a quality score, 0–1, for the SNP assay). Sort the assays in the SNP table according to GenTrain score. SNPs that have a low GenTrain score, below 0.5, should be manually inspected. Replicate and inheritance errors are extremely useful means of identifying problematic clusters. Subsequently, "cluster separation," "heterozygous xs," "call freq," "ChiTestP100," "minor freq," and "AA freq" can be used sequentially to sort the assays and examine the outliers to identify cluster problems. If the pattern of data points cannot be reasonably interpreted either by the Clustering tool or via manual editing, then the SNP should be excluded using the "Zero selected SNP" function (Fig. 4.3). Some SNPs require refinements to the position of the clusters based on manual inspection. This is because the automatic clustering tool is not always able to interpret the data in the best possible way. The shape of the cluster can be altered by dragging the edges of

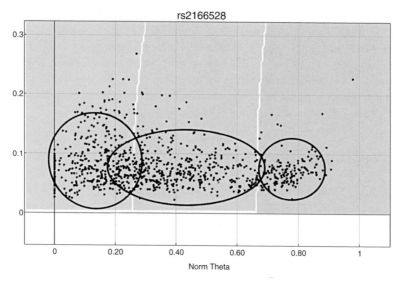

Fig. 4.3 A failed SNP assay. Three distinct clusters could not be defined and the intensity is very low. This assay should be excluded from the analysis as the data are unreliable

the cluster to the most appropriate position and shape. Editing like this often is required for SNPs where the samples do not contain representatives for all three genotypes; that is, one group is missing. In this case, the automatic clustering tool does not always correctly interpret the data. In addition, fine changes to the cluster shape are required to include or exclude samples that lie between groups (Fig. 4.4).

3.4.6 Types of Clusters

The Edited column on the SNP table enables you to track which SNPs have been modified. The Aux Value column also may be used to further track the editing process. You can use the Set Aux Value to record additional information about the types of cluster you reviewed. An example of how you would use this is illustrated next, using a project containing 1,536 custom SNPs with 960 DNAs. The SNPs are given an aux number to represent the type of clustering pattern they display (see Table 4.1).

———➤

Fig. 4.4 (a) Interpretation of data by the automatic clustering tool. Note that a replicate error (different genotypes assigned to two replicates) is indicated by the square. (b) The same data edited by changing the shape of the heterozygous cluster to exclude two samples between the heterozygous and homozygous group on the right. In addition, the replicate error has been removed. It is also interesting to note that, in this case, the replicate error would be removed if the sample were included as part of the homozygous group, but instead it was decided to exclude the two samples, as they appeared almost equidistant between the two groups

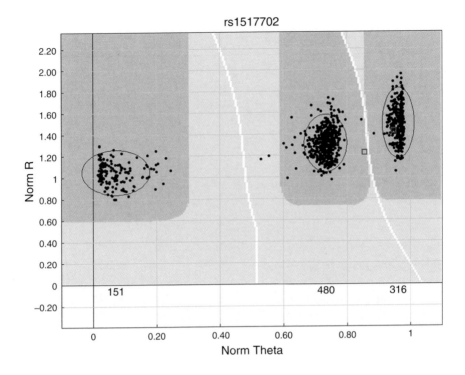

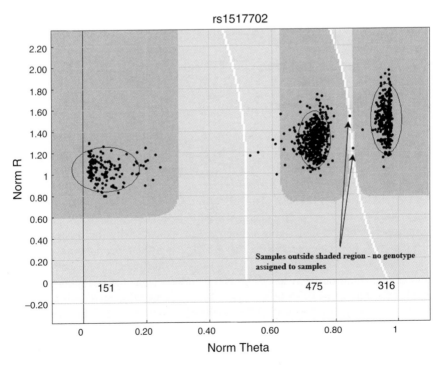

Table 4.1 Example of cluster types produced in a 1,536 custom OPA. The problematic assays have been sorted according to type of problem

Cluster Type	Aux Value	No of SNPs of This Type
Excluded	−1	13
Unaltered	0	1,451
Many inheritance problems	1	4
Three groups not accepted	2	3
Poor separation	3	26
Four groups	4	15
Five groups	5	3
Low intensity	6	2
No third group	7	2
Monomorphic	8	10
Misc		7
		72

3.4.7 Generate the Report

Once the editing is complete, a report can be generated using the Final Report tool within the software. In the output file, it is useful to include the GC score for the genotypes. This provides an indication of confidence in the calling when analyses are performed downstream with the data. The final report file normally exceeds the capacity of Excel, so the data need to be uploaded into a database (or alternative) to allow further work with the data.

3.5 Problems for Analysis

3.5.1 Inheritance Problems

The expectation is that, for each assay, the samples will cluster into three groups, with equal intensity for both alleles and the clusters separated well along the Norm Theta with a distribution that fits with Hardy-Weinberg equilibrium, given the constraints of the sample set. For a given project, a number of assays will not fit this expected cluster pattern. These atypical cluster patterns can be grouped into common types (see Table 4.1). These assays need to be reviewed manually, and subsequent analysis of the underlying sequence can indicate why the assays are not working well. For example, SNP rs10187687 displayed more inheritance errors than other assays in the project. The clusters, in the main, were well defined, although the AA group was spreading and there was an additional group of failed samples for this assay. The primers for this assay were aligned to the genome sequence using "Blat" at the UCSC Genome Bioinformatics Browser (http://genome.ucsc.edu) (Fig. 4.5). We can identify an additional SNP (rs13432838 [A/G]), located adjacent to the variable 3′ base on the ASOs. It can be hypothesized

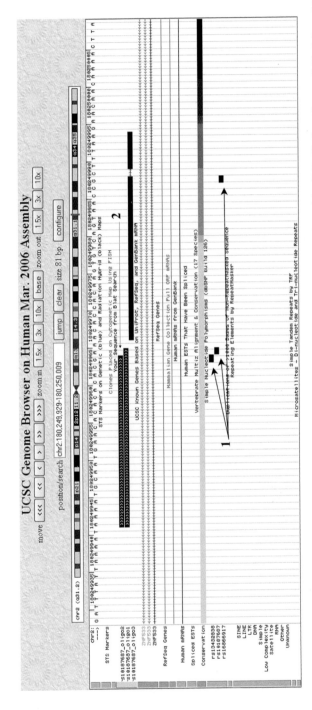

Fig. 4.5 The Blat (UCSC Genome Browser) result for rs10187687. Arrow 1 indicates DNA as the position of the SNPs (indicated as small black boxes) in the sequence. The top left SNP created a mismatch between the allele-specific oligonucleotide and genomic DNA as shown at arrow 2

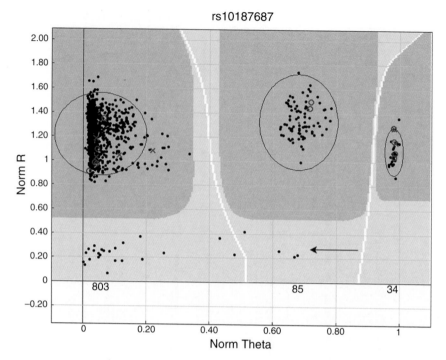

Fig. 4.6 The SNP Graph for rs10187687. Inheritance problems are highlighted, while a number of samples have failed due to low intensity and are excluded from the cluster areas

that a proportion of the samples may contain the A variant, in which case the ASOs do not bind at this base, and subsequently it is likely that the assay will not work. This may explain why there were a high proportion of failed samples for this one assay (Fig. 4.6). In addition, there is another SNP (rs16866917) within the priming sequence of the LSO (22 bases along 5′ to 3′). Adjacent to this base is a cytosine nucleotide in the genomic sequence (according the UCSC Genome Browser) that is not in the sequence of the oligonucleotide. Together, these irregularities in the LSO may reduce its efficiency to specifically hybridize at this site. Finally, there is an incorrect 5′ base for both ASOs, although this is unlikely to have a significant effect on the assay performance. These problems can occur if different builds of the genome sequence are used to design assays and information about the presence of SNP is not available when the assays are designed.

3.5.2 Poor Separation

The problem in genotyping of being left with a large number of samples that cannot be accurately genotyped (illustrated in Fig. 4.7) is caused by the inability to clearly separate clusters. In such a case, it again is important to check the primer sequences

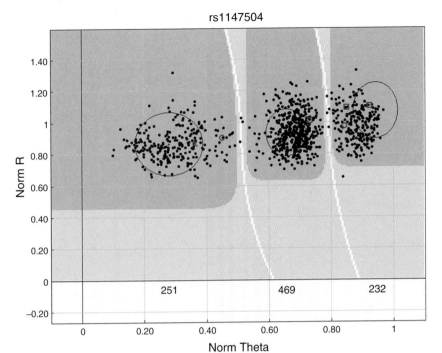

Fig. 4.7 An SNP graph with poorly separated clusters for assay rs1147504

on the genomic sequence using Blat. In the example shown in Fig. 4.8, the ASO had multiple binding sites throughout the genome due to the simple repeat (TGA) in the binding site for ASOs. In addition, there was an additional SNP within the binding site (rs 35947407), which was a deletion insertion polymorphism for [–/TGA]. When the deletion is present, the ASO does not match the sequence at the 5′ end, leaving only 21 bases for specific binding. Hence, the specificity for the desired region of chromosome 7 may be so reduced as to render the assay unusable. Normally, the SNP design output file lists the SNP assays and ranks them with a descending design score. Another indicator is the validation bin from the assay design. It is a good guide for SNP selection, although this is not always a perfect indicator that an assay will work. The SNP in the preceding example was "Goldengate validated" in this OPA pool.

3.5.3 Assay-Specific Problems

These include the presence of four or five groups, the absence of a third group, and low overall intensity. In addition, SNPs may be monomorphic in the population studied.

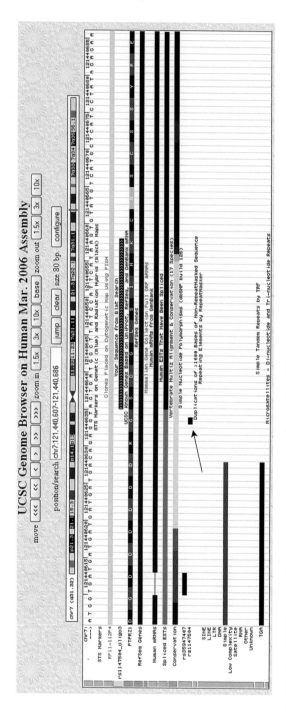

Fig. 4.8 Blat (UCSC Genome Browser) result using the sequence of the allele-specific oligonucleotide indicating an overlap with a deletion/insertion polymorphism at the 5' end (arrowed)

3.5.4 Whole-Genome Amplified DNA

WGA DNA samples generally cluster well with genomic DNA, but on occasion, this sample type may behave differently and form clusters in a different position to gDNA (Fig. 4.9). In such a case, select only the WGA samples and cluster them separately from the gDNA samples.

3.6 Strategy for Problem Solving

1. Review poorly clustered SNPs manually and identify those with high Mendelian inheritance errors.
2. Compare (using Blat) oligonucleotide sequences against the genome sequence to investigate underlying sequence structure. Look for additional SNPs, indels, repeats, low-complexity DNA, and nonspecific binding.
3. Investigate the possibility of a copy number variant (CNV). An example of a bad inheritance problem caused by a CNV is shown in Fig. 4.10. In this example, the inheritance error is seen only for this family within the dataset. The inheritance error occurs when one of the parents has a lower intensity for the

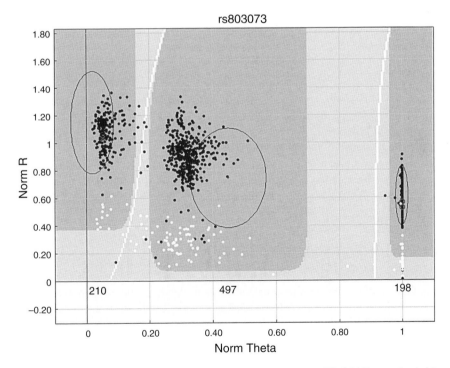

Fig. 4.9 The different clustering behavior of all whole-genome amplified DNA samples (white dots) compared to the genomic DNA samples (black dots)

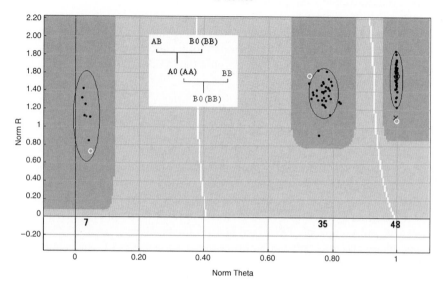

Fig. 4.10 How copy number variants (CNVs) may cause inheritance conflicts. The three family members are shown as circles. The parents appear as AB (heterozygote, middle) and BB (homozygote, right), while the child is genotyped as homozygote AA (left). The possible explanation is shown in the diagram on the left, whereby one parent carries a deletion CNV (represented by O) resulting in a BB genotype and inherited by the child, which would then be genotyped as AA. This is corroborated by the fact that the intensities of the corresponding AA and BB calls are the lowest in their respective clusters, while the heterozygote has an average intensity

homozygous group, appearing lower down on the Norm R axis, due possibly to just one copy of the allele. The offspring similarly shows a lower intensity within the group (Fig. 4.11). Interestingly, the deletion CNV seen in this example can be traced in two generations of the family.

4. If necessary, cluster WGA and gDNA samples separately. As a precaution, it is worth having a good number of either type of DNA (i.e., 96, or more, samples) to enable effective clustering of the separate groups.

3.7 Analysis of Standard Content Panels

Illumina supplies Cluster files (.egt) with their standard content panels (Linkage Panel IV, Infinium products, Mouse Linkage Panels, etc.). These cluster files have been produced by genotyping control sample DNAs and provide a guide for the expected position of the three groupings for each SNP. It is useful to import these and use them to assist the editing process. However, it is necessary to check the fit of the samples to the cluster files. For some experiments, there is very good agreement between samples and cluster positions (e.g., Fig. 4.11). However, occasionally, the agreement is not good (Fig. 4.12). One such case was encountered using

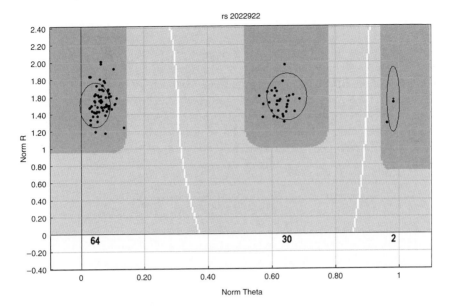

Fig. 4.11 Good fit of data (96 samples) to cluster files provided with Linkage Panel IVb

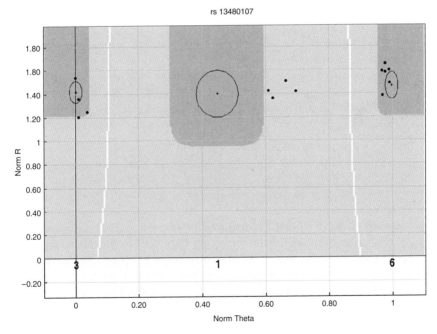

Fig. 4.12 Poor fit of data (14 samples) to Mouse LD Panel cluster files. There is a clear group of heterozygous samples that do not lie in the position predicted by the cluster file

the Mouse LD panel with 16 individuals from a litter of mice that were part of a linkage study. In this case, an alternative approach was taken. Within the group were the parental inbred mouse lines Bl6 and C3H. Rather than using the cluster files provided with the panel, the data were clustered using the clustering tool. The genotypes of the Bl6 and C3H individuals then were checked against the expected genotypes for these mouse lines (files supplied by Illumina, Mouse_LD_Linkage_ Inbred_Genotypes), and 100% concordance was found. It then was straightforward to analyze the cluster positions and adjust according to the groupings for this sample set. The heterozygous group were clearly defined in all cases. Normally, known genotypes are not available within an experiment, and so the typical process that one would follow for the analysis of SNP panels where Illumina clusters are available would be to check the fit of the experimental data to the Illumina clusters using a Gen-Call threshold set to 0.15 (rather than 0.25 for the GoldenGate assay). The fit of the data is good where the call-rate for a samples exceeds 0.98 (*see* **Note 10**). In addition, the control probes can be checked for technical anomalies in the sample processing and for consistency across samples processed in separate batches.

4   Notes

1. For the BeadArray genotyping assay to be performed effectively there are several laboratory requirements:
   - Separate pre-PCR room.
   - Clean benches with 10% bleach immediately before commencing laboratory work.
   - Wear separate laboratory coats in pre- and post-PCR areas.
   - Do not transfer items from post-PCR into pre-PCR areas.
   - Have dedicated multichannel and single channel pipettes for pre- and post-PCR areas.

2. It was not possible to set our Hettich Rotanta 460 centrifuge to the low speed required in the protocol ($8 \times g$, 1 min). Instead, we used a setting of $50 \times g$ for 6 s. The speed does not exceed $20 \times g$ during this short run. Within our laboratory, this has been found to produce good results, removing the liquid from the wells without dislodging the pellet.

3. The washing step is the rate-limiting step for the protocol, as it is important that the beads are not allowed to dry during the step. We have found that the maximum number of plates that one person can process at a time is limited to 2×96 well plates. During the washing step, the beads are resuspended by vortexing. Sometimes, the beads are difficult to resuspend, especially if any drying has occurred; and in this case, we often extend the vortex time by 1–2 min to ensure that all beads are resuspended.

4. While the addition of UDG is optional, we always choose to add it to the reactions.

5. We prefer to seal the plates using heat sealing rather than using Microseal "A" PCR plate sealing film. This reduced evaporation during the PCR step.

6. The prepared hybridization plate has to be centrifuged once the GS#-UB2 and samples have been added. The protocol recommends that this is performed at 3,000 × g for 4 min at 25 °C. However, at this speed, we found the plate is liable to crack, so we reduced the speed to 2,000 × g to overcome this problem.

7. If BeadChips are used, and therefore less than 96 samples are being processed, for the preceding steps, the volumes of the reagents used need to be reduced in proportion with the numberof samples being processed at one time. In this case, when the reagents are defrosted, aliquot them according to the batch numbers that you intend to process, thus avoiding repeated freeze/thaw cycles for each batch.

8. Whenever the SAMs are placed into an OmniTray, ensure that the level of liquid in the tray is sufficient to cover the bottom of the fiber bundles, checking that the level is uniform across the SAM.

9. We do not use the Grip 'n Strip tool but instead remove the cover seal by hand. It is possible to improvise and use an alternative (e.g., other small tips) to the gel-loading tip specified (if these tips are not available) when loading reagent GS#-AC1 onto the BeadChip.

10. Illumina modified and extended their product range, and as of beginning 2007, the Linkage Panel and other whole genome panels (240 K, 300 K, 370 K, 450 K, 550 K, 650 K, 1 M) are based on the Infinium assay, for which cluster files are provided. The principles for reviewing and editing cluster files as described in this chapter are applicable to the Infinium data but the large number of SNPs within the panels makes it unfeasible to carry out manual editing. A number of alternative calling algorithms are being developed (www.illumina.com/illuminaconnect) that may be more appropriate for calling such large data sets, especially where there is a poor fit to the cluster files provided.

Acknowledgments We thank Anna Richardson for the PicoGreen DNA quantification protocol and Laura Winchester for technical assistance. The work was funded by the Wellcome Trust.

References

1. Lander ES, Linton LM, Birren B, Nusbaum C, Zody MC, Baldwin J, Devon K, Dewar K et al (2001) Initial sequencing and analysis of the human genome. International Human Genome Sequencing Consortium. Nature 409:860–921

2. Altshuler D, Brooks LD, Chakravarti A, Collins FS, Daly MJ, Donnelly P (2005) A haplotype map of the human genome. Nature 437:1299–1320

3. Zeggini E, Rayner W, Morris AP, Hattersley AT, Walker M, Hitman GA, Deloukas P, Cardon LR, McCarthy MI (2005) An evaluation of HapMap sample size and tagging SNP performance in large-scale empirical and simulated data sets. Nat Genet 37:1320–1322

4. Syvanen AC (2005) Toward genome-wide SNP genotyping. Nat Genet 37 Suppl:S5–S10

5. Oliphant A, Barker DL, Stuelpnagel JR, Chee MS (2002) BeadArray technology: Enabling an accurate, cost-effective approach to high-throughput genotyping. Biotechniques Suppl:56–58, 60–61

6. Gunderson KL, Steemers FJ, Lee G, Mendoza LG, Chee MS (2005). A genome-wide scalable SNP genotyping assay using microarray technology. Nat Genet 37:549–554
7. Steemers FJ, Chang W, Lee G, Barker DL, Shen R, Gunderson KL (2006) Whole-genome genotyping with the single-base extension assay. Nat Methods 3:31–33
8. Ragoussis J (2006) Genotyping systems for all. Drug Discovery Today: Technologies 3:115–122

Chapter 5
Microsatellite-Based Candidate Gene Linkage Analysis Studies

Cathryn Mellersh

Abstract As the number of mammalian genomes to be completely sequenced continues to grow, researchers attempting to identify mutations responsible for inherited disease increasingly take advantage of studies undertaken in alternative species, usually humans, to identify and investigate previously identified "candidate genes" in their own species. This chapter describes the identification of microsatellites closely associated with candidate genes and their subsequent analysis in cohorts of samples segregating a disease or trait of interest for an association between the candidate genes and the condition under investigation.

Keywords candidate gene, microsatellite, linkage, association, canine, genetics

1   Introduction

As the number of mammalian genomes to be completely sequenced increases, seemingly daily, so do our opportunities to use interspecies comparison to understand how genomes evolve and the relationship between genotype and phenotype. The technical and analytical resources available with which to identify genes associated with various traits are now highly sophisticated and genomewide studies aimed at identifying genes associated with both simple Mendelian and complex traits are commonplace, at least in the field of human genetics. However, large-scale projects that involve the analysis of hundreds of thousands of single nucleotide polymorphism markers (SNPs) in thousands of DNA samples are extremely costly, usually involve the collaboration of many groups and consortia, and therefore, often are beyond the means of most modest research groups. However, investigators who study a species other than man can take advantage of extensive (and expensive) studies that have already been conducted on humans to identify genes associated with corresponding traits in their own species. Phenotypically similar conditions that afflict humans and other species may be caused by mutations in the same gene, and

From: Methods in Molecular Biology, Vol. 439: *Genomics Protocols: Second Edition*
Edited by Mike Starkey and Ramnath Elaswarapu © Humana Press Inc., Totowa, NJ

the investigation of these "candidate genes" for their role in an inherited condition under investigation can provide a shortcut to mutation identification. In our own field, that of canine genetics, many pathogenic mutations responsible for inherited conditions have been identified by simply investigating genes first associated with homologous conditions in man. Examples include the mutations that cause canine leukocyte adhesion deficiency (CLAD), a fatal immunodeficiency disease [1]; cystinuria, an inherited renal and intestinal disease [2]; and fucosidosis, a lysosomal storage disease [3, 4]. For a more extensive summary of hereditary disorders characterized at the molecular level, interested readers should read the reviews in references 5 and 6.

Unfortunately, not all candidate gene studies are successful; sometimes, conditions that appear to be homologous across species turn out to be distinct at the molecular level. In the absence of a whole genome sequence, the task of investigating a gene for its role in a disease involves complex and time-consuming cloning and sequencing stages with DNA from normal individuals to first determine the wild-type gene sequence before the gene can be sequenced in affected individuals. Testament to the effort involved to investigate a candidate gene for its role in a particular disease, prior to the genome era, are the host of papers that exclude a gene or genes for their role in a given inherited disease. In the dog, some of many available examples include the exclusion of the canine photoreceptor specific cone-rod homeobox (CRX) gene as a candidate for early onset photoreceptor diseases in the dog [7], exclusion of urate oxidase as a candidate gene for hyperuricosuria in the Dalmatian dog [8], and exclusion of the PAX3 gene as a candidate for deafness [9]. In summary, a candidate gene approach to mutation identification in the absence of a genome sequence and without any additional evidence to implicate the gene under investigation can be considered a risky strategy [10].

The availability of a genome sequence changes the situation dramatically. For any species that has had its genome sequenced, it is relatively straightforward to investigate any gene included in the sequence assembly for its involvement with a particular inherited disease for which adequate case and controls samples are available. Typically, and briefly, polymorphic markers located close to or within the candidate gene are first identified, using suitable Internet resources, and then genotyped, either in samples from a pedigree segregating the condition or in a cohort of unrelated cases and control samples. Appropriate linkage or association analysis is undertaken subsequently with the genotyping data to test for an association between the gene and the disease. If the selected candidate gene is not involved with development of the disease, the candidate gene-associated marker(s) and the disease mutation will segregate independently from one another, and therefore, be no association will be found between the two, in which case the gene can be simply excluded. If, however, statistically significant association is detected between a polymorphic marker and a disease, this provides evidence that the marker, and hence its linked gene, are located in a region associated with the disease; the gene therefore is a candidate gene for the disease.

Armed with such evidence, it is justifiable for the gene to be investigated further by using the wild-type gene sequence, available from the whole genome sequence, to design assays to determine the sequence of the gene in affected individuals and compare the affected and unaffected genes for pathogenic mutations. By the judicious use of linked markers to initially obtain linkage evidence for the involvement of a candidate gene, unnecessary and expensive sequencing experiments are avoided.

We recently used this strategy to investigate a small number of genes associated with various forms of inherited cataracts in humans for their possible role in the development of hereditary cataract (HC) in dogs. The approach lead to the successful identification of the first mutation associated with HC in the dog: a single nucleotide insertion in exon 9 of HSF4 that causes a very early onset and progressive form of HC in Staffordshire Bull Terriers and Boston Terriers [11].

2 Materials

2.1 Software or a Browser

Software or an Internet browser are needed for accessing genome sequence databases and oligonucleotide primer selection software.

2.2 DNA Extraction

2.2.1 DNA Extraction from Blood

1. Nucleon DNA Extraction Kit (Tepnel Life Sciences).
2. Freshly prepared nucleon reagent A: 10 mM Tris-HCl, 320 mM sucrose, 5 mM MgCl$_2$, 1 % (v/v) Triton X-100; pH to 8.0 with 40 % (w/v) NaOH.
3. 15 mL and 50 mL polypropylene centrifuge tubes.

2.2.2 DNA Extraction from Buccal Swabs

1. QIAamp DNA Mini Kit (Qiagen).
2. Freshly prepared lysis master mix, comprising 20 μL of protease (Qiagen), 400 μL of phosphate-buffered saline, 400 μL of AL lysis buffer (Qiagen).
3. 2 mL polypropylene microcentrifuge tubes.
4. ND-100 spectrophotometer (Nanodrop Technologies).

2.3   Materials for PCR Amplification of Microsatellites from Genomic DNA Samples

1. Genomic DNAs (10 ng/µL).
2. Amplitaq Gold DNA polymerase (with 10×PCR Gold Buffer and $MgCl_2$; Applied Biosystems).
3. 4×100 mM dNTPs (GE Health Care).
4. Fluorescently labeled tail primer (5′-TGACCGGCAGCAAAATTG-3′).
5. 96-well thermal cycler plates and sealing film (Alpha).
6. Peltier thermal cycler.

2.4   Materials for Fractionation of Microsatellites Alleles

1. Thermo-Fast 96, nonskirted plates (Abgene).
2. Genescan 400HD[Rox] size standard (Applied Biosystems).
3. Genescan 500HD[Rox] size standard (Applied Biosystems).
4. Hi-Di formamide (Applied Biosystems).
5. ABI 3100 Plate Assembly Kit (Applied Biosystems).
6. 3100 Genetic Analyzer (Applied Biosystems).

2.5   Software

GeneMapper® software for automated microsatellite allele scoring is available from Applied Biosystems.

2.6   Software for Linkage or Association Analyses

1. For linkage analysis, Linkage [12], Fastlink [13], VITESSE [14], Genehunter [15], and Allegro [16, 17].
2. Case-control association studies: e.g. Haploview [18], plink (http://pngu.mgh. harvard. edu/~purcell/plink/); a comprehensive list is provided by a recent review [19].

3   Methods

3.1   Selection of Candidate Genes

Candidate genes for any inherited disease or condition are usually genes known to be associated with similar diseases in other species, genes implicated by their known normal activity or distribution among relevant cells or tissues, or genes located in a genomic region previously shown to be associated with the disease under investigation. If the disease under investigation has not been mapped to a particular region of the publicly available genome databases (e.g., Ensembl, http://www.ensembl.org/index.html), PubMed (http://www.ncbi.nlm.nih. gov/entrez/query.fcgi? db=PubMed) can be used to search the scientific literature for published reports of genes associated with similar diseases in other species. Alternatively, genes known to be expressed in the tissue or organ affected by the disease can be accessed by similar methods. Another useful information source is the Online Mendelian Inheritance in Man database (http://www.ncbi.nlm.nih.gov/ entrez/query.fcgi?db=OMIM), which is a catalogue of human genes and genetic disorders. The database contains textual information, references, and copious links to MEDLINE; sequence records in the Entrez system; and links to additional related resources at the National Center for Biotechnology Information (NCBI) and elsewhere. Entrez is the integrated, text-based search and retrieval system used at NCBI for the major databases, including PubMed, Nucleotide and Protein Sequences, Protein Structures, Complete Genomes, Taxonomy, and others (http://www.ncbi.nlm.nih.gov/gquery/gquery.fcgi). For diseases already mapped to a genomic region, it is a rational to select candidate genes within the linked region on the basis of their role, structure, or transcript distribution and the databases just detailed provide the necessary resources for that or links to alternative databases.

3.2 Identification of Microsatellites Closely Flanking Candidate Gene(s)

Once a short list of candidate genes has been compiled, the next step is to identify polymorphic markers within, or closely flanking, each gene. Suitable markers include both SNPs and microsatellites. This chapter focuses on microsatellites, because at the time of writing, the technology to genotype microsatellites generally is more accessible to modest sized labs, especially those working in nonhuman biology, than SNP technology. This situation is likely to change in the coming years, with large numbers of characterized SNPs becoming available for an increasing number of species. However, until that time arrives, microsatellites represent perfectly suitable markers with which to analyze tens, or even hundreds, of candidate genes simply and efficiently. If sufficient resources are available, two microsatellites that flank each candidate gene should be selected. If the budget is limited or large numbers of candidate genes are being investigated, it may be more appropriate to initially investigate a single microsatellite per gene and confirm preliminary associations by analyzing additional markers in follow-up investigations.

Several databases can be used to download sequence from within or surrounding candidate genes, but we focus on the use of Ensembl (http://www.ensembl.org/index.html), a software system that produces and maintains automatic annotation on selected eukaryotic genomes.

The first step is to select the genome of choice from the Ensembl home page and enter the symbol of the gene into the search box. Alternatively, start with the human genome (http://www.ensembl.org/Homo_sapiens/index.html), search for the gene of interest, and then follow the link to the appropriate predicted orthologue from the species under investigation. User-specified amounts of species-specific sequence surrounding the candidate gene can be exported, via Ensembl, in a variety of formats, the simplest of which is as a FASTA format text file. Additional export options include the option to export only coding, peptide, cDNA, 5'UTR, 3'UTR, or genomic sequence and whether the output format is HTML, text, or compressed text. The exported sequence then can be accessed as a Word file, for example, and searched for microsatellite repeat motifs. Suitable microsatellite motifs include di-, tri-, and tetranucleotides repeats; we recommend dinucleotides, because they occur more frequently than other microsatellites in most genomes and, in the canine genome at least, are more stable than tetranucleotides [20]. Microsatellites should consist of an uninterrupted run of at least 12 (CA) repeats and be located as close to the gene as possible to maximize the likelihood that the microsatellite, the gene, and therefore any disease-associated mutations will be in linkage disequilibrium with one another. Linkage disequilibrium varies among species; in the dog, it varies among breeds but averages around 2 MB [21], so we typically select microsatellites within 500 kb of each candidate gene if they are available.

3.3   PCR Primer Design

Once microsatellites that flank each candidate gene have been identified, it is necessary to design a pair of primers with which to amplify each microsatellite. Primers can be designed manually, but it usually is more convenient to use specialized software, such as Primer 3 ([22]; http://frodo.wi.mit.edu/primer3 /input.htm), software that incorporates user-specified criteria into its automated primer design. Microsatellite sequence data is simply pasted into the text box and the microsatellite motif, which the primers need to flank, is surrounded by square brackets []. Primer length, melting temperature, GC content, and amplification product length all can be specified, although for most purposes the default values are adequate. Amplification products should be a maximum of 400–450 bp in length. We amplify microsatellites using unlabeled forward and reverse primers and a third, fluorescently labeled primer, identical to an 18 nucleotide extension (TGACCGGCAGCAAAATTG) that is added to the 5' end of each forward primer. This method has been described elsewhere [23] and routinely is used in our laboratory. It has the advantage over conventional dual-primer PCR that only unlabeled forward and reverse primers need be designed and purchased for each microsatellite, which are much less expensive than labeled primers, and the PCR products are labeled via a third, universal, labeled primer used to amplify all microsatellites.

Once primers have been designed, it is necessary to screen the genome of the species from which they were designed to ensure they are sequence-specific and do not hybridize within any repeat elements. Short, interspersed elements (SINES), for example, are between 10- and 100-fold more common within the canine genome than in humans [24], and primers that are inadvertently designed to hybridize to such repeat elements will bind nonspecifically throughout the genome and fail to produce a single, specific PCR product. This screening process can be achieved in silico using Ensembl BLASTView (http://www.ensembl.org/ Homo_sapiens/ helpview?se=1&kw=blastview), which provides access to the WU-BLAST and SSAHA sequence similarity search algorithms via a single interface.

3.4 DNA Extraction

3.4.1 DNA Preparation from 2–10 ml of Blood Using a Nucleon Kit

1. In a 50-mL polypropylene tube, combine blood and nucleon reagent A to a total volume of 45 mL (red blood cell lysis).
2. Mix by inversion and incubate for 4 min at room temperature. Centrifuge at $1{,}300 \times g$ for 4 min at room temperature. Discard the supernatant without disturbing the pellet (white blood cells) (*see* **Note 1**).
3. Add 2 mL of nucleon reagent B to the pellet, vortex, and incubate at 37 °C for at least 10 min (white blood cell lysis) (*see* **Note 2**).
4. Transfer to a 15-mL polypropylene tube and add 500 μL of sodium perchlorate. Mix by inversion at least seven times (protein precipitation).
5. Add 2 mL of chloroform and mix by inversion at least seven times. Centrifuge at $1{,}300 \times g$ for 3 min.
6. Transfer the top aqueous phase to a new 15-mL tube, avoiding contact with the protein layer at the interface between the upper and lower phases (*see* **Note 3**).
7. Add 2 mL of chloroform and mix by inversion at least seven times. Add 300 μL of nucleon resin without remixing the phases. Centrifuge at $1{,}300 \times g$ for 3 min.
8. Transfer the top aqueous phase to a new 15-mL tube, avoiding contact with the resin layer.
9. Centrifuge at $1{,}300 \times g$ for 3 min and transfer the supernatant to a fresh tube without disturbing the resin pellet.
10. Add 5 mL of ice-cold 100% (v/v) ethanol and slowly invert, until the DNA is fully precipitated (*see* **Note 4**).
11. Using a sealed glass hook made from a glass Pasteur pipette, hook out the DNA and allow to air dry for 10 min.
12. Wash the DNA off the hook with 200–400 μL of ultrapure H_2O, collecting in a 1.5-mL microcentrifuge tube. Incubate at 37 °C for up to 60 min to assist the DNA to dissolve.
13. Measure the DNA concentration ($OD_{260\,nm}$) and purity ($OD_{260\,nm}/OD_{280\,nm}$).

3.4.2   DNA Extraction from Buccal Swabs

1. Snap or cut off a swab head and collect it in a 2-mL microcentrifuge tube.
2. Add 820 μL of lysis master mix and vortex for 15 s.
3. Incubate at 56 °C for 10 min. Pulse the centrifuge tube to remove droplets from the lid.
4. Add 400 μL of ethanol and vortex for 10 s to mix.
5. Apply 600 μL of sample to a QIAamp column.
6. Centrifuge at 6,000 × g for 1 min. Discard the flow through.
7. Repeat steps 6 and 7 to apply the remaining sample onto the QIAamp column.
8. Apply 500 μL of Qiagen wash buffer AW 1 and centrifuge at 6,000 × g for 1 min. Discard the flow through.
9. Apply 500 μL of Qiagen wash buffer AW 2 and centrifuge at 20,000 × g for 3 min. Discard the flow through.
10. Centrifuge at 20,000 × g for 1 min and discard the flow through.
11. Place the column inside a 1.5-mL microcentrifuge tube add 100 μL of Qiagen elution buffer AE (preheated to 37 °C) to the column filter. Incubate at room temperature for 2 min. Elute the DNA by centrifuging at 6,000 × g for 1 min.
12. Measure the DNA concentration ($OD_{260\,nm}$) and purity ($OD_{260\,nm}/OD_{280\,nm}$).

3.5   PCR Amplification of Microsatellites from Genomic DNA Samples (see Note 5)

1. Amplify DNA samples in 12-μL reactions comprising 2 μL of 10 ng/μL genomic DNA, 4×0.2 mM dNTPs, 1×PCR buffer, 1.5 mM $MgCl_2$, 0.17 μM forward primer, 0.42 μM reverse primer, 0.5 μM fluorescently labeled universal primer, 1.2 units of AmpliTaq Gold DNA polymerase. If a microsatellite is being amplified in multiple DNA samples, prepare a "master mix" containing all the PCR ingredients except the genomic DNA, allowing for an additional 10% of each ingredient. Vortex to mix briefly and centrifuge very briefly in a microcentrifuge.
2. Aliquot 10 μL of PCR mix into separate wells of a 96-well thermal cycler plate and add 2 μL of 10 ng/μL DNA to each well. Seal the plate with sealing film.
3. Centrifuge the plate briefly, transfer to thermal cycler, and amplify using the following PCR program: 94 °C, 4 min; (94 °C, 1 min; annealing temperature—the annealing temperature can be adjusted as appropriate; a typical starting temperature is 56 °C—1 min; 72 °C, 1 min)×30 cycles; (94 °C, 1 min; 50 °C, 1 min; 72 °C, 1 min)×8 cycles; 72 °C, 30 min.

3.6   Fractionation of Fluorescently Labeled Microsatellite Alleles on a 3100 Genetic Analyzer

1. Dispense 1 μL of each PCR product into a separate well of a 96-well plate.
2. Add 10 μL of Hi-Di formamide mix (0.04 μL of size standard and 9.96 μL of Hi-Di formamide) to each well (for PCR products up to 350 bp, use the 400 Rox

size standard; and for PCR products up to 450 bp, use the 500 Rox size standard).

3. Denature the PCR products at 95 °C for 1 min and place on ice for 2 min.
4. Load the plate into the 3,100-plate assembly kit and then onto the ABI 3100 Genetic Analyzer, and select the relevant run module.

3.7 Genotyping Microsatellites

Automated microsatellite genotyping is carried out using GeneMapper software v4.0 (Applied Biosystems). This genotyping software package provides automated genotyping of microsatellite data generated using any of the compatible Applied Biosystems electrophoresis-based genotyping systems and Data Collection Software. The software reads data from the *.fsa files generated during fractionation of the fluorescently labeled amplification products, detects peaks, makes size calls, and finally makes automated genotyping calls with associated quality scores. The sizing and genotyping algorithms used by the GeneMapper software are determined by a collection of user-defined settings, which include information such as the microsatellite repeat length, dye color, and allele size range of the markers being analyzed and the size standard used.

3.8 Linkage and Association Analysis (see Note 6)

The choice of the most appropriate method with which to analyze microsatellite genotyping data for an association with the disease under investigation depends on a wide variety of factors. The design and statistical interpretation of linkage and association studies is an extremely complex area, and a exhaustive discussion of the subject therefore is outside the scope of this review; readers that are new to either linkage or association studies are well advised to familiarize themselves with the fundamental principles first by reading, for example, the following basic text on the subject [25].

Whether the samples derive from individuals closely related to one another or more distantly related or unrelated subjects determines whether a pedigree-based linkage analysis or a population-based association study is more appropriate. In addition, the mode of inheritance of the disease also needs to be taken into account. The analysis of conditions or traits that exhibit a straightforward mode of inheritance, be it dominant or recessive, are straightforward to analyze using computer programs such as Linkage [12], Fastlink [13], and VITESSE [14], all of which can calculate multipoint lod scores for a few markers, although execution time increases rapidly with the number of markers. In contrast to genomewide screens, candidate genes approaches usually do not involve the analysis of large numbers of markers, and for this reason, these programs are likely to be adequate for most studies involving modest numbers of markers. Multipoint linkage calculations involving many markers is possible with the program Genehunter [15], which can implement both parametric and nonparametric analyses, although the run time of Genehunter grows exponentially with pedigree size, making it unsuitable for large pedigrees. Another computer program capable of both parametric and nonparametric multipoint analyses is Allegro, which is considerably faster than Genehunter [16, 17].

An equally formidable range of software is available with which to carry out association studies. A recent review [19] comprehensively details software currently available for association analyses. A wide-ranging list of computer software available for both genetic linkage and association analysis (and much more) is available from the North Shore LJJ Research Institute (http://www.nsljj-genetics.org/soft/#c). Various texts also cover a comprehensive range of relevant statistical methods [26, 27].

4   Notes

1. For old or frozen blood, it may be necessary to perform 2 × nucleon reagent A–effected red blood cell lysis treatments (steps 1 and 2).

2. The DNA preparation can be frozen and stored at the end of step 3.

3. If the aqueous phase is cloudy at step 6, transfer it to a new tube and repeat steps 5 and 6.

4. If the DNA does not precipitate in 5 mL of ethanol, centrifuge it at $5,000 \times g$ for 15 min to pellet the DNA. Discard the supernatant, wash the pellet with 2 mL of 70% (v/v) ethanol, and recentrifuge. Discard the supernatant and dry the DNA at 37 °C until all the ethanol has evaporated. Resuspend the DNA in 100 µL of ultrapure H_2O.

5. The efficiency and cost effectiveness of microsatellite analysis can be increased by amplifying multiple microsatellites simultaneously in a single multiplex PCR. Two or three microsatellites routinely can be amplified simultaneously in a single PCR; to amplify more than this number might involve adjusting the concentrations of individual primers. Care must be taken to ensure the amplification products of microsatellites amplified together are sufficiently different for alleles of individual markers to be distinguished. For dinucleotide microsatellites, we ensure a margin of at least 50 bp between the allele ranges of markers amplified together. Microsatellites also can be amplified individually and fractionated together. This method had the advantage that different dyes can be used to label the PCR products of different microsatellites, and the difference in allele size range just discussed is not an issue.

6. The extent of linkage disequilibrium in the species under investigation must be considered when analyzing the data from candidate gene–associated microsatellites. An association between a particular microsatellite and a disease merely indicates that the microsatellite, and hence the associated candidate gene, are located in an area of the genome that is in linkage disequilibrium with the disease. In the dog, for example, linkage disequilibrium extends for around 2 Mb [21], and so all the genes within approximately this distance of a microsatellite

that displays association with the disease being studied, by definition, must be considered candidates for the disease, although obviously some genes are better candidates than others.

References

1. Kijas JM, Bauer TR, Jr., Gafvert S, Marklund S, Trowald-Wigh G, Johannisson A, Hedhammar A, Binns M, Juneja RK, Hickstein DD, Andersson L (1999) A missense mutation in the beta-2 integrin gene (ITGB2) causes canine leukocyte adhesion deficiency. Genomics 61:101–107
2. Henthorn PS, Liu J, Gidalevich T, Fang J, Casal ML, Patterson DF, Giger U (2000) Canine cystinuria: Polymorphism in the canine SLC3A1 gene and identification of a nonsense mutation in cystinuric Newfoundland dogs. Hum Genet 107:295–303
3. Occhiodoro T, Anson DS (1996) Isolation of the canine alpha-L-fucosidase cDNA and definition of the fucosidosis mutation in English Springer Spaniels. Mamm Genome 7:271–274
4. Skelly BJ, Sargan DR., Herrtage ME, Winchester BG (1996) The molecular defect underlying canine fucosidosis. J Med Genet 33:284–288
5. Starkey MP, Scase TJ, Mellersh CS, Murphy S (2005) Dogs really are man's best friend— Canine genomics has applications in veterinary and human medicine! Brief Funct Genomic Proteomic 4:112–28
6. Giger U, Sargan DR, McNiel EA (2006) Breed-specific heriditary diseases and genetic screening. In: Ostrander EA, Giger U, Lindblad-Toh K (eds)The dog and its genome. Cold Spring Harbor Laboratory Press, New York, pp. 249–289
7. Akhmedov NB, Baldwin VJ, Zangerl B, Kijas JW, Hunter L, Minoofar,KD, Mellersh C, Ostrander EA, Acland GM, Farber DB, Aguirre GD (2002) Cloning and characterization of the canine photoreceptor specific cone-rod homeobox (CRX) gene and evaluation as a candidate for early onset photoreceptor diseases in the dog. Mol Vis 8:79–84
8. Safra N, Ling GV, Schaible RH, Bannasch DL (2005) Exclusion of urate oxidase as a candidate gene for hyperuricosuria in the Dalmatian dog using an interbreed backcross. J Hered 96:750–754
9. Brenig B, Pfeiffer I, Jaggy A, Kathmann I, Balzari M, Gaillard C, Dolf G (2003) Analysis of the 5′ region of the canine PAX3 gene and exclusion as a candidate for Dalmatian deafness. Anim Genet 34:47–50
10. Aguirre-Hernandez J, Sargan DR (2005) Evaluation of candidate genes in the absence of positional information: A poor bet on a blind dog! J Hered 96:475–484
11. Mellersh CS, Pettitt L, Forman OP, Vaudin M, Barnett KC (2006) Identification of mutations in HSF4 in dogs of three different breeds with hereditary cataracts. Vet Ophthalmol 9:369–378
12. Lathrop GM, Lalouel JM, Julier C, Ott J (1984) Strategies for multilocus linkage analysis in humans. Proc Natl Acad Sci USA 81:3443–3446
13. Cottingham RW Jr, Idury RM, Schaffer AA (1993) Faster sequential genetic linkage computations. Am J Hum Genet 53:252–263
14. O'Connell JR, Weeks DE (1995) The VITESSE algorithm for rapid exact multilocus linkage analysis via genotype set-recoding and fuzzy inheritance. Nat Genet 11:402–408
15. Kruglyak L, Daly MJ, Reeve-Daly MP, Lander E S (1996) Parametric and nonparametric linkage analysis: A unified multipoint approach. Am J Hum Genet 58:1347–1363
16. Gudbjartsson DF, Thorvaldsson T, Kong A, Gunnarsson G, Ingolfsdottir A (2005) Allegro version 2. Nat Genet 37:1015–1016
17. Gudbjartsson DF, Jonasson K, Frigge ML, Kong A (2000) Allegro, a new computer program for multipoint linkage analysis. Nat Genet 25:12–13
18. Barrett JC, Fry B, Maller J, Daly MJ (2005) Haploview: Analysis and visualization of LD and haplotype maps. Bioinformatics 21:263–265

19. Almasy L, Warren DM (2005) Software for quantitative trait analysis. Hum Genomics 2:191–195
20. Francisco LV, Langston AA, Mellersh CS, Neal CL, Ostrander EA (1996) A class of highly polymorphic tetranucleotide repeats for canine genetic mapping. Mamm Genome 7:359–362
21. Sutter NB, Eberle MA, Parker HG, Pullar BJ, Kirkness EF, Kruglyak L, Ostrander EA (2004) Extensive and breed-specific linkage disequilibrium in Canis familiaris. Genome Res 14:2388–2396
22. Rozen S, Skaletsky HJ (2000) Primer3 on the WWW for general users and for biologist programmers. In: Krawetz S, Misener S (eds) Bioinformatics methods and protocols: Methods in molecular biology. Humana Press, Totowa, NJ, pp. 365–386
23. Oetting WS, Lee HK, Flanders DJ, Wiesner GL, Sellers TA, King RA (1995) Linkage analysis with multiplexed short tandem repeat polymorphisms using infrared fluorescence and M13 tailed primers. Genomics 30:450–458
24. Kirkness EF, Bafna V, Halpern AL, Levy S, Remington K, Rusch DB, Delcher AL, Pop M, Wang W, Fraser CM, Venter JC (2003) The dog genome: Survey sequencing and comparative analysis. Science 301:1898–1903
25. Strachan T, Read A (2003) Human molecular genetics. Garland Publishing, New York
26. Barnes M R, Gray IC (2003) Bioinformatics for geneticists. John Wiley and Sons, New York
27. Koenig IR, König IR, Ziegler A (2006) A statistical approach to genetic epidemiology. Wiley-VCH, Hoboken, NJ

Chapter 6
Full Complexity Genomic Hybridization on 60-mer Oligonucleotide Microarrays for Array Comparative Genomic Hybridization (aCGH)

Alexei Protopopov, Bin Feng, and Lynda Chin

Abstract Recurrent structural alterations, such as amplification, deletion, or translocation, are hallmark features of the cancer genome. Mapping of these DNA copy number aberrations, using approaches such as comparative genomic hybridization (CGH), has enabled the discovery of bona fide tumor suppressor genes and oncogenes. With the emergence of high-density oligo-based microarray platforms, array-based CGH has become a powerful technology that can facilitate the accurate mapping and rapid identification of novel cancer genes. Here, we describe the optimized technical protocol for comparative genomic hybridization with full-complexity genomic DNA on 60-mer oligonucleotide microarrays.

Keywords Cancer genomics, array-CGH (aCGH), oligo microarrays, genomic profiling, copy number changes

1 Introduction

Comparative genomic hybridization (CGH) is a method of molecular cytogenetics for the analysis of copy number changes, such as gains and losses, in the DNA content of tumor cells. Unlike expression, which can be influenced by many factors that are not relevant to cause of cancer (e.g., time of day), the ground truth of the copy number generally is known: a diploid with two copies. Availability of germline normal can further exclude copy number polymorphisms. Therefore, one can think of copy number alterations as a way to focus on more likely causal cancer-related genes when combined with expression information.

With advance in microarray spotting technology, Pinkel et al. pioneered the adaptation of CGH to microarray format with large probe elements such as BACs [1]. However, genomewide BAC array-CGH was still limited in resolution (3–1 Mb). The oligomer microarrays offer significant advantages, including in situ array fabrication, flexible design, accurate annotation, and ultimate flexibility and programmability. In addition, unlike large DNA elements such as cDNA and

From: Methods in Molecular Biology, Vol. 439: *Genomics Protocols: Second Edition*   87
Edited by Mike Starkey and Ramnath Elaswarapu © Humana Press Inc., Totowa, NJ

BAC, the computational design of oligomer probes should minimize cross-hybridization artifacts. With technical optimization of the protocols and implementation of statistical modeling for copy number transition, we successfully established a robust array-CGH platform, which offered a median resolution of less then 10 Kb across the mouse and human genomes [2].

2   Materials

2.1   Clinical Biospeciments and Cell Lines

1. Specimens should have minimal 80% tumor:normal ratio and are snapped-frozen in LN2 within 20 min of excision. In addition to the confirmation of a histological diagnosis by at least two pathologists, tumor DNA quality must be superb.
2. For human genomic DNA, male DNA G147A (Promega, Madison, WI).
3. For human genomic DNA, female G152A (Promega).
4. For mouse modeling of cancer, determination of a "normal" number of genomic copies is a challenging task; therefore, as a reference to tumor sample, we use the normal tissue from the host animal (tail DNA, for example).

2.2   Special Equipment

1. Microarray Scanner G2565BA (Agilent, Palo Alto, CA) or equivalent.
2. Hybridization Set. These instructions assume the use of either Agilent set of hybridization equipment or the BioMicro MAUI System (BioMicro, Salt Lake City, UT). Agilent: hybridization oven G2545A supplemented with hybridization oven rotator G2530-60029 and hybridization chambers G2534A; hybridization chamber gasket slides G2534-60003 or G2534-60002. BioMicro: MAUI Hybridization System; MAUI hybridization chambers FL or DC.
3. N_2 purge desiccator for slide storage (Terra Universal, Fullerton, CA).
4. Ozone destruct unit NT-70 (Ozone Solutions, Sioux Center, IA).
5. At least three glass 20 slides staining dish with removable rack 900200 (Fisher, Hampton, NH) or equivalent. All should be dedicated for array-CGH and not substituted.

2.3   DNA Isolation and Digestion

1. *Alu*I (10 U/μL) (Promega).
2. *Rsa*I (10 U/μL) (Promega).

3. QIAquick PCR Purification Kit 28104 (QIAGEN, Valencia, CA).
4. Microcon YM-30 centrifugal filters (Millipore, Billerica, MA).

2.4 Random-Prime Labeling for Array-CGH

1. Primer/reaction buffer (2.5X): 750 ng/µL random octamers, 125 mM Tris-HCl, pH 6.8, 12.5 mM MgCl, 25 mM 2-mercaptoethanol. Store in aliquots at −20 °C.
2. Cy3-dNTP Mix (5X): 1.0 mM dATP, 1.0 mM dGTP, 1.0 mM dTTP, 0.25 mM dCTP (Invitrogen, Carlsbad, CA), 1.0 mM Cyanine 3-dCTP (PerkinElmer, Wellesley, MA) in 5 mM Tris-HCl, pH 8, 0.5 mM EDTA. Store in aliquots at −20 °C.
3. Cy5-dNTP mix (5X): 1.0 mM dATP, 1.0 mM dGTP, 1.0 mM dTTP, 0.25 mM dCTP (Invitrogen), 1.0 mM Cyanine 5-dCTP (PerkinElmer) in 5 mM Tris-HCl, pH 8, 0.5 mM EDTA. Store in aliquots at −20 °C.
4. Exo-Klenow Fragment (40 U/µL) (Invitrogen). Keep at −20 °C.
5. Stop buffer: 0.5 M EDTA.
6. TE buffer (1X): 10 mM Tris-HCl, pH 7.5, 1 mM EDTA.
7. Nuclease-free water (Ambion, Austin, TX).

Labeling mixtures may be purchased in kit form from several manufacturers (Enzo Life Science, Invitrogen, Agilent). The CGH Labeling Kit 42671 (Enzo Life Science, Farmingdale, NY) has been used most successfully.

2.5 Hybridization and Wash

1. Cot-1 DNA 1 mg/mL (Invitrogen). Store in aliquots at −20 °C.
2. Blocking agent: 2 mg/mL yeast tRNA (Invitrogen), 100 µg/mL poly(dA)-poly(dT) (Sigma, St. Louis, MO) in nuclease-free water. Store in aliquots at −20 °C.
3. 2X Hyb buffer: 7.8X SSC, 0.5% SDS (v/v).
4. Wash 1: 0.5X SSPE, 0.005% N-lauryl sarcosine (v/v) (Sigma) in MilliQ H_2O. Filter using a 1-liter 0.22 µm filter unit 430015 (Corning, Corning, NY) or equivalent.
5. Wash 2: 0.1X SSPE, 0.005% N-lauryl sarcosine (v/v) (Sigma) in MilliQ H_2O. Filter using the filter unit 430015 (Corning). Use only disposable plasticware for the wash buffers.
6. 3 M sodium acetate pH 5.8.
7. Ethanol 100 and 70% in nuclease-free water (v/v).
8. Lintless wipes DURX 670 (Berkshire, Great Barrington, MA) or equivalent.

Hybridization and wash mixtures may be purchased from Agilent.

3   Methods

3.1   Environmental Parameters

1. Equip the array-CGH laboratory with the NT-70 ozone interceptor, which provides an ozone-free "blanket" of air around the operational area (*see* **Note 1**).
2. Monitor the humidity in the room and use a dehumidifier to maintain the moisture level below 60% (*see* **Note 2**).
3. Use the nitrogen purge box to store the open packages of microarray chips in an oxygen-free atmosphere.

3.2   Genomic DNA Preparation

CGH requires equal amounts of test and reference probes. For most cases the best-suited reference is commercial human male genomic DNA prepared by pulling samples from multiple healthy donors. The protocol begins with 2250 ng of a DNA double digested with *Alu*I and *Rsa*I (the choice of enzymes depends on the array). DNA quality controls take two aliquots of about 250 ng DNA combined. Therefore, taking into the account some variability in the digestion yield, it is reasonable to start with at least 5 µg of intact genomic DNA of each, the test and reference channels (*see* **Note 3**).

1. Use any method yielding a nondegraded pure genomic DNA. For example, the DNeasy Tissue Kit (QIAGEN) has been used successfully to isolate high-quality DNA from cell lines and fresh and frozen primary tissue samples (*see* **Note 4**).
2. Determine the concentration of sample DNA and level of contaminating proteins, RNA, phenol, and other molecules by measuring for OD at 280, 260, and 230 nm. Use a minimum A_{260}/A_{280} ratio of 1.8 and a minimum A_{260}/A_{230} ratio of 1.9 for a DNA sample to be considered of satisfactory for array-CGH (*see* **Note 5**). The lowest DNA concentration acceptable for the following digestion is 58 ng per 1 µL of nuclease-free water.
3. Set a heat block to 37 °C.
4. Digest 5 µg of each test DNA by adding 10 µL 10X Buffer C (supplied with *Rsa*I), 1.5 µL *Alu*I, 1.5 µL *Rsa*I, and water to make 100 µL final volume. Repeat with the corresponding amount of reference DNA. Incubate overnight at 37 °C (*see* **Note 6**).
5. Clean up the digested fragments using the QIAquick PCR Purification Kit. Elute with 50 µL of water (add water to the center of the column and allow to sit for 1 min before centrifugation).
6. Quantitate using the NanoDrop. Please note that all DNA samples must have a concentration of 100 ng/µL. Concentrate or dilute the DNA samples if necessary. For concentration use a Vacufuge set to 45 °C. We consider a DNA is sufficiently pure, if the A_{260}/A_{280} ratio is above 1.8 and the A_{260}/A_{230} ratio is above 2.0.

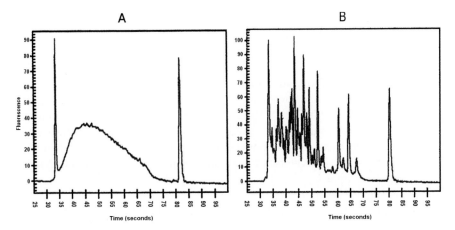

Fig. 6.1 An example of profiles for cleaved DNA of a good (A) and unacceptable (B) quality

7. To determine whether the DNA is appropriately fractionated and how much degradation it contains, examine the *AluI/RsaI* digested samples on Agilent 2100 Bioanalyzer using a DNA 12000 chip. An example of profiles is shown in Fig. 6.1. All samples should be compared to this image to make sure DNA is not degraded.

3.3   Synthesis of Cy3- and Cy5-Labeled Targets

The current protocol describes enzymatic-based labeling of DNA with Klenow DNA polymerase, random primers, and cyanine-labeled nucleotides. For a single polarity array-CGH, consider labeling each test/reference pair as follows: test DNA with cyanine-3 (Cy3) and the reference with cyanine-5 (Cy5)(*see* **Note 7**).

1. Set the hybridization system to 65 °C; heat blocks to 98 °C and 37 °C; refrigerating bath to 2 °C.
2. Create a table for the following information being collected: unique sample name, polarity of labeling, dye incorporation, labeling yield, array ID.
3. Thaw at room temperature and protect from light the primers/reaction buffer, Cy3-dNTP mix, and Cy5-dNTP mix. Keep the Klenow enzyme at −20 °C. Thaw and dissolve DNA at 37 °C (up to 15 min).
4. Count the number of Cy3/Cy5 pairs and mark the corresponding number of 1.5-mL microcentrifuge reaction tubes with their unique sample names. Mark the same number of reaction tubes for reference probes. The reference name may be the same.
5. Make ready the following materials required for the processing: stop buffer, nuclease-free water.
6. Add 20 μL of the primer/reaction buffer to each reaction tube.

7. Add 20 µL DNA (2 µg) to the corresponding tubes. Mix by pipetting four times. Quick spin the reaction tubes.

8. Place reaction tubes on 98 °C heat block for 10 min and chill in the 2 °C bath for 2 min. Keep at 2 °C.

9. Mark one new 1.5-mL microcentrifuge tube with "Cy3" and another with "Cy5" during the incubation.

10. Combine the Cy3-dNTP mix 10 µL and the Klenow 1 µL per reaction into the "Cy3" tube. Mix thoroughly by gentle flicking four times. Combine 10 µL Cy5-dNTP mix and 1 µL Klenow per reaction into the "Cy5" tube. Mix thoroughly by gentle flicking four times. Quick spin both the Cy3 and Cy5 tubes. Protect from light.

11. Quick spin the reaction tubes after the 2 °C incubation.

12. Add 11 µL of the appropriate nucleotide-enzyme mixture per tube: Cy3 to the reactions with test samples and Cy5 to references; mixing thoroughly by pipetting four times. Quick spin all tubes.

13. Transfer the reaction tubes to the 37 °C heat block. Incubate at 37 °C for 3 h.

14. During the 3 h incubation, for each Cy5/Cy3 pair, carefully mark one new 1.5-mL microcentrifuge and one supplied Microcon collector tube (*see* **Note 8**).

15. Stop the reaction by adding 5 µL stop buffer to each tube after 3 h incubation. Quick spin all tubes.

16. Combine paired Cy3 and Cy5 reactions in the tube that has a unique sample name. The total volume of the target mix is 110 µL.

17. Purify the target using Microcon YM-30 filters. Add 400 µL TE to the stopped labeling reaction. Lay onto a YM-30 filter/collector. Centrifuge 10 min at 8 000 g. Discard flow-through and place column back into a collector tube. Add 450 µL water to the column and repeat spin. Discard flow-through. Again add 450 µL water to the column and repeat spin. Invert filter into the premarked 1.5-mL microcentrifuge tube and spin 1 min at 8,000 g to elute.

18. Bring the volume to 153 µL with water. Determine the yield and incorporation as follows.

19. Take 1.5 µL of each target to measure for OD at 260, 550, and 650 nm. Measure samples in duplicate. Use the following formulas: $A_{550}/0.15$ = pmoles Cy3/µL; $A_{650}/0.25$ = pmoles Cy5 /µL; $A_{260 \times 33}/0.1$ = ng DNA /µL (*see* **Note 9**).

20. Proceed only if Cy3 readings exceed 3.9 pmol/µL; Cy5 readings exceed 2.6 pmol/µL; DNA reading exceed 80 ng/µL. This is a critical quality control step, as it is the final one prior to hybridization onto a microarray, the most expensive component in array-CGH (*see* **Note 10**).

21. Labeled probes can be stored up to 3 d at −20 °C in the dark (*see* **Note 11**).

3.4   Preparation of Labeled Target and Hybridization

Make ready the following materials required for the hybridization: nuclease-free water, blocking agent, 2X Hyb buffer. Thaw Cot-1 DNA at 37 °C. The total volume of the hybridization mix is specific to a hybridization set being used and can vary from 50 μL to 500 μL. It is easy to adjust the volume using the procedure in Sect. 3.4.2 (*see* **Note 12**).

3.4.1 Agilent Hybridization Set

1. Add the following to the 150 μL labeled target: 50 μL Cot-1 DNA, 50 μL blocking agent (*see* **Note 13**). Add 250 μL 2X Hyb Buffer. Quick spin.
2. Denature the hybridization mix by placing in 98 °C heat block for 3 min. Transfer tubes to 37 °C heat block.
3. Incubate for 30 min at 37 °C (preannealing). Centrifuge for 1 min at ~18,000 g (*see* **Note 14**). The target is ready for immediate hybridization.
4. Assemble hybridization chambers and gasket as directed by manufacturer for this type of arrays. Record the array ID for future reference.
5. Transfer the entire 500 μL for hybridization to a slide-size array.
6. Hybridize at 65 °C in a rotating hybridization oven for 40 h at 20 rpm.

3.4.2 MAUI Hybridization Set

1. Add the following to the 150 μL labeled target: 50 μL Cot-1 DNA, 50 μL blocking agent, 25 μL 3 *M* sodium acetate. Mix thoroughly by pipetting four times. Precipitate the DNA by the addition of 750 μL ethanol. Incubate at −60 °C (dry ice) for 5 min.
2. Pellet the precipitated DNA by centrifugation for 10 min at ~18,000 g. Carefully remove supernatant from tubes with a pipette. Rinse the pellets with 400 μL of 70% ethanol. Vortex briefly to dislodge pellet. Recentrifuge and carefully remove supernatant with a pipette. Centrifuge samples without the caps at ~5,000 g for 3 min to make sure samples are dried completely.
3. Add 27 μL water to the pellet. Gently tickle the tube until the DNA dissolves.
4. Add 27 μL 2X Hyb buffer. Mix thoroughly by gentle flicking four times. Centrifuge at maximal speed for 1 min.
5. Denature the hybridization mix by placing in 98 °C heat block for 3 min. Transfer tubes to 37 °C heat block.
6. Incubate 30 min at 37 °C. Centrifuge at ~18,000 g for 1 min. The targets are ready for immediate hybridization.
7. Place the array in the MAUI hybridization chamberand use as directed by manufacturer. Record the array ID for future reference.
8. Check the integrity of the MAUI assembly. Slowly preload the assembly with 2X Hyb buffer diluted 1:1 with water, without the target. Briefly, insert the end of the tip into the fill port, wiggling the tip gently and applying a slight downward pressure to ensure a good seal. Keeping the tip perpendicular to the slide, inject the blank hybridization buffer into the fill port. When the solution begins to emerge from vent port, tilt tip to relieve pressure. In 30 s check for leakage (*see* **Note 15**).

9. Slowly load 50 µL of the target into the fill port, pushing the blank buffer out. Avoid bubbles.
10. Lock the ports as directed by manufacturer.
11. Hybridize at 65 °C in the MAUI Hybridization System for 18h using Mix Mode B.

3.5 Take Down

Wash arrays using glass slide staining dishes and slide trays, as outlined next. Wash in batches of no more than five arrays, and use fresh wash buffers for each batch. Use forceps only for handling the arrays. Rinse wash dishes and all accessories in water only, never use any detergent (*see* **Note 16**).

1. Make ready the following materials required for the take down. Wash 1, two forceps, lintless wipes, water bath for magnetic stirrers at 37 °C. Using the water bath, equilibrate Wash 2 to 37 °C.
2. Prepare the Bath 1 to be used for disassembly of the hybridization sandwich. Add 250 mL of Wash 1 to a staining dish at room temperature.
3. Prepare Bath 2: Place a stir bar and glass slide rack in the second staining dish. Add 250 mL of Wash 1 room temperature. Place on a stir plate, at a stir setting of 6.
4. Prepare Bath 3: Add 250 mL of heated Wash 2 to the third staining dish. Add a stir bar. Place the dish in the water bath for magnetic stirrers heated to 37 °C. Place on the stir plate, at a stir setting of 6. The wash "station" is set (*see* **Note 17**).
5. Place the hybridization sandwich into Bath 1. Using forceps, disassemble. Be wary not to scratch the array.
6. Immediately after disassembly, place the array in the glass slide holder of Bath 2. Repeat steps 5 and 6 for each of five arrays to be washed together.
7. Once all the arrays are in the slide holder rack in Bath 2, wash covered for 5 min at room temperature.
8. Transfer the slide holder rack from Bath 2 to Bath 3 and wash covered for 1 min at 37 °C.
9. Remove each slide individually, bar-code-free side first, using the slow pull from Bath 3 to dry the arrays. Arrays are now ready to scan. Refill baths for the washing of next set of arrays.

3.6 Array Scanning

Scan arrays as directed by manufacturer of the microarray scanner, setting the resolution to 5 µm.

3.7 First-Pass Analyses

The protocols and software described here were tested on the Agilent's feature extraction output and also can be used on tab-delimited text format of other scanners. Array map files were generated based on NCBI build 36 (hg18) for human and NCBI build 36 (mm8) for mouse. Map files contain genomic positions for all probes on the array. All map files and analysis software could be downloaded from Lynda Chin's website (http://genomic.dfci.harvard.edu). We developed an R package, aCGHNorm, which generates normalized log2 ratio using LOWESS (locally weighted regression) from feature extraction text output.

1. Download R package from http://r-project.org and install it on your computer.
2. Download aCGHNorm package from Lynda Chin's website (http://genomic. dfci.harvard.edu) and install it according to the manual.
3. Put Agilent's feature extraction text output files into one folder with an annotation file according to the manual in aCGHNorm package.
4. Call doNorm2 function (see the aCGHNorm manual) to finish the normalization step. This call generates two text files (masterResults.xls and QA.xls) and some graphs showing the quality of the experiment. The masterResults.xls combines all experiments to report an average of the dye swaps. All ratios are ordered by each probe's genomic position from chromosome 1 to chromosome Y. Figure 6.2 is a graphic representation of Male/Female hybridization.

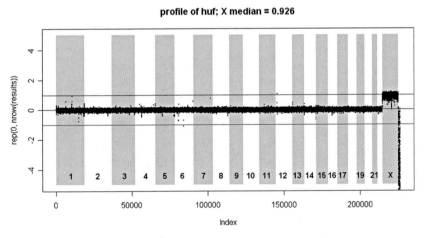

Fig. 6.2 Genomic profiles of female versus male normal genomic DNAs. Array-CGH profiles of pooled female DNA against pooled male DNA (as reference) with x-axis coordinates representing oligo probes ordered by genomic map positions and y-axis coordinates representing each oligo probe's log2 ratio

3.8   QA Assessment

During the normalization process, some figures and a QA summary file (QA.xls) were generated to provide a statistical overview of the quality of the experiments. Some of the QA summary columns are explained here:

goodSpots:
The percentage of good spots (average of hyb and swap's log2 ratios) after considering Agilent feature extraction flags.

hybSpots:
The percentage of good spots after considering Agilent feature extraction flags for one hybridization.

swapSpots:
The percentage of good spots after considering Agilent feature extraction flags for dye swap.

agiFlag.hyb, agiFlag.swap:
The number of bad spots flagged by Agilent's feature extraction software.

faint.hyb, faint.swap:
The number of bad spots due to the weak intensity level for both Cy5 and Cy3 channels.

dsd:
A measurement of the noise level of the good spots. Derivative standard deviation (dsd) is defined to estimate noise level of profiles with large regional gain/loss. Let $X1, \ldots, Xn$ be the log2 ratio of probes which are ordered by chromosome map positions.

$$Yi = Xi + 1 - Xi \text{ while } i \in (1\!:\!n\text{-}1)$$

$$DSD = sd(Y)/sqrt(2)$$

dsd.hyb:
Derivative standard deviation calculated using hyb spots.

dsd.swap:
Derivative standard deviation calculated using swap spots.

rep.cons:
The number of spots dropped due to the inconsistency between hyb and swap. The difference between hyb and swap was calculated for each feature to form a vector rep, where $2 \times$ sd (rep) is used as the threshold to filter out inconsistent probes.

cor:
A measurement of consistency of hyb and swap. It calculated as Pearson correlation between hybSpots and swapSpots.

Cy5.hyb.med: The median raw intensity of hyb's Cy5 channel.
Cy3.hyb.med: The median raw intensity of hyb's Cy3 channel.
Cy5.swap.med: The median raw intensity of swap's Cy5 channel.
Cy3.swap.med: The median raw intensity of swap's Cy3 channel.

Different facilities and different platforms have different ranges for these QA parameters. The most important parameters are goodSpots, hybSpots, swapSpots,

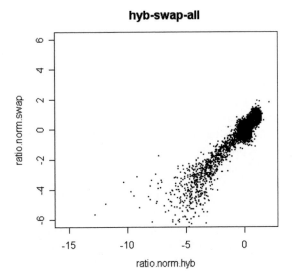

Fig. 6.3 Performance consistency on dye-swap hybridization. A scatter plot of the raw log2 ratio of the pairwise dye-swap hybridization profiles from one tumor cell line sample

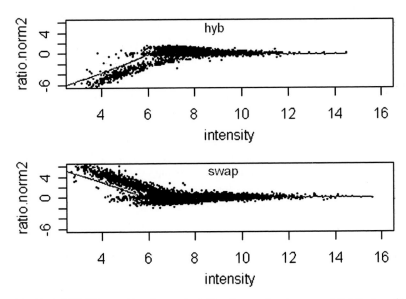

Fig. 6.4 The LOWESS smoothing figures for hyb and swap showing a good distribution of log2 ratios for all intensity probes

dsd, dsd.hyb, and dsd.swap. The thresholds for failing experiments are set arbitrarily. Users should set thresholds based on the range of the parameters for a collection of experiments and correlate these parameters with the profiles figures.

Figures generated during normalization process indicate experiment's quality (Figs. 6.3 and 6.4 are examples).

The analytical software may also be purchased from Agilent Technologies.

4 Notes

1. Ozone level as little as a 5 ppb can affect the Cy5 signal and compromise array-CGH performance. During some summer days, we detected ozone levels as high as 60 ppb in the laboratory.
2. High air humidity affects the autofocus of the microarray scanner.
3. Although low-input protocols with as little as 200 ng of genomic DNA have been successful for purposes of oligo array-CGH profiling, our standard protocol requires 2 μg of DNA per hybridization. A single hybridization quite often is sufficient to generate high-quality data. However, in 25% of the samples, we found that rehybridization and averaging of two datasets lowered the profile noise below the passing threshold (dsd<0.3). We believe that this category of experiments relates to marginal DNA quality or minor technical errors, therefore, benefiting from repeat hybridization [2]. If the latter is suspected, we recommend pairwise dye-swap hybridizations be carried out on the same day. In case of dye-swap hybridization, the recipes of DNA preparation, labeling, and hybridization should be scaled up accordingly.
4. The quality of the DNA is the most important factor in determining quality of the profile. In addition to the amount of contaminating DNA from a source other than tumor cells (e.g., stromal cells, normal to tumor ratio), purity of the genomic DNA and size range of fragments for input are major determinants of successful labeling reactions.
5. When using the NanoDrop, select Sample Type to be DNA-50. A low A_{260}/A_{230} ratio may indicate a guanidine salt carryover (Trizol or the single-column based isolations) that can interfere with downstream digestion and labeling. Additionally, the presence of guanidine will lead to higher 260 nm absorbance. This means that the DNA quantity may be overestimated. Use Microcon YM-30 centrifugal filters to desalt the DNA sample.
6. There is an optimal size range for labeling and control of DNA quality. We previously tested and compared performance of DNase digestion, mechanical shearing, and digestion with different restriction enzymes for this purpose and decided that the restriction enzyme digest provided the most controlled and consistently reproducible fractionation profile. Although we used *Dpn*II or *Hinc*II or *Alu*I/*Rsa*I, the Agilent oligo-arrays are designed for the latter and, therefore, *Alu*I/*RSa*I are current protocol.
7. The Cy5 incorporation is more difficult than Cy3, so the trick is to balance the incorporation rates by using a better quality DNA (usually the reference) for Cy5 labeling. Labeling efficiency is a major determinant of the DNA input requirement and overall signal and noise of the array. More efficient dye incorporation reduces the DNA input and potentially lowers the overall reagent cost.
8. Use bar codes when possible.
9. When using the NanoDrop, select the MicroArray mode ssDNA-33.
10. Both the yield of labeling and the dye incorporation are predictive of hybridization success. Dye incorporation below the cutoff is considered a labeling failure.

The source of failure can be traced to labeling reagents or technical procedures by readings on the reference DNA (conducted in parallel) versus DNA quality. Dye incorporation readings above 8 pmol/µL strongly suggess dNTP carryover and, therefore, is considered a failed labeling reaction. The amount of labeling product from 2 µg of input DNA should exceed 80 ng/µL in 150 µL reaction.

11. The time frame for the labeling following the start of hybridization is one day. About 12 individual hybridizations can be set up in one day by one person.

12. The hybridization system affects the DNA input requirement and cost of array-CGH, because the hybridization volume affects the amount of labeled product required to generate a specific level of signals. Dynamic hybridization systems greatly enhance performance. However, among the various dynamic platforms we tested, including Agilent hybridization oven, BioMicro MAUI, and Advalytix ArrayBooster, we found that hybridization volume is critical. With a large-volume system, such as Agilent's hybridization oven, mixing is sufficient with rotation and performance is consistent. With a small-volume system, such as the BioMicro and Advalytix, the mixing mechanics are a common source of technical failure. However, with the smaller volume, such as BioMicro's MAUI system, which requires 50 µL instead of 500 µl in the Agilent oven, spot intensity typically is twofold higher (~600–700 scanner units on average compared to ~350). Therefore, adaptation of a smaller-volume protocol would permit a 50% reduction in DNA input or labeling reagent (e.g., Cy3/Cy5 dyes). This improved reaction kinetics, leading to higher signals of hybridization, means lower noise. We had extensive experience with the MAUI system, as we developed the original protocol that enabled sufficient signals of detection on cDNA and unoptimized gene-expression oligo arrays. Despite the better signal performance, we adapted Agilent's hybridization protocol as our standard on the optimized CGH arrays for the following reasons: The lid manufacturer for the MAUI system was unable to deliver consistently high-quality hybridization chamber lids, so we suffered a significant failure rate due to leaky or warped lids. We have a commitment from BioMicro that it will work to improve the lid quality and minimize leakage. In parallel, we will develop screening procedures to cull leaky lids. Moreover, the smaller hybridization volume and higher signals allows a shorter hybridization time. This reduces the array-CGH turnaround time from 40 h to 18 h, translating into increased array-CGH cycle per workweek, thus reducing personnel cost.

13. Cot-1 DNA solution and blocking agent can be prepared as a master mix.

14. Historically, preannealing of Cot-1 DNA has been used for hybridization of a repeat-rich probe with high-complexity targets, like BAC or cDNA arrays. For modern oligo-microarrays, the utility of this step is questionable.

15. In case of leaks, insert the assembly into the Jig immersed to water. Holding the Jig firmly in place, grasp the removal tab on the lid, and slowly peel the lid away from the slide. Bend the lid as you peel it away from slide. Place the slide into the slide staining dish. Wash for 5 min. Dip the slide in isopropanol for 10 s, and air dry. Repeat the assembling and the hybridization buffer preloading.

16. The residual detergent may cause green backgrounds on the arrays. Try to remove the green haze by washing arrays in acetonitrile (Sigma). Do acetonitrile washes in a vented fume hood.
17. For high-throughput experiments, consider automated wash stations for the microarrays (Advalytix, Tecan).

Acknowledgments We acknowledge significant contributions of Craig Cauwels, Melissa Donovan, and Ilana Perna in the development and optimization of these technical protocols and Dr. Cameron Brennan for customization of the analysis tools. This work has been performed in the Arthur and Rochelle Belfer Cancer Genomics Center, made possible by a gift from the Belfer Foundation. Additional support from the NIH (RO1CA99041, PO1CA095616, P50CA093683) and the Goldhirsh Foundation are acknowledged.

References

1. Pinkel D, Segraves R, Sudar D, Clark S, Poole I, Kowbel D et al (1998) High resolution analysis of DNA copy number variation using comparative genomic hybridization to microarrays. Nat Genet 20:207–211
2. Brennan C, Zhang Y, Leo C, Feng B, Cauwels C, Aguirre AJ et al (2004) High-resolution global profiling of genomic alterations with long oligonucleotide microarray. Cancer Res 64:4744–4748

Chapter 7
Detection of Copy Number Changes at Multiple Loci in DNA Prepared from Formalin-Fixed, Paraffin-Embedded Tissue by Multiplex Ligation-Dependent Probe Amplification

Minoru Takata

Abstract With the increasing knowledge on the genetic alterations associated with various cancers, molecular analysis of these alterations on formalin-fixed paraffin-embedded tissue is of increasing importance. Multiplex ligation-dependent probe amplification (MLPA) is a novel technique to measure the copy number of up to 45 nucleic acid sequences in a single reaction. This method relies on sequence-specific probe hybridization to genomic DNA followed by multiplex-PCR amplification of the hybridized probe, and semi-quantitative analysis of the resulting PCR products. This method is easy to perform; requiring as little as 50 ng of DNA extracted from formalin-fixed, paraffin-embedded (FFPE) tissue.

Keywords Chromosome aberrations; FFPE tissue; DNA imbalances; melanocytic neoplasm

1 Introduction

With the increasing knowledge about genetic alterations associated with various cancers, molecular analysis of these alterations on formalin-fixed, paraffin-embedded (FFPE) tissue is of increasing importance in diagnostic pathology as well as cancer research. Different methods are available for examining routinely processed material for copy number changes, including comparative genomic hybridization (CGH), fluorescence in situ hybridization (FISH), and loss of heterozygosity (LOH) analysis. However, most of these techniques have technical limitations. With FISH and LOH analysis, only one to several loci can be investigated. Although CGH is able to screen copy number changes throughout the whole genome, it requires a relatively large amount of tissues to obtain sufficient DNA (at least 500 ng for a single experiment).

Multiplex ligation-dependent probe amplification (MLPA) is a novel technique to measure the copy number of up to 45 nucleic acid sequences in one reaction [1]. This method relies on sequence-specific probe hybridization to genomic DNA followed by multiplex-PCR amplification of the hybridized probe and semi-quantitative analysis of the resulting PCR products. The method is easy to perform; requiring as little as 50 ng of DNA. A further advantage of this method is that it can be performed on partially degraded DNA, which makes this technique very suitable for analysis of routinely processed paraffin-embedded sections. A couple of reports, published recently, utilize this method for the analysis of copy number changes in melanocytic tumors [2, 3].

2 Materials

2.1 DNA Extraction from FFPE Tissues

1. Proteinase K solution: 2 mg/mL proteinase K in 25 mM Tris-HCl, pH 8.2 (stock solution kept at −20 °C). The stock is diluted 10-fold in 25 mM Tris-HCl, pH 8.2 at 25 °C. Make fresh dilutions.
2. 1 M NaSCN: 81 g sodium thiocyanate (Sigma S7757)/ L in demineralized water. Store at room temperature.
3. TE buffer: 10 mM Tris-HCl, pH 8.2, 1 mM EDTA.
4. Xylol (Sigma).
5. Heating block for slides.
6. Ethanol series comprising successively 75, 96, and 100% ethanol.
7. Inverted microscope.

2.2 MLPA Reaction

1. The SALSA human chromosome aberration kits, P005, P006 and P007 (MRC Holland, Amsterdam, the Netherlands).
2. Each kit contains the following:

   Probe mix.
   MLPA buffer (1.5 M KCl, 300 mM/L Tris-HCl, pH 8.5, 1 mM/L EDTA).
   Ligase-65.
   Ligase-65 buffer A.
   Ligase-65 buffer B.
   PCR buffer.
   FAM-labeled PCR primers including dNTPs.
   Polymerase.
   Enzyme dilution buffer.
3. Ligase-65 mix: This should be made less than 1 h before use and stored on ice. Mix 3 μL Ligase-65 buffer A, 3 μL Ligase-65 buffer B, and 25 μL water. Add 1 μL Ligase-65 and mix again.

4. Polymerase mix: This should be made less than 1 h before use and stored on ice. Mix 2 μL SALSA PCR-primers, 2 μL SALSA enzyme dilution buffer, and 5.5 μL water. Add 0.5 μL SALSA polymerase. Mix well, but do not vortex.
5. Thermal cycler with heated lid.

2.3 Gel Electrophoresis

1. ROX-labeled internal size standard (ROX-500 Genescan, Applied Biosystems, Foster City, CA).
2. Formamide, deionized (nuclease and protease tested, Sigma).

3 Methods

3.1 DNA Extraction from FFPE Tissues (see Note 1)

1. Prepare 10-μm thick sections of normal and tumor tissues that are mounted on glass slides (see Notes 2 and 3).
2. Heat a slide for 15 min at 75 °C to melt the paraffin. Then place the slide in xylol for 5 min. Repeat this step to dissolve the paraffin as much as possible.
3. Immerse the slide sequentially in 100% ethanol for 30 s, 96% ethanol for 30 s, and 75% ethanol for 30 s, followed by a final wash in tap water for 30 s.
4. Incubate the slide in 1 M NaSCN at 37 °C overnight (see Note 4).
5. Rinse the slide with tap water and finally in TE and let it dry.
6. Manually microdissect tumor areas under an inverted microscope, referring to a corresponding hematoxylin and eosin-stained pathology slide (see Note 5).
7. Pour a few drops (20–40 μL) of proteinase K solution onto the tissue and scratch it off into a 1.5-mL tube. Add more proteinase K solution to make a final volume of 50–100 μL.
8. Incubate at 37 °C overnight.
9. The next morning, heat the tube for 20 min at 80 °C to inactivate the proteinase K.
10. Centrifuge for 10 min at 12,000 g.
11. Transfer the supernatant to a clean tube. Use 2–5 μL for each MLPA reaction.

3.2 MLPA Reaction

All the reactions should be carried out in a thermal cycler equipped with a heated lid (see Note 6). Use 200-μL PCR tubes for the following reactions.

3.2.1   DNA-Denaturation and Hybridisation of the SALSA-Probes

1. Dilute the DNA sample (20–500 ng DNA) with TE buffer to 5 µL.
2. Heat for 5 min at 98 deg;C and cool to 25 °C before opening the thermal cycler.
3. Add 1.5 µL of SALSA probe mix and 1.5 µL of MLPA buffer. Mix with care.
4. Incubate for 1 min at 95 °C followed by 16 h at 60 °C.

3.2.2   Ligation Reaction

1. Reduce temperature of the thermal cycler to 54 °C. While at 54 °C, add 32 µL of Ligase-65 mix to each sample and mix well.
2. Incubate at 54 °C for 10–15 min, then heat to 98 °C for 5 min.

3.2.3   PCR Reaction

1. Add 4 µL of 10X SALSA PCR buffer to 26 µL of water and 10 µL MLPA ligation reaction.
2. While tubes are in the thermal cycler at 60 °C, add 10 µL polymerase mix to each tube and start the PCR reaction.
3. Cycle the reactions with the following program: 95 °C for 30 s, 60 °C for 30 s, and 72 °C for 60 s, for 33 cycles. Follow this by final extension at 72 °C for 20 min (*see* **Note 7**).

3.3   Gel Electrophoresis

The amount of MLPA PCR reaction is quantified by capillary electrophoresis or sequence gel electrophoresis, depending on the instrument and fluorescent label used. The electrophoresis conditions for the ABI 310 and ABI3100 Genetic Analyzer we use are shown.

3.3.1   ABI-310 (One Capillary)

1. Mix: 0.75 µL of the PCR reaction, 0.75 µL of water, 0.5 µL of ROX-labeled internal size standard, and 12 µL of Hi-Di formamide. Volume is based on the use of a 0.5-mL genetic analyzer sample tube (for a 48-well tray).
2. Incubate the mixture for 2 min at 94 °C and cool on ice.
3. Capillaries: 310 capillaries, 47 cm × 50 µm. Polymer: POP-6. Settings: Run temperature 60 °C. Injection time 5–10 s. Run voltage 15 kV. Run for 30 min. Filter set D (select the correct dye set for using other internal standard).

3.3.2 ABI-3100 (16 Capillaries)

1. Mix: 1 µL of the PCR reaction, 0.5 µL of ROX-labeled internal size standard, and 8.5 µL Hi-Di formamide. Volume is based on the use of a 96-well plate (for a 96-well tray).
2. Incubate the mixture for 2 min at 94 °C and cool on ice.
3. Capillaries: 3100 capillary array, 36 cm. Polymer: POP-4 or POP-6. Settings: Run temperature 60 °C. Cap fill volume: 184 steps. Prerun voltage: 15 kV. Prerun time: 180 s. Injection voltage: 3.0 kV. Injection time 10–30 s. Run voltage 15 kV. Run time 1,500 s. Filter set D (select the correct dye set for using other internal standard).

3.4 Data Analysis

3.4.1 Data Quality Control

DNA quantity control fragments of 64, 70, 76, and 82 bp are included in each SALSA probe mix. These generate amplification products much smaller than the probe amplification products. The data quality control fragments present a warning when the amount of sample DNA used was lower than the 20 ng human DNA required for reliable MLPA results. The four data quality control fragments generate amplification products when ligation is omitted or when no sample DNA is present. The four 64–82 bp amplification products present a much smaller peak size than the 94 bp amplification product that is sample DNA and ligation-dependent control product, when 50 ng human DNA or more is used in the MLPA reaction (Fig. 7.1, top). If the 64, 70, 76, and 82 bp amplification products have similar or larger peak sizes than the 94 bp fragment and the 130–472 bp MLPA probe amplification products, either the ligation reaction failed or the amount of sample DNA was less than 20 ng (Fig. 7.1, bottom). In either case, the results obtained may not be reliable and should not be used for further analysis.

3.4.2 Calculations of the Relative Probe Ratios

1. Export the gene scan data of sizes and peak areas of multiplex PCR products to an Excel or SPSS file. A data flow manual can be downloaded from the MRC-Holland website [4].
2. Remove nonspecific amplification products and primer-dimer peaks. They are either low peaks or of short length.
3. Normalize all the expected MLPA products by dividing each peak area by the combined peak area of all peaks in that lane (relative peak area).

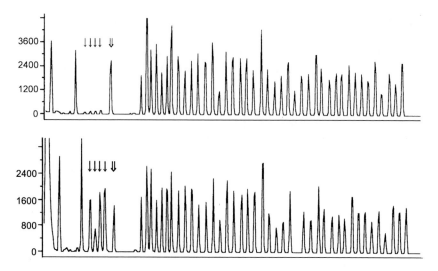

Fig. 7.1 Representative examples of the MLPA electrophoresis profiles using a SALSA P006 Chromosome Aberration Kit. (Top) An example of good experiment. The 64-, 70-, 76-, and 82-bp quality control peaks (black arrows) have a much smaller peak size than the 94 bp amplification product (a white arrow), which are sample DNA and ligation-dependent control product. (Bottom) An example of bad experiment with insufficient DNA, which should not be used for copy number calculations. The 64, 70, 76, and 82 bp amplification products (black arrows) have similar or larger peak sizes than the 94 bp fragment (a white arrow) and the 130–472 bp MLPA probe amplification products

4. Calculate the relative copy number for each probe, which is expressed as a ratio of the relative peak area for each locus of the sample to that of the relative peak area of the normal sex-matched control (*see* **Note 8**).
5. Perform similar calculations for peak heights. Normally, there are no significant differences in the outcome of normalized areas and heights.
6. Assess the copy number change for each locus (*see* **Note 9**).

4   Notes

1. A number of protocols and kits are available for DNA extraction from FFPE tissues. A critical point in choosing the DNA extraction method is to reduce impurities, which would inhibit ligation and PCR reactions, as much as possible. The protocol described in the text is simple and, as far as we know, the best for an MLPA reaction. Other methods could be used, but the DNA extraction protocol should be identical between control and tumor samples.
2. Sufficient amounts of DNA can be obtained from a 5 mm × 5 mm × 10 μm-thick section. Do not use more starting material than necessary, to reduce the contamination of impurities that could influence the polymerase activity.

3. Control samples of normal male and female DNA should be prepared, respectively, from tissues fixed and processed in the same way as the tumor samples. We use normal lymph node tissues for control samples.

4. This incubation step in the chaotropic NaSCN solution removes impurities from the slides and improves the efficacy of proteolytic digestion [5].

5. This step can be skipped if the section consists of tumor cell populations only. For small tumors or tumors with marked infiltration of nonneoplastic cells, laser microdissection may be used to obtain pure tumor cell populations.

6. The following program of a thermal cycler covers the complete MLAP reaction:

    98 °C for 5 min, 25 °C pause.
    95 °C for 1 min, 60 °C pause.
    54 °C for 15 min.
    98 °C for 5 min, 4 °C pause.
    60 °C pause.
    95 °C for 30 s, 60 °C for 30 s, 72 °C for 1 min for 33 cycles.
    72 °C for 20 min, 4 °C pause.

7. PCR cycles can be increased up to 40.

8. As the MLPA analysis is based on a PCR reaction, the specificity and reproducibility of this method needs to be critically evaluated. When the specificity of MLPA was examined by calculating the ratio for each MLPA robe for normal DNA sample, false-positive results of 7.3% were found. When ratios were calculated for test samples using normal samples from other MLPA experiments instead of reference samples included in the same experiment, the false-positive results rose to 20%. Therefore, normal DNA samples should be included in each experiment to decrease the number of false positives [2]. Alternatively, the average values of pooled data of normal DNA samples obtained from a number of experiments can be used for calculations. Duplicate or triplicate experiments for one sample decreases the ratio of false-positive results.

9. The copy number ratio, which is considered significant, varies among studies. One study considered a probe ratio below 0.8 or above 1.2 to represent DNA copy number losses and gains, respectively [2], while another study considered ratios less than 0.7 as loss and higher than 1.3 as gain calculated from both peak areas and peak heights [3]. It is recommended that a stricter criteria be adopted to reduce false-positive results.

References

1. Schouten JP, McElgunn CJ, Waaijer R, Zwijnenburg D, Diepvens F, Pals G (2002) Relative quantification of 40 nucleic acid sequences by multiplex ligation-dependent probe amplification. Nucleic Acids Res 30:e57

2. van Dijk MC, Rombout PD, Boots-Sprenger SH, Straatman H, Bernsen MR, Ruiter DJ, Jeuken, JW (2005) Multiplex ligation-dependent probe amplification for the detection of chromosomal gains and losses in formalin-fixed tissue. Diagn Mol Pathol 14:9–16

3. Takata M, Suzuki T, Ansai S, Kimura T, Shirasaki F, Hatta N, Saida T (2005) Genome profiling of melanocytic tumors using multiplex ligation-dependent probe amplification (MLPA): Its usefulness as an adjunctive diagnostic tool for melanocytic tumors. J Dermatol Sci 40:51–57
4. Available at http://www.mlpa.com/pages/support_mlpa_analysis_infopag.html
5. Hopman AH, van Hooren E, van de Kaa CA, Vooijs PG., Ramaekers FC (1991) Detection of numerical chromosome aberrations using in situ hybridization in paraffin sections of routinely processed bladder cancers. Mod Pathol 4:503–513

Chapter 8
Application of Microarrays for DNA Methylation Profiling

Axel Schumacher, Andreas Weinhäusl, and Arturas Petronis

Abstract Comprehensive analyses of the human epigenome may be of critical importance in understanding the molecular mechanisms of complex diseases, development, aging, tissue specificity, parental origin effects, and sex differences, among other systemic aspects of human biology. However, traditional DNA methylation methods allowed for screening of only relatively short DNA fragments. The advent of microarrays has provided new possibilities in DNA methylation analysis, because this technology is able to interrogate a very large number of loci in a highly parallel fashion. There are several permutations of the microarray application in DNA methylation profiling, and such include microarray analysis of bisulfite modified DNA and also the enriched unmethylated or hypermethylated DNA fractions using methylation-sensitive restriction enzymes or antibodies against methylated cytosines. The method described in detail here is based on the analysis of the enriched unmethylated DNA fraction, using a series of treatments with methylation-sensitive restriction enzymes, adaptor ligation, PCR amplification, and quantitative mapping of unmethylated DNA sequences using microarrays. The key advantages of this approach are the ability to investigate DNA methylation patterns using very small DNA amounts and relatively high informativeness in comparison to the other restriction-enzyme- based strategies for DNA methylation profiling [1].

Keywords DNA methylation, microarrays, epigenetic profiling, epigenetic biomarkers, whole genome approach, epigenetics

1 Introduction

The key principles and some technical details of this method are described in our recent article [1] and the strategy for enrichment of unmethylated portions of the genome is presented in Fig. 8.1. Briefly, genomic DNA (gDNA) is

From: Methods in Molecular Biology, Vol. 439: *Genomics Protocols: Second Edition*   109
Edited by Mike Starkey and Ramnath Elaswarapu © Humana Press Inc., Totowa, NJ

digested with methylation-sensitive restriction enzymes, such as *Hpa*II and *Hin*6I. Whereas methylated restriction sites remain unaltered, the sites containing unmethylated CpGs are cleaved by the enzymes, and DNA fragments with 5′-CpG protruding ends are generated. In the next step, the double-stranded adapter CG-1 is ligated to the CpG overhangs. At this point, it is expected that most of the relatively short (<1.5 kb) and amplifiable DNA fragments derive from the unmethylated DNA regions. Some ligation fragments, however, still may contain methylated cytosines. A large proportion of these fragments are eliminated by treatment with McrBC, thereby increasing the specificity of the enrichment of the unmethylated DNA fraction. McrBC cleaves DNA containing methylcytosine on one or both strands, recognizing two half sites of the form (G/A)^{m}C; these half sites can be separated by up to 3 kb, but the optimal separation is 55–103 base pairs. The remaining pool of unmethylated DNA fragments then is enriched by aminoallyl-PCR amplification that uses primers complementary to the adapter CG-1. An important advantage of using protruding ends in the adapter-ligation step is that degraded gDNA fragments will not be ligated and amplified and, therefore, will not interfere with DNA methylation analysis (which is especially useful when analyzing tissues with relatively long postmortem interval or paraffin-embedded samples). The enriched unmethylated DNA fractions then are labeled with fluorescent dyes and hybridized to microarrays. Several different types of microarrays can be used for epigenetic analysis; for example, oligonucleotide arrays of individual genes or microarrays containing relatively larger DNA fragments of gene regulatory regions, such as CpG islands [1]. Yet, it is evident that epigenetic profiling should be performed in a systematic, unbiased fashion and not limited to the traditionally preferable regions, such as CpG islands. Numerous other genomic loci exist that may be sites for important epigenetic modification, including enhancers, imprinting control elements, and the regions that encode regulatory RNA elements. It therefore is beneficial to use high-density tiling arrays that can cover entire chromosomes and even entire genomes represented by millions of oligonucleotides on glass chips. Whole genome tiling arrays are already available for several species and will soon be available for the entire human genome [1–3].

2 Materials

2.1 *Adapter Design*

1. Adapter storage buffer ST (100 mL): 10 mM Tris-HCl, pH 8.5, 50 mM NaCl.
2. Primer for preparation of universal adapter CG-1:

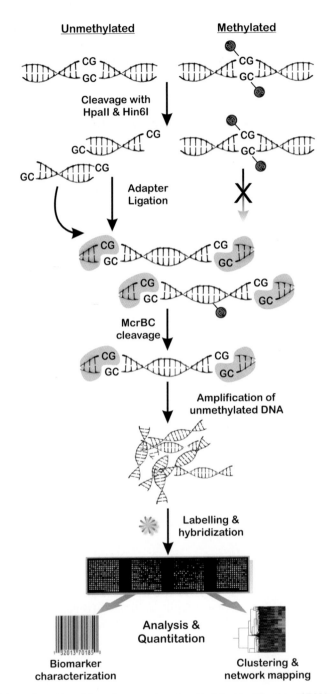

Fig. 8.1 Schematic outline of the microarray-based method for identification of DNA methylation differences in genomic DNA. Samples are cleaved by methylation-sensitive restriction endonucleases, such as *Hpa*II and *Hin*6I, ligated to the CpG-overhang specific adapter, then cut by McrBC to eliminate residual methylated DNA fragments. The resulting unmethylated DNA fragments are selectively enriched by adapter-specific aminoallyl-PCR, labeled, and hybridized to microarrays

primer CG-1a: 5′-CGTGGAGACTGACTACCAGAT-3′
primer CG-1b: 5′-AGTTACATCTGGTAGTCAGTCTCCA-3′.

2.2   Methylation-Sensitive Cleavage of DNA

1. *Hpa*II 10 U/µL (Fermentas).
2. *Hin*6I 10 U/µL (Fermentas, *see* **Note 1**).
3. Spike DNA (optional, see Sect. 3, Methods).

2.3   Ligation

1. ATP stock-solution (NEB). Make 10 µM stock solution, and store as aliquots at −20 °C.
2. T4 DNA Ligase (Fermentas).

2.4   McrBC Digestion

1. McrBC enzyme (NEB).
2. Guanosine triphosphate (GTP, supplied with the enzyme): Make 10X aliquots, since GTP is highly unstable.
3. Agarose (Peqlab).
4. Ethidium bromide (10 mg/ml solution, Bio-Rad).
5. Gel electrophoresis apparatus.

2.5   CpG-Specific Adapter Amplification

1. 100 mM nucleotide stock solutions (Fermentas).
2. 50 mM aminoallyl-dUTP (Ambion).
3. Labeling nucleotides (20 mM): Thaw one vial aminoallyl-dUTP. Add 16.5 µL of H$_2$O to this stock tube (contains 50 µL aa-dUTP). To this, add 16.6 µL dTTP, 41.6 µL dGTP, 41.6 µL dATP, and 41.6 µL dCTP (from 100 mM stock nucleotide tube). Mix and store at −20 °C.
4. Taq polymerase, 5 U/µL (NEB).
5. 200 pmol/µL CG-1b-primer (see Sect. 2.1).

2.6   Purification of Amplification Products

1. Microcon YM-50 columns (Millipore).
2. Spectrophotometer.

2.7 Labeling

1. Na_2CO_3 (Sigma). Prepare $0.1\,M$ solution in water.
2. FluoroLink monofunctional dyes, Cy3 and Cy5 (Amersham Biosciences).
3. DMSO (Sigma).
4. Hydroxylamine (Sigma) (10 mL): $4\,M$ hydroxylamine. Make aliquots and store at $-20\,^{\circ}\mathrm{C}$.
5. Sodium bicarbonate buffer ($0.1\,M$, pH 9.0): Add 0.42 g of sodium bicarbonate to a flask. Add 50 mL of water, and mix with a magnetic stirrer until it dissolves completely. The pH should be around 8.5. Adjust the pH to 9.0 by adding ~2.5–3.0 mL of the $0.1\,M$ Na_2CO_3 solution. Monitor the pH change carefully, and do not exceed pH 9.3. Filter sterilize the buffer (prepare as fresh as possible). Aliquot and store at $-20\,^{\circ}\mathrm{C}$. Do not refreeze the buffer. Use each aliquot only once.
6. Prior to coupling, prepare the dyes. The Cy dyes come in packages, which contain five vials. Dissolve one vial of dye in 72 μL DMSO. Aliquot 4.8 μL in 15 light-protected amber reaction tubes, dry immediately in a speedvac, and store at $-20\,^{\circ}\mathrm{C}$ protected from light (*see* **Note 2**).
7. MinElute PCR-purification kit (Qiagen).
8. $3\,M$ Sodium acetate solution (pH 5.2).

2.8 Hybridization

1. SlideHyb glass array hybridization buffer 2 (Ambion).
2. Yeast tRNA (Sigma).
3. Bovine serum albumin (Sigma).
4. Cot-1 DNA (Roche Diagnostics).
5. Oligonucleotides poly(dA)-poly(dT).
6. Isopropanol (>99.8%, Sigma).
7. Wash solution I: 2X SSC, 0.5 % SDS. Filter-sterilize the buffer (*see* **Note 3**).
8. Wash solution II: 0.5X SSC, 0.5 % SDS. Filter-sterilize the buffer.
9. Coverslips: Hybri-Slips (Sigma).

3 Methods

Here, we present a complete protocol for DNA methylation profiling using the unmethylated fraction of the genome. This protocol works with most microarray types, as exemplified here with CpG island and oligonucleotide microarrays. A detailed description of how specific "epigenetic" microarrays are designed and how the slides are processed can be found elsewhere [1]. DNA samples can be either interrogated as pairs (tester and control) on one array, as described in this protocol, or hybridized independently, with only one fluorescent dye (e.g., for Affymetrix tiling arrays). This protocol provides two examples: (1) a comparison of DNA derived from human control fibroblast cells, with the same cells treated with either staurosporine (STS) or retinoic acid (RA), and (2) lymphocyte DNA from Prader-Willi syndrome (PWS) patients compared with healthy controls and Angelman syndrome (AS) patients.

3.1 Adapter Design

In the CpG-specific adapter design, the following aspects must be taken into account:

1. It must contain a CpG overhang, which fits to the restriction site of the enzymes used (Fig. 8.2a).
2. The two nucleotides next to the CpG overhang have to be different from the nucleotides within the recognition sequence of all the enzymes used in the restriction

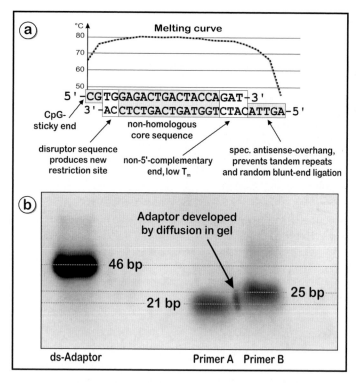

Fig. 8.2 Preparation of the CpG-specific adapter. (a) CG1-adapter components. (b) Annealing of the CG primers produces a double-stranded adapter that can be identified by electrophoresis as a strong band (200 pmol loaded; 2.5 % agarose gel)

digests. This ensures that, after ligation of the adapters, the old restriction site is disrupted and the sequence cannot be recut by the enzymes.

3. The melting temperature (T_m) should be about equal throughout the adapter and decreasing at the 3′ end opposite to the CpG overhang.

4. The core sequence of the adapter should be nonhomologous to any sequence in the gDNA. Additionally, it is advisable to avoid palindromic sequences, which can lead to mispriming.

5. If you want to cut the adapter from the target fragments, include a very specific, new recognition site within the adapters core sequence (close to CpG overhang), which does not cut frequently in the genome you analyze. Use a rare eight-mer (e.g., SdaI (5′-CC↑TGCA↓GG) for sticky ends or MssI (GTTT↓AAAC) for blunt ends.

6. Use a specific, long antisense overhang, which prevents the forming of tandem repeats and random blunt-end ligation to the genome (e.g., through degraded DNA).

7. To avoid improper annealing between the two primers, choose a non-5′-complementary end (see Fig. 8.2a).

The adapter preparation involves the following program:

1. Dissolve the nonphosphorylated ssDNA oligonucleotides in ST buffer with a concentration of approximately 800 pmol. The presence of some salt is necessary for the oligos to hybridize into a double-stranded (ds) adapter. Keep some of the CG-1b oligo, as it will be later used as a primer to amplify the unmethylated DNA fraction.

2. Measure the concentration in a spectrophotometer and adjust the concentration to 400 pmol (see **Note 4**).

3. Mix together equal molar amounts of each complementary oligonucleotide to a final concentration of 200 pmol, and heat for 5 min in a thermalcycler at 80 °C. Cool down at 1 °C/min until the mixture reaches room temperature (RT). The longer the cooling period, the lower is the risk of hairpin structures.

4. Check a small amount of the ds adapter on a 2.5% agarose-gel (see Fig. 8.2b), which should generate a strong band. Run the single-stranded primers next to the adapter as a control.

5. Store the adapters at −20 °C.

3.2 Methylation-Sensitive Cleavage of DNA

Over 260 methylation-sensitive restriction enzymes (MSREs, including isoschizomers) are available; however, not all enzymes are useful and informative for DNA methylation profiling. Informative MSREs are defined by the number of cleavage fragments that can be ligated to adapters, efficiently amplified, and are not lost during column-purification steps [3]. Some enzymes, although they cut frequently in the genome, produce fewer informative fragments than enzymes that do not cut as frequently. For example, the nonpalindromic *Aci*I (5′-CCGC-3′) recognizes more than twice as many CpG sites in CpG island regions than *Hpa*II but, on the other hand, produces fewer fragments in the size range that can be detected by PCR or amplified fragment-length polymorphism (AFLP) methods (Table 8.1). Different requirements for the enzyme reduces the list of potentially useful and informative MSREs to about 17, which would cover up to 85% of all CpG island CpG dinucleotides, but less than 50% of all CpG dinucleotides in other genomic regions [3]. Here, we demonstrate the technology with a double digestion using *Hpa*II and *Hin*6I enzymes (*see* **Note 5**). Before the test DNA can be processed, it is advantageous to add "spiking" DNA to the gDNA samples. The spiking DNA consists of "alien" (exogenous) DNA, which allows monitoring of each step of the experiment (*see* **Note 6**); as with any microarray experiment, controls are crucial for monitoring experimental variability. Suitable spiking DNA consists of sequences that are frequently spotted on commercially available microarrays, such as *Arabidopsis*, RNA spikes, or artificial sequences.

1. 400 ng of genomic DNA is digested with 10 U *Hpa*II plus 10 U *Hin*6I for 6 h at 37 °C. The DNA should be highly concentrated to keep the total volume of the reaction below 20 μL.
2. The samples are then incubated for 10 min at 65 °C to inactivate the enzymes. Let the sample cool down at 1 °C/min. This step ensures reannealing of small DNA fragments that can denature during heat inactivation.
3. Immediately proceed with the ligation reaction. It is important that the digested DNA is processed as soon as possible to avoid degradation of the CpG overhangs.

3.3 Ligation

1. Take 200 ng of the digested DNA for a standard ligation. If digested DNA is stored too long before ligation, ligation efficiency decreases.
2. Adjust the buffer concentration to 1X, taking into account that the unpurified restriction product still contains salts from the previous DNA cleavage reaction. That means that the buffer has to be supplemented for only the extra volume added to the DNA.
3. Add ATP to a final concentration of 1 m*M*.
4. Add 120 pmol of the double-stranded CG adapter and mix (*see* **Note 7**).
5. Adjust the volume of the reaction with water to 18 μL, mix, then heat the sample for 10 min at 45 °C in a water bath or thermal cycler (*see* **Note 8**).
6. Take the sample out of the cycler and chill on ice immediately.
7. Add quickly 2 μL (10 U) of the ligase to a final volume of 20 μL, mix well without introducing bubbles and place the tubes back into the cycler. Incubate for 16 h, followed by a short, 5 min, inactivation step at 65 °C. To favor reannealing of short denatured DNA fragments, let the sample cool down at 1 °C/min until room temperature is reached (*see* **Note 9**).
8. Store the ligation mixture at −20 °C (4 °C if the reaction is processed at the same day).

Table 8.1 Methylation-sensitive restriction enzymes

MSRE	Cut site 5′–3′	Approx. % of CpGs in CpG islands[*]	Approx. % of CpGs in non-CpG islands[*]	Fragments/kb in CpG islands[*]	Fragments/kb in non-CpG islands[*]
AciI (SsiI)	CCGC, GCGG	30.60	17.36	3.23	1.79
Hin6I[**] (HinP1I)	GCGC	14.40	5.05	3.98	0.61
HpaII (BsiSI)	CCGG	11.70	9.33	3.98	1.18
HinI1 (BsaHI)	GRCGYC	2.6	0.9	1.92	0.11
HpyCH4IV (TaiI)	ACGT	1.66	6.73	1.24	0.97
Bsp119I (AsuII)	TTCGAA	0.11	0.13	0.11	<0.02
Bsu15I (ClaII)	ATCGAT	<0.05	0.39	<0.02	0.02

Note: Methylation-sensitive restriction enzymes. All the MSREs produce sticky ends that can be ligated to the CG-1 adapter for high throughput microarray-based DNA methylation profiling. Other MSREs that fit to the CG-1 adapter are Psp1406I, XmiI, BstBI or NarI; however, these enzymes have very few restriction sites in the human genome or are too expensive to be used in the required amount.

[*] Indicates the number of 75 bp, 2 kb long, that is, "informative," fragments, derived from CpG-island and non-CpG-island sequences in the human genome [3]. R = A/G; Y = C/T.

[**] See **Note 1**.

3.4   McrBC Digestion

1. Perform the reaction with NEB buffer 2, 1X BSA (10X), and 2X GTP (2 mM).
2. Add buffer only for the extra volume of the reaction. The ligation already took place in a restriction enzyme buffer.
3. ATP inhibits the reaction; therefore, make sure that the DNA sample was either purified before McrBC cleavage or the present ATP was degraded by heating the sample for some time during the "cool-down" phase of the ligation procedure.
4. Digest the DNA with 10 U/μg of McrBC for 8 h at 37 °C (see **Note 10**).
5. Heat inactivate the enzyme for 20 min at 65 °C. Store the ligation mixture at −20 °C (4 °C if the reaction is processed at the same day).

3.5   CpG-Specific Adapter Amplification

The protocol is based on aminoallyl (aa) nucleotide incorporation followed by coupling to N-hydroxysuccinimide (NHS) functionalized dyes. This indirect labeling method is advantageous compared to direct incorporation of dye labeled nucleotides in gDNA. The aminoallyl-nucleotide is better tolerated by the Taq polymerase than other fluorescent nucleotides analogues (see **Note 11**). While direct incorporation of dye labeled nucleotide protocols typically have incorporation efficiencies in the range of 2–5 dye molecules per 1,000 nucleotides, the aminoallyl protocol can achieve frequencies of incorporation in the range of 10–20 dye molecules per 1,000 nucleotides. Furthermore, the reagent costs for the aminoallyl-NHS protocols are lower. Aminoallyl-dUTP contains a reactive amino group on a two-carbon spacer attached to the methyl group on the base portion of dUTP. During the coupling reaction, after DNA synthesis, this amino group reacts with the NHS ester of the monoreactive Cy3 and Cy5 dyes.

1. For each 200 ng DNA, prepare the PCR-mixture as follows (final volume 100 μL):

Volume	Component	Final concentration
x μL	dd-H_2O	
10 μL	10X buffer	1X
14 μL	$MgCl_2$ (25 mM)	3.5 mM
1.25 μL	Allyl-dNTP mix (20 mM)	250 μM
1 μL	CG1b primer (200 pmol)	~1 pmol for each ng DNA
x μL	Adapter ligation	200 ng
3 μL	Taq polymerase (5 U/μL)	15 U (depending on amount of starting DNA)

2. Depending on the thermalcycler, the PCR reaction may have to be split into several tubes. Run the following program:

Step	Time	Temp	Cycles	Notes
1	5 min	72 °C	1	Fills in 3′ recessed ends[*]
2	1 min	95 °C	1	Starting denaturation
3	30 s	93 °C	24	Denaturation
4	2 min + 3 sec/cycle	68 °C		Annealing and extension
5	5 min	72 °C	1	Final extension
6	hold	4 °C		Storage

[*]This amplicon-PCR does not work with a hot-start procedure.

3. Check the size of the amplified DNA by separating 8 μL product on a 1% agarose gel (the size of DNA should range from approximately 150 to 2000 bp; see Fig. 8.3).

3.6 Purification of Amplification Product

The overall yield of the amplicon PCR is relatively high (20–100 μg). However, most commercially available column-based PCR cleanup kits cannot handle this large amount of DNA without clogging. Therefore, our laboratory uses YM-50 Microcon columns for cleanup, but other systems may work as well (*see* **Note 12**).

1. Add 500 μL of nuclease-free water to the empty columns and spin ~6 min at ~9,000 g until the column is dry (the columns contain small amounts of Tris on their membrane that can inhibit the subsequent labeling reaction. Therefore, the columns have to be washed prior to adding the allyl-PCR product).
2. Insert the Microcon sample reservoir into a 1.5 mL vial. Pipette the whole PCR product into the reservoir. Spin as per guidelines (approximately 1 min at ~9,000 g).

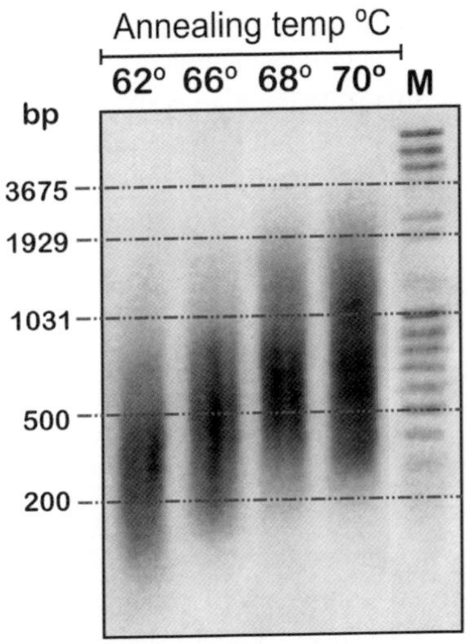

Fig. 8.3 The adapter-amplification results in a typical smear in the size range of 150–2,000 bp. To some extent, the lengths of the amplified fragments depend on the primer annealing temperature of the PCR reaction. Usually, an increased annealing or elongation temperature produces larger DNA fragments

The exact spin time depends on temperature and rotor and may have to be elucidated empirically. Be sure to align the strap of the collection tube cap toward the center of the rotor. Following the spin, the membrane still should be slightly wet.

3. Add 500 μL of dd-H$_2$O and spin again for 6 min at ~9,000 g. The membrane should look wet but not contain a visible amount of liquid. If there is liquid, spin it down for 1 min.
4. Add additional 90 μL of nuclease-free dd-H$_2$O (*see* **Note 13**).
5. Place the vial with the column into the Eppendorf shaker and shake at 400 rpm for 1 min (this helps to elute the entire DNA, which may be bound to the membrane). If you have no Eppendorf shaker, vortex the column slightly for several seconds.
6. Uncap the Microcon unit. Separate the sample reservoir from the filtrate cup, and place the sample reservoir upside down into a new vial. Spin for 3 min at ~1,500 g in invert spin mode to elute DNA.
7. Remove the column. Cap the vial with the eluate (~100 μL) for storage at −20 °C.
8. Use 5 μL of the eluate to measure the DNA concentration in a spectrophotometer.

3.7 Labeling

3.7.1 Coupling

The amount of unmethylated DNA fraction to be hybridized to an array depends on the surface area of the microarray. In our experience, 2 μg of the eluate is enough for a standard slide.

1. For all steps, shield all samples thath contain fluorescent dyes from light!
2. Dry the DNA sample (2 μg) in a speedvac.
3. Resuspend the DNA pellet in 9 μL of the 0.1 M sodium bicarbonate buffer. Add 3 μL DMSO. Let sample sit for 10–15 min.
4. Denature the sample for 2 min at 100 °C. In our experience, single-stranded DNA incorporates the Cy dyes better than native DNA.
5. Quickly spin down and add the sample to a dry aliquot of monofunctional Cy3 or Cy5 dye. Resuspend the dye pellets by pipetting carefully; try not to introduce air bubbles (if it happens, centrifuge the sample for a few seconds, this will get rid of the bubbles).
6. Incubate for 2 h at 30 °C in the dark.

3.7.2 Quenching

After the coupling step, reactive groups on the dyes must be quenched to prevent cross reactivity or exchange of dye molecules between active aminoallyl nucleotides. This protocol uses hydroxylamine to quench.

1. Add 4.5 µL of 4 M hydroxylamine. Mix well by pipetting up and down.
2. Incubate for 15 min at room temperature with slight agitation (e.g., on an Eppendorf shaker, ~400 rpm) in the dark.
3. Combine the Cy3 and C5 samples.

3.7.3 Cleanup of Product

To remove unincorporated dyes, columns have to be applied that efficiently purify fragments in the size range of ~50–2,000 bp. For this protocol, we are using the Qiagen MinElute PCR-Purification Kit.

1. Add 25 µL of dd-H$_2$O to the sample.
2. Add 3 µL of 3 M sodium acetate (pH 5.2) to ensure a low pH of the mixture. It is important to keep in mind the DNA binding curve for silica, on which this kit is based, is favorable at low pH but falls off precipitously around pH 8.0. Therefore, it is essential that the pH of the reaction be below pH 7.5.
3. Add 275 µL (5X vol.) of buffer PB and mix. Apply to column and centrifuge at 10,000 g for 1 min.
4. Discard the flow-through and wash the DNA five times with 750 µL buffer PE at 10,000 g for 1 min. A critical step for minimizing the background is to effectively wash away all unbound dye molecules during the final cleanup; small amounts of unincorporated dye can have great absorbance and thus give misleading results. If the color from the dyes still is visible in the wash, continue running wash buffer through the column until the eluate is clear.
5. Centrifuge the column for an additional 1 min at 10,000 g.
6. Place the column in a new, clean 1.5 mL microcentrifuge tube and elute by adding 25 µL prewarmed (~50 °C) elution buffer (EB) to the center of the membrane. Let it stand for 1 min at room temperature and spin at 10,000 g for 1 min.
7. Repeat the elution step with an additional 25 µL (final volume 50 µL).

3.7.4 Determination of Incorporated Dyes

1. Place the entire 50 µL sample in a clean microcuvette. Measure the absorbance of the Cy3/Cy5 sample in a spectrophotometer at 260 nm (DNA), 550 nm (Cy3 fluorescence), and 650 nm (Cy5 fluorescence). Be very careful about contamination in the cuvette as you will be recovering the sample for hybridization. Wash carefully after each measurement.
2. Calculate the amount of recovered DNA: $A_{260} \times 50 \times$ total volume of sample (µL) = ng of target.
3. Calculate the frequency of incorporation (FOI). For Cy3 incorporation, $86.5 \times (A_{550}/A_{260})$; and for Cy5 incorporation, $51.9 \times (A_{650}/A_{260})$. The results are expressed as the number of Cy-dCTP incorporated per 1,000 nucleotides of DNA (see **Note 14**). Optimal frequencies of incorporation are 15–20 and higher, although anything higher than 10 gives satisfactory results. Using targets with an FOI less than 6 may give high background or very weak signals.

3.8 Hybridization

Before starting to hybridize, it is recommended that the coverslips already are cut, the hybridization chambers ready, and all solutions prepared and adjusted to the correct temperature. To prepare the prehybridization buffer, combine 2 mL of SlideHyb glass array hybridization buffer 2, 80 µL of 20 µg/µL yeast tRNA, and 200 µg bovine serum albumin. As the hybridization buffer, use SlideHyb buffer 2 (*see* **Note 15**). Add 50 µL of 20 µg/µl yeast tRNA and 200 µL COT-1 DNA (1 µg/µL). The COT fraction of human placental gDNA consists mainly of repetitive DNA elements. COT DNA is expensive and may be omitted if the microarray used does not contain larger amounts of repetitive sequences. If using cDNA arrays, 10–20 µg poly(dA)-poly(dT), which blocks hybridization to polyA tails of cDNA, may be added. Also, scan one array prior to hybridization to determine basal level of background on the array (*see* **Note 16**).

3.8.1 Prehybridization

1. Briefly heat the prehybridization solution in an amber reaction tube to 72 °C and let it cool down to room temperature.
2. To locate the area on the microarray that has to be hybridized, place the dry microarray on a hybridization template that shows the dimensions of the slide and the area where the grids are printed.
3. Add water into the hybridization chamber to prevent drying of the samples.
4. Pipette the prehybridization solution to the designated area of the slide. Avoid surface contact with objects such as pipette tips. If any bubbles appear, try to remove these with a syringe needle.
5. When bubble free, apply the coverslip carefully (use coverslips larger than the grid area). Use tweezers.
6. Put the slide into a hybridization chamber and prehybridize the array for 1 h at 45 °C.
7. After incubation, move the slides to a 45 °C jar containing dd-water and remove the coverslip carefully, without using force.
8. Transfer the array into a new jar containing 45 °C dd-water and leave them there for 2 min, moving the slide up and down several times. Repeat this step two more times.
9. Transfer the slide into a new jar containing room-temperature isopropanol and repeat the washing by moving the slide up and down several times (~1 min).
10. Immediately blow dry the slide with pressurized air. The prehybridized arrays can now be hybridized or stored in a dark place for later usage.

3.8.2 Hybridization

1. Place the complete 50 µL DNA sample in a speedvac (shield from light) until the solution is reduced to about 5 µL. Do not allow the labeled DNA to dry completely.

2. Preheat the hybridization solution prior to use for 5 min at 72 °C. Make sure the hybridization solution is thoroughly resuspended. Centrifuge for 1 min at 10,000 g to get rid of particulate material. Pipette the needed amount (~60 μL for a 12,200 feature CpG- island array) of the hybridization solution and keep it at room temperature for 10 min.

3. Add the DNA (~5 μL) to the hybridization mix, denature it for 5 min at 80 °C, and briefly spin down.

4. Add water into the hybridization chamber to prevent drying of the samples.

5. Pipette the hybridization solution to the designated area of the slide.

6. Apply the coverslip carefully to the bubble-free hybridzation solution. Again, avoid any contact with the array surface.

7. Place the slide quickly but carefully into a hybridization chamber to prevent renaturation of the probe and evaporation of the hybridization solution. Make certain that the array surface is level.

8. Protect the hybridization chambers from light, and hybridize for approximately 12–16 h at 50 °C (oligo arrays may require a lower temperature, whereas higher temperatures are needed for BAC/PAC arrays).

3.8.3   Washing

1. Fill three jars with Wash I and two jars with Wash II and heat them to 50 °C.

2. Remove the coverslip by rinsing the slide carefully in Wash I (jar 2). Take extreme care when handling the array. Allow coverslip to float off the surface. Do not attempt to force off the coverslip, as this will damage the array.

3. Place the array slide quickly into the second Wash I (jar 2) for 15 min (carry out all wash steps in the dark), moving the slide up and down from time to time.

4. Transfer the slides to a new jar (jar 3) and repeat the wash step for 15 min.

5. Place the array slide into Wash II (jar 4) for 15 min, moving the slide up and down from time to time.

6. Transfer the slides to a new Wash II (jar 5) and repeat the wash step (15 min).

7. Rinse the slide for 2 min within another jar in dd-water (jar 6) to remove all traces of SDS and SSC.

8. Transfer the slide into a new jar with isopropanol (jar 7) for 30 s, moving it up and down (*see* **Note 17**).

9. Blow dry the slide as quickly as possible using pressurized air. Any buffer left on the array will appear as haze on array and result in a strong green background.

10. Scan the array in the Cy3 and Cy5 channel (532 nm and 635 nm).

3.9   Examples

The preceding protocol, in combination with other epigenetic profiling methods, may help identify interindividual variation in genomewide methylation patterns as well as epigenetic changes that arise during tissue differentiation or through the influence of various environmental factors. Figure 8.4 demonstrates how the technology can be used to detect abnormal DNA methylation patterns in genomic imprinting diseases, such as Prader-Willi syndrome. In approximately 25% of subjects affected with PWS, both copies of chromosome 15 are inherited from the mother (maternal disomy or UPD) [4]. However, the missing chromosome 15 from the father, which contains the paternally expressed genes, is required for normal development. The maternal genes are inactive due to methylation of the chromosomal

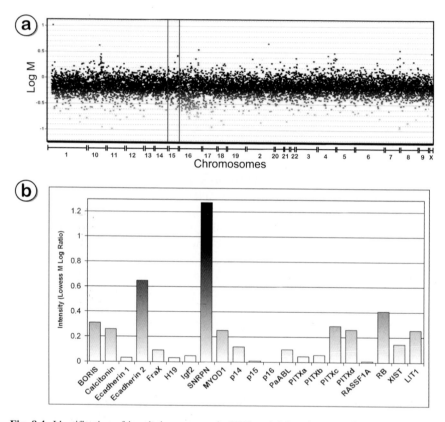

Fig. 8.4 Identification of imprinting patterns in PWS and AS patients. (a) Comparison of two PWS patients with healthy controls using a CpG-island array displays numerous interindividual methylation differences throughout the genome. Additionally, a proportion of the outliers may be caused by single nucleotide polymorphisms (SNPs; *see* **Note 18**). Chromosome 15 data points are highlighted. (b) Comparison of the unmethylated fraction of two PWS and two AS UPD patients (mean values) using an oligonucleotide-microarray. The AS patients have two paternal chromosomes; hence, they should exhibit no methylation of *SNRPN* (default methylated only on the maternal allele). The PWS patients have two maternal chromosomes, resulting in a fully methylated *SNRPN* gene (methylated only on the maternal allele). Of the tested genes, only *SNRPN* showed significant methylation differences (log ratio >1) between the PWS and AS cases

region. In contrast, in UPD cases of Angelman syndrome, the imprinting is the other way around: Both copies of chromosome 15 are inherited from the father and unmethylated. As can be seen in Fig. 8.4a, many epigenetic differences between PWS patients and healthy controls could be observed. Most detected methylation differences are due to interindividual variation and are not disease specific. However, a comparison of the *SNRPN* gene on chromosome 15 with other imprinted genes on other chromosomes and nonimprinted control genes revealed that the different methylation levels at *SNRPN* could be identified (see Fig. 8.4b). Another example for the application of the presented protocol is shown in Fig. 8.5, where fibroblast cells were either treated with the neurogenic substance, staurosporine, or retinoic acid; several cell lines are known to undergo neuronal and partly cardiomyocyte

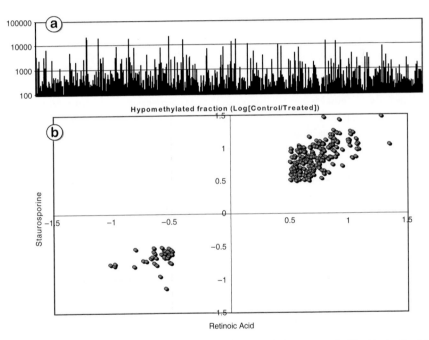

Fig. 8.5 Influence of STS- and RA-treatment on DNA methylation patterns in fibroblast cells. (a) Methylation differences detected with CpG-island microarrays after staurosporine treatment. Each bar indicates the absolute methylation difference at one CpG island reflected in intensity differences of fluorescent dyes on the array (x-axis absolute fluorescent differences between the Cy3 and Cy5 channels). (b) Cosegregation of methylation changes in STS- and RA-treated fibroblast cells. Shown are only the CpG-islands sequences that changed in both STS- and RA-treated cells (log[control/treated] >0.5) compared to untreated controls. Some of the genes that displayed a methylation difference after treatment of both substances, for example, were the postsynaptic density scaffolding protein (PDSP1), the roundabout homolog 2 and 3 precursor (receptor for SLIT2, and probably SLIT1, which are thought to act as molecular guidance cue in cellular migration, including axonal navigation and projection of axons to different regions during neuronal development), NT-3 growth factor receptor precursor, nexin 9, Galectin-1, and MAX interacting protein 1

differentiation in vitro after induction of these substances. Some genes change methylation after induction of staurosporine and retinoic acid, possibly due to common pathways [5].

4   Notes

1. *Hha*I is an isoschizomer of *Hin*6I; however, the *Hha*I restriction enzyme is not suitable for adapter-based methylation analyses, since this enzyme does not produce a CpG overhang that can be ligated to the adapter.
2. Alternatively, water also can be used to dissolve Cy dyes; however, the mono reactive ester is labile in water and not recommended. Do not use Dioxane, which was suggested elsewhere as an alternative to DMSO. It appears that Cy5 is not fully soluble in dioxane. DMSO on the other hand is hygroscopic and will absorb moisture from the air, which reacts with the NHS ester of the dye and significantly reduces the coupling reaction efficiency. Therefore, keep the DMSO supplied in an amber screw-capped vial at −20 °C, and let the vial warm to room temperature before opening to prevent condensation. While the dyes usually are good when first opened, they are sensitive to moisture, with a short half-life in aqueous environments. Therefore, take care to store aliquots of them desiccated, if possible under vacuum.
3. Wash buffer impurities can cause a grainy background; SDS will fluoresce in the Cy5 channel, while salt will fluoresce in the Cy3 channel. To eliminate this problem use high-quality SDS and SSC for preparation of wash buffers and filter sterilize. Insufficient washing also may contribute to a grainy background, therefore, increasing the wash times or slight agitation may help decrease the background.
4. Quantify in spectrophotometer. Remember the $OD_{260} = 1$ of single-stranded DNA is equivalent to 33 µg/mL not 50 µg/mL. This relationship, however, can be inaccurate for short fragments of DNA, such as oligonucleotides. Base composition and even linear sequence affect optical absorbance; hence, the precise value of the OD to mass relationship is unique for each oligo. For example, 1.0 OD_{260} of CCCCCCCCCCCC (homopolymeric deoxycytidine) equals 39 µg while 1.0 OD_{260} of AAAAAAAAAAAA (homopolymeric deoxyadenosine) equals only 20 µg. Do not believe the amount of primer as indicated by the manufacturer. Perform the appropriate calculation (800 pmol primer = 0.0008 × MW oligo).
5. To get the most out of restriction analyses, it is crucial to choose the right enzyme combination for the targets to be interrogated. For example, some MSREs cut relative frequently in CpG islands but rarely recognize a sequence outside of a CpG-island region, as is the case for *Hin*6I (5′-GCGC-3′) or *Bsp143*II (5′-PuGCGCPy-3′). In contrast, enzymes such as *Hpy*CH4IV (5′-ACGT-3′) cut predominantly outside of CpG-island sequences and are less useful in the interrogation of CpG islands, for instance, in CpG-island microarray-based studies [3]. Several other methods rely on the specific methylation-sensitive

cleavage of the rare cutter *Not*I (5′-GCGGCCGC-3′), for example, RLGS and AFLP methods and a couple of microarray approaches [6, 7]. However, *Not*I-sites are not well represented in the genome and provide only a very rough overview of methylation patterns. Hence, it is not advisable to include *Not*I in genomewide analyses of complex diseases.

6. Additionally, spike DNA can be used to test the methylation-sensitive cleavage reaction, ligation efficiency, McrBC digestion, and PCR reaction. Spiking DNA is added to the reactions with concentrations equivalent to the template concentration (1 genome equivalent, GE, assuming 3×10^9 bp in the human genome). For example, 16.2 pg of λ plasmid is added to 1 μg of template DNA. Make 12X GE stock solutions in 1 m*M* EDTA; aliquot into small volumes and store at −20 °C. To prepare a McrBC digestion control, digest pUC57 with *Hpa*II, ligate the CG adapter, and premethylate with SssI methylase. McrBC controls are added to the McrBC cleavage step as premethylated plasmid. Other suitable controls are ΦX174 to test forPCR bias or pBR322 to test the ligation efficiency [1]. The ΦX174 DNA has to be digested separately with *Hpa*II/*Hin*6I then ligated to the CG adapter. Complementary spiking oligonucleotides have to be representative sequences of *Hpa*II and *Hin*6I cleavage fragments.

7. The amount of CG adapter depends directly on the amount of restriction enzymes used. For each enzyme add ~0.3 pmol of the double-stranded CG adapter per ng DNA. For instance, in a triple digest (*Hpa*II/*Hin*6I/*Aci*I) of 500 ng DNA, it is advised to add 500×0.3 pmol $\times 3 = 450$ pmol of the adapter.

8. All ligation components initially are combined without the T4-ligase. Since the CpG overhang of the adapter is complementary to itself, it could form adapter dimers during the ligation. Therefore, a heating step is introduced before the ligase is added to the mixture. In this way, adapters are dissociated to bind specifically to the DNA overhangs produced by the restriction enzymes.

9. It is important to reduce the amount of ATP in the mixture, since ATP interferes with the McrBC digestion by competing with GTP for a binding site in McrBC. Since ATP cannot be utilized by the enzyme to perform the enzymatic reaction, McrBC binds to the DNA without cleavage. The best way to decrease the amount of ATP is heat degradation, since free ATP is highly unstable.

10. McrBC makes one cut between each pair of half sites, cutting close to one half site or the other, but cleavage positions are distributed over several base pairs approximately 30 base pairs from the methylated base. Therefore, the enzyme does not produce defined DNA ends on cleavage. Also, when multiple McrBC half sites are present in DNA (as is the case with cytosine-methylated genomic DNA), the flexible nature of the recognition sequence results in an overlap of sites, so a smeared rather than a sharp banding pattern is produced. GTP is more labile than most other nucleotides, so it is recommended to aliquot the 100 m*M* solution supplied. McrBC cuts the DNA (both strands) even if the methylated cytosines are on only one strand (hemi-methylated).

11. This protocol works only with standard Taq polymerases. Other, high-processing polymerases usually cannot incorporate modified nucleotides efficiently enough to produce PCR fragments of the desired length.

12. Centrifugation of Microcon YM-50 columns forces liquid through the low-binding cellulose membrane. Solutes larger than the nominal cutoff of the units are retained: YM-50 columns have a double-stranded nucleotide cutoff of 100 bp. This includes about 46 bp from the adapters on both sides of the amplification products. Therefore, all methylation-sensitive restriction products below ~54 bp and unincorporated adapters and primers should be removed during cleanup. Recovery of amplification products generally is above 90%. Millipore columns also are available as plates, which could be suitable for high-throughput analysis. Recently, we also used Qiagen columns for purification; the overall amount that can be loaded is lower, however, the cleanup still produces enough material for several hybridizations with high-quality DNA.

13. The following aminoallyl/N-hydroxysuccinimide reaction requires that no free amines be present in the coupling reaction buffer. This means that no Tris buffer can be used during any steps, including the DNA preparation.

14. The numbers 86.5 and 51.9 are conversion factors calculated by using the average molecular weight of dNTPs (324.5 g/mole), the absorption coefficient of Cy3 and Cy5 ($150,000/M^{*}cm$ and $250,000/M^{*}cm$), and the concentration of Cy-labeled allyl-PCR products that absorb 1 AU of 260-nm light (50 μg/mL). When measuring ss-DNA, the concentration of Cy-labeled ss-DNA that absorbs 1 AU of 260-nm light is 37 μg/mL. The conversion factors changes then to 117 for Cy3 and 70.2 for Cy5. Note that sample volume is not factored in; this is because volumes cancel out when DNA and cyanine absorbance are measured simultaneously. Frequently, the FOI for the different dyes may differ. In our experience, Cy5 is not as well incorporated as Cy3, however, it really depends on several factors (humidity, ozone concentration, etc.) and that is why it is recommended to do reciprocal labeling.

15. Other hybridization buffers may be used. The optimal buffer depends on several experimental variables, such as type of glass slide used, the length and GC content of the attached nucleic acid targets, and the manufacturing protocol employed.

16. As arrays age, some surface chemistries, particularly poly-L-lysine, show an increase in autofluorescence in the Cy3 channel. Aged microarrays also may show a higher affinity for salt, which will fluoresce in the Cy3 channel. To minimize these effects, use arrays as fresh as possible and store them properly (in the dark and under vacuum, if possible).

17. Salt from the final wash buffer may precipitate out of solution onto the array surface on submersion into isopropanol. This is likely a result of a higher than intended concentration of salt in the wash buffer; accidentally switching the wash buffers can cause this phenomenon. A brief rewash in water followed by another isopropanol wash likely will remove the salt from the array.

18. SNPs within the recognition sites for *Hpa*II and *Hin*6I may simulate epigenetic differences [8]. To exclude the impact of DNA sequence variation, check the available SNP databases and identify the DNA sequence variation within the restriction sites of the used enzymes. From CpG-island microarray studies, the estimate is that 10–30% of methylation variation detected in among individuals, in fact,

could be due to DNA sequence variation [1]. For comparison, in a study for the Human Epigenome Project (HEP), interrogation of 3,273 unique CpG sites on chromosome 6 revealed that 101 CpGs overlapped with known SNPs (3%) [9]. Another way to differentiate DNA sequence effects from genuine epigenetic differences consists of performing an identical microarray experiment on a DNA sample that has been stripped of all methylated cytosines [1]. To eliminate all methylation, the whole genome has to be amplified (thereby replacing all ^{met}C with C), for example, using the Phi29 DNA polymerase. Amplified DNA samples then are subjected to the same steps as depicted in Fig. 8.1 and cohybridized on the microarrays. In this experiment, all the outliers must be a result of DNA sequence variations within the restriction sites of the enzymes used and can be eliminated from further epigenetic analyses.

Acknowledgments We thank Professor Bernhard Horsthemke for providing DNA samples of Prader-Willi and Angelman syndrome patients and Dr. Jon Mill and Carolyn Ptak for their critical reading of the manuscript.

References

1. Schumacher A, Kapranov P, Kaminsky Z, Flanagan J, Assadzadeh A, Yau P, Virtanen C, Winegarden N, Cheng J, Gingeras T, Petronis A (2006) Microarray-based DNA methylation profiling: technology and applications. Nucleic Acids Res 34:528–542
2. Cheng J, Kapranov P, Drenkow J, Dike S, Brubaker S, Patel S, Long J, Stern,D, Tammana H, Helt G, Sementchenko V, Piccolboni A, Bekiranov S, Bailey DK, Ganesh M, Ghosh S, Bell I, Gerhard DS, Gingeras TR (2005) Transcriptional maps of 10 human chromosomes at 5-nucleotide resolution. Science 308:1149–1154
3. Schumacher A, Petronis A (2006) Epigenetics of complex disease: From general theory to laboratory praxis. Curr Top Microbiol Immunol 310:81–115
4. Schumacher A (2001) Mechanisms and brain specific consequences of genomic imprinting in Prader-Willi and Angelman syndromes. Gene Funct Dis 1:7–25
5. Schumacher A, Arnhold S, Addicks K, Doerfler W (2003) Staurosporine is a potent activator of neuronal, glial, and "CNS stem cell-like" neurosphere differentiation in murine embryonic stem cells. Mol. Cell Neurosci. 23:669–680
6. Yamamoto F, Yamamoto M (2004) A DNA microarray-based methylation-sensitive (MS)-AFLP hybridization method for genetic and epigenetic analyses. Mol Genet Genomics 271:678–686
7. Li J, Protopopov A, Wang F, Senchenko V, Petushkov V, Vorontsova O, Petrenko L, Zabarovska V, Muravenko O, Braga E, Kisselev L, Lerman MI, Kashuba V, Klein G, Ernberg I, Wahlestedt C, Zabarovsky ER (2002) NotI subtraction and NotI-specific microarrays to detect copy number and methylation changes in whole genomes. Proc Natl Acad Sci USA 99:10724–10729
8. Schumacher A, Friedrich P, Schmid J, Ibach B, Eisele T, Laws SM, Foerstl H, Kurz A, and Riemenschneider M (2006) No association of chromatin-modifying protein 2B with sporadic frontotemporal dementia. Neurobiol Aging Sep 14 (e-pub ahead of print)
9. Rakyan VK, Hildmann T, Novik KL, Lewin J, Tost J, Cox AV, Andrews TD, Howe KL, Otto T, Olek A, Fischer J, Gut IG, Berlin K, Beck S (2004) DNA methylation profiling of the human major histocompatibility complex: A pilot study for the human epigenome project. PLoS Biol 2:e405

Chapter 9
Genomewide Identification of Protein Binding Locations Using Chromatin Immunoprecipitation Coupled with Microarray

Byung-Kwan Cho, Eric M. Knight, and Bernhard Ø. Palsson

Abstract Interactions between *cis*-acting elements and proteins play a key role in transcriptional regulation of all known organisms. To better understand these interactions, researchers developed a method that couples chromatin immuno-precipitation with microarrays (also known as ChIP-chip), which is capable of providing a whole-genome map of protein-DNA interactions. This versatile and high-throughput strategy is initiated by formaldehyde-mediated cross-linking of DNA and proteins, followed by cell lysis, DNA fragmentation, and immu-nopurification. The immunoprecipitated DNA fragments are then purified from the proteins by reverse-cross-linking followed by amplification, labeling, and hybridization to a whole-genome tiling microarray against a reference sample. The enriched signals obtained from the microarray then are normalized by the reference sample and used to generate the whole-genome map of protein-DNA interactions. The protocol described here has been used for discovering the genomewide distribution of RNA polymerase and several transcription factors of *Escherichia coli*.

Keywords ChIP-chip, chromatin immunoprecipitation, microarray, RNA polymer-ase, transcription factor, transcription factor binding

1 Introduction

In the postgenomic era, systematic and high-throughput technologies allow us to enumerate biological components on a large scale. As one of the approaches, chromatin immunoprecipitation coupled with microarrays (ChIP-chip) has been used to explore the genomewide interactions between proteins and *cis*-acting elements, such as a comprehensive identification of transcriptional regulatory regions of the human genome [1] and in other organisms [2–8]. Maps of genomewide protein-DNA interactions are essential in understanding many fundamental biological components, such as the logic of regulatory networks, chromosome structure, DNA replication, DNA repair, heat shock response, and

From: Methods in Molecular Biology, Vol. 439: *Genomics Protocols: Second Edition*   131
Edited by Mike Starkey and Ramnath Elaswarapu © Humana Press Inc., Totowa, NJ

metabolism [9–14]. Due to a dramatic improvement in microarray technologies, it is possible to create high-resolution maps of genomewide protein-DNA interactions [1, 3].

Although many variations of the ChIP-chip protocol exist, the essential steps begin with an in vivo fixation of protein-DNA complexes mediated by formaldehyde. The cells are then lysed and the DNA fragmented to a desired size using sonication. The protein bound DNA is then enriched through immunoprecipitation by the specific antibody against the protein of interest (or against epitope-tag flanked with the target protein), then purified of the protein through a heat-mediated reversal of the cross-links. The DNA then is purified, amplified, and hybridized to a microarray [15–20]. Several features make the ChIP-chip protocol difficult for all applications. First, not every available antibody is efficiently applicable to the immunoprecipitation of the cross-linked protein-DNA complexes [17]. This limitation probably is due to antibody specificity and an affinity against the target protein and epitope masking by formaldehyde-mediated cross-linking. To address this limitation, epitope-tagging methods have been developed to use in ChIP-chip of yeast and *E. coli* [4, 21]. A second limitation is amplifying the ChIP DNA, which often is required to obtain sufficient amounts of DNA for labeling and hybridization without introduction of bias. Generally, two PCR-based methods have been widely used to amplify the ChIP DNA. The first method uses a degenerate oligo that randomly anneals to DNA [22], while the other uses ligation-mediated PCR (LM-PCR), and achieves roughly 100- to 1,000-fold amplification [2]. Finally, current microarray platforms are fairly expensive, but these costs are beginning to decrease significantly as new technologies are developed and competition among suppliers increases. This chapter describes the ChIP-chip protocol used for mapping the RNA polymerase binding in *Escherichia coli* along with several transcription factors.

2 Materials

2.1 Cell Culture and Cross-Linking

1. 10X M9 salt stock (M9 minimal medium): Dissolve 60 g Na_2HPO_4, 30 g KH_2PO_4, 5 g NaCl, and 10 g NH_4Cl in dH_2O; adjust to 1 L final volume; and sterilize by autoclaving for 20 min at 15 psi (1.05 kg/cm^2) on a liquid cycle. Store at room temperature.
2. 10X glucose stock: Dissolve 20 g glucose in dH_2O, adjust to 1 L final volume, and sterilize by passing the solution through a 0.22-μm filter. Store at room temperature.
3. 500X $MgSO_4$ stock (1 M): Dissolve 22.85 g $MgSO_4 \cdot 6H_2O$ in dH_2O, adjust the volume to 0.1 L, and sterilize by autoclaving for 20 min at 15 psi (1.05 kg/cm^2) on a liquid cycle. Store at room temperature.

4. 1000X $CaCl_2$ stock (0.1 M): Dissolve 1.47 g $CaCl_2·2H_2O$ in dH_2O, make up the volume to 0.1 L, and sterilize by autoclaving for 20 minutes at 15 psi (1.05 kg/ cm^2) on a liquid cycle. Store at room temperature.
5. 100X trace element solution: Dissolve 16.67 g $FeCl_3·6H_2O$, 0.18 g $ZnSO_4·6H_2O$, 0.12 g $CuCl_2·2H_2O$, 0.12 g $MnSO_4·H_2O$, 0.18 g $CoCl_2·6H_2O$, 0.12 g Na_2MoO_4, and 22.25 g $Na_2EDTA·2H_2O$ in dH_2O; adjust to 1 L final volume; and sterilize by passing the solution through a 0.22-μm filter. Store at room temperature.
6. 37% formaldehyde solution (Fisher Scientific, F79-500). Store in the chemical hood.
7. 2.5 M glycine solution: Dissolve 187.68 g glycine in dH_2O, make up the final volume to 1 L, and sterilize by passing the solution through a 0.22-μm filter. Store at room temperature.
8. Tris-buffered saline (Sigma, T5912). Store at 4 °C.

2.2 Cell Lysis, Preparation of Chromatin Complexes, and Immunoprecipitation

1. Lysis buffer: 10 mM Tris-HCl, pH 7.5, 100 mM NaCl, and 1 mM EDTA. Store at 4 °C.
2. Lysozyme (Epicentre, R1810M). Follow storage instructions provided by supplier.
3. Protease inhibitor cocktail (Sigma, P8465): Dissolve at 200 mg/mL in DMSO and dilute in four volumes of dH_2O. Make fresh prior to use and keep cold on ice.
4. IP buffer: 100 mM Tris-HCl, pH 7.5, 200 mM NaCl, 2 mM EDTA, and 2% Triton X-100. Store at 4 °C.
5. Misonix 3000 sonicator equipped with microtip.
6. Antibody: Anti-RNAP mouse antibody (Neoclone, W0001, W0002), anti-myc mouse antibody (Santa Cruz Biotechnology, sc-40), Mouse IgG (Upstate, 12–371). Follow storage instructions provided by supplier.
7. Bead washing buffer: Dissolve BSA (Sigma, A7906) at 5 mg/mL in PBS. Make fresh prior to use and keep cold on ice.
8. Dynabeads: Pan mouse IgG (Invitrogen, 112.05). Follow storage instructions provided by supplier.

2.3 Immunoprecipitation and Reverse Cross-Linking

1. IP washing buffer 1 (W1 buffer): 50 mM Tris-HCl, pH 7.5, 100 mM NaCl, 1 mM EDTA, and 1% Triton X-100. Store at 4 °C.
2. IP washing buffer 2 (W2 buffer): 50 mM Tris-HCl, pH 7.5, 500 mM NaCl, 1 mM EDTA, and 1% Triton X-100. Store at 4 °C.

3. IP washing buffer 3 (W3 buffer): 10 mM Tris-HCl, pH 8.0, 250 mM LiCl (Sigma, L7026), 1 mM EDTA, and 1% Triton X-100. Store at 4 °C.

4. TE buffer: 10 mM Tris-HCl, pH 8.0, and 1 mM EDTA. Store at 4 °C.

5. IP Elution buffer: 10 mM Tris-HCl, pH 8.0, 1 mM EDTA, and 1% SDS. Store at room temperature.

2.4   Purification, qPCR and Amplification of DNA

1. TE buffer: 10 mM Tris-HCl, pH 8.0, and 1 mM EDTA.

2. RNaseA solution, 10 mg/mL (Qiagen, 19101).

3. Protease K solution, 20 mg/mL (Invitrogen, 25530-049).

4. 3 M sodium acetate, pH 5.2.

5. Qiagen PCR Purification Kit (Qiagen, 28106).

6. 2X SYBR green PCR master mix (Qiagen, 204145). Follow storage instructions provided by supplier.

7. iCycler real-time PCR detection system (Bio-Rad, CA).

8. Sequenase (USB, 70775Z), 5X Sequenase buffer (supplied with Sequenase), and Sequenase dilution buffer (supplied with Sequenase).

9. 0.1 M DTT (USB, 70726).

10. 0.5 mg/mL BSA (diluted from 10 mg/mL stock from NEB, B9001S).

11. Rand 9-Ns primer: 5′ TGGAAATCCGAGTGAGTNNNNNNNNN 3′.

12. Rand univ primer: 5′ TGGAAATCCGAGTGAGT 3′.

13. dNTP mix (Takara, 4030).

14. *pfu* turbo polymerase (Stratagene, 600135) and 10X *pfu* buffer (supplied with polymerase).

2.5   Microarray Hybridization and Scanning

1. Cy3-labeled nine-mers (TriLink Biotechnologies, N46-0001-50).

2. Cy5-labeled nine-mers (TriLink Biotechnologies, N46-0002-50).

3. Random nine-mer buffer: 125 mM Tris-HCl, pH 8.0, 12.5 mM MgCl$_2$, and 0.175% β-mercaptoethanol.

4. 50X dNTP mix solution: 10 mM Tris-HCl, pH 8.0, 1 mM EDTA, 10 mM dNTP.

5. 100 U Klenow fragment (NEB, M0212M).

6. MAUI hybridization unit (BioMicro Systems, Utah).

7. NimbleGen custom microarrays (Design ID: 1881, *Escherichia coli* whole-genome tiling array consisting of 371,034 oligonucleotides spaced 25 bp apart across the whole genome).

8. NimbleGen Array Reuse Kit 40 (NimbleGen, KIT001-2).

9. Axon scanner, model 4000B.

10. Cy3 CPK6 50-mer (IDT, Custom oligo synthesis).

11. Cy5 CPK6 50-mer (IDT, Custom oligo synthesis).
12. 20X SSC (Sigma, S6639).
13. 0.1 M DTT.
14. Wash I: 250 mL ddH$_2$O, 2.5 mL 20X SSC, 5 mL 10% SDS, and 250 μL 0.1 M DTT.
15. Wash II: 250 mL ddH$_2$O, 2.5 mL 20X SSC, and 250 μL 0.1 M DTT.
16. Wash III: 250 mL ddH$_2$O, 625 μL 20X SSC, and 250 μL 0.1 M DTT.

2.6   Normalization and Peak Identification

SignalMap (www.nimblegen.com), Matlab ver 7.0.4 with bioinformatics toolbox (www.mathworks.com), Microsoft Excel (www.microsoft.com), Mpeak (the complete program is available from http://www.stat.ucla.edu/~zmdl/mpeak/).

3   Methods

3.1   Cell Culture and Cross-Linking

1. Add 2.8 mL of 37% formaldehyde solution directly to each 100 mL culture that contains the number of cells used for a ChIP-chip experiment (*see* **Note 1**). Continue to incubate with gentle shaking for 20 min at room temperature (*see* **Note 2**).
2. Add 5 mL of 2.5 M glycine solution directly to each 100 mL sample followed by incubation for 5 min at room temperature.
3. Centrifuge at 4,700 g for 5 min at 4 °C and pour off the supernatant. Wash each pellet three times with one volume of ice-cold TBS (*see* **Note 3**) and resuspend the cell pellet in the TBS remaining after decanting the supernatant. Transfer the sample to a new 1.5-mL tube and centrifuge at the maximum speed (15,800 g) for 1 min.
4. Remove all supernatant using a pipette and store the cell pellet at −80 °C until use.

3.2   Cell Lysis, Preparation of Chromatin Complexes and Immunoprecipitation

1. Completely resuspend the cell pellet in 0.5 mL of lysis buffer. Add 40 μL of protease inhibitor cocktail and 0.5 μL of lysozyme solution. Incubate the sample for 30 min at 37 °C on a rocker, and then add 0.5 mL of IP buffer and 40 μL of protease inhibitor cocktail. Continue to incubate on ice until the lysate is cleared (*see* **Note 4**).

2. Shear the lysate by sonicating for four 20 s pulse with a Misonix microtip sonicator at output setting 2 (*see* **Note 5**). To avoid overheating the sample, keep the
sample on ice at least 1 min between cycles. Centrifuge at 15,800 g for 10 min at
4 °C to clarify the chromatin solution. Take 10 μL of the chromatin solution to
use as "total DNA (tDNA)" sample and store at −20 °C for further use.
3. Split the chromatin solution into two 0.5-mL aliquots. Add 1 μg of specific antibody to one aliquot, and 1 μg of mouse IgG to the other (*see* **Notes 6 and 7**).
Incubate samples overnight at 4 °C on a rocker.

3.3   *Immunoprecipitation and Reverse Cross-Linking*

1. For each sample, wash 50 μL of Dynabeads Pan mouse IgG beads three times with
1 mL of bead washing buffer. Add the incubated samples ("with specific antibody
[iDNA]" and "with mock antibody [mDNA]") to the washed magnetic beads (*see*
Note 8). Continue incubation for at least 6 h at 4 °C on a rocker at 8 rpm.
2. Collect the magnetic beads using an MPC magnet and remove the supernatant
by aspiration. If needed, save the supernatant for the "unbound fraction sample."
Sequentially, wash the beads twice with 1 mL of W1 buffer, once with 1 mL of
W2 buffer, once with 1 mL of W3 buffer, and once with 1 mL of TE buffer (*see*
Notes 9 and 10).
3. Resuspend the beads in 200 μL of IP elution buffer and add 190 μL of that solution to the tDNA sample. Continue to incubate overnight at 65 °C to reverse
cross-links.

3.4   *Purification, qPCR and Amplification of DNA*

1. Pull down the magnetic beads using an MPC magnet and transfer 200 μL of
supernatant to a new tube. Add 200 μL of TE buffer and 8 μL of RNaseA solution (10 mg/mL) to each sample. Continue to incubate for 2 h at 37 °C. Add 4 μL
of protease K solution (20 mg/mL) to each sample and continue to incubate for
2 h at 55 °C. Purify DNA using the Qiagen PCR Purification Kit and elute with
50 μL of ddH$_2$O (*see* **Note 11**). At this point, qPCR can be done using the iDNA,
mDNA, and tDNA to confirm the enrichment fold required to run a microarray
(*see* **Note 12**).
2. Set up the round A reaction mix on ice as described in Table 9.1 (*see* **Note
13**).
3. In a PCR tube, mix 7 μL of the iDNA or mDNA, 2 μL of 5X Sequenase buffer,
and 1 μL of 40 μM Rand 9-Ns primer. Cycle to 94 °C for 2 min then cool to
10 °C. Add 5.05 μL of round A mix. Ramp up from 10 °C to 37 °C over 8 min,
hold at 37 °C for 8 min, heat to 94 °C for 2 min, then cool to 10 °C. Add 1.2 μL
mixture of 0.9 μL of Sequenase dilution buffer and 0.3 μL of Sequenase enzyme.

Table 9.1 Round A reaction mix

		Number of Samples		
Component	Stock	1	6	12
Sequenase buffer	5X	1.0	6.0	12.0
dNTP mix	2.5 mM	1.5	9.0	18.0
BSA	0.5 mg/mL	1.5	9.0	18.0
DTT	0.1 M	0.75	4.5	9.0
Sequenase	13 U/μL	0.3	1.8	3.6
Total		5.05	30.3	60.6

Table 9.2 Round B reaction mix

		Number of Samples		
Component	Stock	1	6	12
ddH$_2$O		63	378	756
pfu buffer	10X	10	60	120
dNTP mix	2.5 mM	10	60	120
Rand univ primer	100 μM	1	6	12
pfu polymerase	5 U/μL	1	6	12
Total		85	510	1,020

Ramp up from 10 °C to 37 °C over 8 min, hold at 37 °C for 8 min, then cool to 4 °C. Dilute the samples by addition of 45 μL of ddH$_2$O.

4. Set up the round B reaction mix on ice as described in Table 9.2. Transfer 15 μL of the diluted template into a new PCR tube and add 85 μL of the round B reaction mix to each tube. Prepare four tubes per sample to achieve enough DNA for microarray hybridization. Cycle to 94 °C for 30 s, 40 °C for 30 s, 50 °C for 30 s, and 72 °C for 2 min (25 cycles) (*see* **Note 14**). Purify the amplified DNA using a Qiagen PCR Purification Kit and elute with 120 μL EB buffer supplied with the kit. Use one purification column per two reactions and combine two elutions in a new tube. The total volume per sample should be 240 μL. Add 24 μL of ice-cold 3 M sodium acetate (pH 5.2) and 700 μL of ethanol. Continue to incubate overnight at −20 °C.

3.5 Labeling, Hybridization, and Scanning

1. Centrifuge at 37,000 g at 4 °C for 30 min and wash the pellet with cold 80% ethanol. Dry the pellet and dissolve the pellet in 9 μL (iDNA) and 7 μL (mDNA) of ddH$_2$O, respectively. Dilute 1 μL of the sample in 99 μL of EB buffer and measure the DNA quantity and quality using a spectrophotometer. The DNA yields range from 5 to 10 μg and A$_{260/280}$ should be between 1.8 and 2.0.

2. Dilute Cy3 and Cy5 dye-labeled nine-mers to 1 OD in 42 μL random nine-mer buffer. Aliquot to 40 μL individual reaction volumes in 0.2 mL thin-walled PCR tubes. Add 1 μg of iDNA and mDNA to the Cy5 and Cy3 tubes, respectively, and bring the final volume to 80 μL using ddH$_2$O. Denature the samples in a thermocycler at 98 °C for 10 min and then quickly chill in an ice-water bath. Add 20 μL of 50X dNTP mix solution and mix well by pipetting at least 10 times.

3. Incubate at 37 °C for 2 h in a thermocycler (light sensitive) and then stop the reaction by the addition of 10 μL of 0.5 M EDTA. Transfer the reaction to a 1.5-mL tube. Precipitate the labeled samples by adding 11.5 μL of 5 M NaCl and 110 μL of isopropanol to each tube. Vortex and incubate for 10 min at room temperature in the dark.

4. Centrifuge at 37,000 g for 10 min and remove the supernatant with a pipette. Rinse the pellet with 500 μL of 80% ice-cold ethanol and centrifuge again at 37,000 g for 2 min. Remove the supernatant with a pipette and dry the pellet for 5 min in a SpeedVac using low heat and protection from light. Rehydrate the dried pellets in 25 μL of ddH$_2$O. Vortex for 30 s and quick spin to collect contents at bottom of tube. Measure the A$_{260}$ in each sample to determine DNA concentration. Typical yields range from 10 to 30 μg per reaction.

5. Set MAUI hybridization unit to 42 °C and allow time for the temperature to stabilize. Combine 13 μg of the both iDNA (Cy5) and mDNA (Cy3) into a single 1.5 mL tube (*see* **Note 15**). Dry the combined contents in a SpeedVac on low heat. Resuspend the sample in 10.9 μL of ddH$_2$O and vortex to completely dissolve the sample. Spin down the tube briefly to collect the contents in the bottom.

6. Using the NimbleGen Array Reuse Kit, add 19.5 μL of 2X hybridization buffer, 7.8 μL of hybridization component A, 0.4 μL of Cy3 CPK6 50-mer oligo, and 0.4 μL of Cy5 CPK6 50-mer oligo to each sample (*see* **Note 16**). Mix the tube briefly, and then spin down to collect the contents in the bottom and place at 95 °C for 5 min.

7. Immediately transfer the tube to the MAUI 42 °C sample block and hold at this temperature until ready for sample loading (*see* **Note 17**). Place the MAUI mixer SL hybridization chamber on the array using the provided assembly/disassembly jig and carefully follow MAUI setup instructions. Use the braying tool to remove all air bubbles from the adhesive gasket around the outside of the hybridization chamber.

8. Load the sample using the pipette supplied with the MAUI station, following manufacturer's instructions. During loading, a small amount (3–7 μL) of the sample may flow out of the outlet port. Confirm that there are no bubbles in the chamber.

9. Place the loaded array into one of the four MAUI bays and let it equilibrate for 30 s. Wipe off any sample leakage at the ports and adhere MAUI stickers to both ports. Close the bay clamp and select mix mode B. Hold down the mix button to start mixing. Confirm that the mixing is in progress before closing the cover. Hybridize the sample overnight.

10. Remove chip from MAUI hybridization station, load it back into the MAUI assembly/disassembly jig, and immerse in the shallow 250 mL Wash I (*see* **Note 18**). While the chip is submerged, carefully peel off the mixer. Gently agitate the chip in Wash I for 10–15 s (*see* **Note 19**).

11. Transfer the slide into a slide rack in the second dish of Wash I and incubate 2 min with agitation. Transfer to Wash II and incubate 1 min with agitation. Rock the dish to move the wash over the tops of the arrays. Transfer to Wash III and incubate for 15 s with agitation.

12. Remove the array and spin dry in an array-drying unit for 1 min. Store the dried array in a dark desiccator and proceed immediately the scanning of the arrays.

3.6 Data Normalization and Peak Identification

1. Conceptually, the normalization approach using sum of intensity of each channel is the simplest way, whose assumption is that the total DNA used is same for both channels. Calculate the sum of each Cy5 and Cy3 channel and the ratio between the total intensity of Cy5 and that of Cy3. Multiply the ratio (N) to each data point (G_k, R_k) and then calculate $\log_2$ ratio (R) of each point (Eq. 9.1–Eq. 9.3). $\log_2$ ratio can be used for the peak identification step (*see* **Note 20**).

$$N = \frac{\sum_{k=1}^{N\mathrm{array}} R_k}{\sum_{k=1}^{N\mathrm{array}} G_k} \tag{9.1}$$

$$G'_k = NG_k, \; R'_k = NR_k \tag{9.2}$$

$$R = \log_2\left(\frac{R'_k}{G'_k}\right) \tag{9.3}$$

2. The binding sites should appear in the data as runs of consecutive points with enhanced amplitude shown in Fig. 9.1. Several peak identification methods have been developed using an error model to compute p-values for single array probes [2], a sliding window approach with Gaussian error function [23], double-regression model [1], a percentile approach [24], a hierarchical empirical Bayes model [25], a joint-binding deconvolution method [26], a tiled model-based analysis of tiling-arrays method [27], and a variance stabilization approach [28]. Of those methods, we used the double-regression method to identify the protein binding sites across genome.

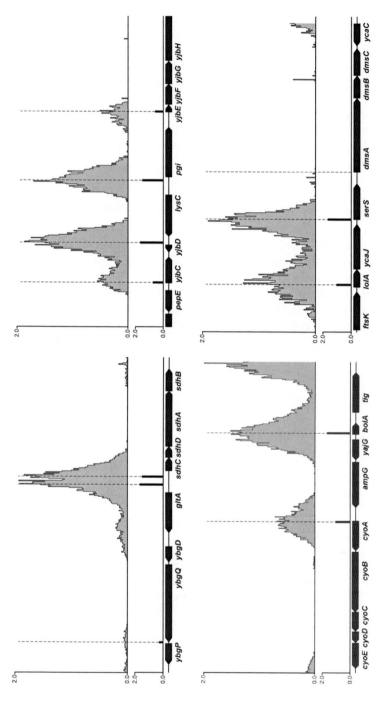

Fig 9.1 Typical ChIP-chip results from the NimbleGen whole-genome tiling arrays

4   Notes

1. The protocol describes the growth of *E. coli* MG1655 in minimal media to the mid exponential phase. Because the protein-DNA interactions are very sensitive to the physiological state of the cell, it is very important to control the growth conditions as tightly as possible. Generally, each immunoprecipitation requires $5 \times 10^7 \sim 1 \times 10^8$ cells (approximately, OD_{600} 0.4 ~ 1.0). To find out all the potential promoters of *E. coli*, the cells were treated with rifampicin for 30 min [3].

2. The cross-linking time should be empirically optimized to each protein-DNA-antibody combination. We found that the cross-linking time (20 min) and formaldehyde concentration (1%) described in this protocol are generally applicable to RNAP (β and β′ subunits) and several transcription factors of *E. coli*. However, theoretically insufficient cross-linking would result in the inability to capture protein-DNA complexes. On the other hand, overcross-linking can make giant protein-DNA complexes and cause the epitope-masking problem. Formaldehyde should be used with appropriate safety measures, such as protective gloves, glasses, and clothing and adequate ventilation. Formaldehyde waste should be disposed of according to regulations for hazardous waste.

3. PBS can be used to wash cell pellets instead of TBS.

4. Cell lysis condition varies depending on the target cell type. For *E. coli*, the lysis method using lysozyme works very well. If the lysozyme is not available, alternative methods, such as French press or glass-bead-based lysis, can be used.

5. The sonication step should be optimized to achieve the optimum fragmentation of DNA (300–1,000 bp). One way to optimize the sonication step is to take 20 μL of the chromatin solution from each sonication cycle and determine the average size of the fragmented DNA in the chromatin solution. Add 80 μL of IP elution buffer to each 20 μL of chromatin solution and continue to incubate at least 6 h (or overnight) at 65 °C to reverse cross-link the chromatin complexes. Add 100 μL of TE and 4 μL of RNaseA solution (10 mg/mL) and continue to incubate for 2 h at 37 °C. Add 2 μL of protease K solution (20 mg/mL) and continue to incubate for 2 h at 55 °C. Purify the DNA with a Qiagen PCR Purification Kit and elute using 30 μL of EB buffer supplied with the kit. The average size of DNA then is analyzed on a 2% agarose gel (Fig. 9.2). Under- and overfragmentation result in a loss of resolution of binding events and more noise in the microarray analysis, respectively (16).

6. Not every available antibody is efficiently applicable to the ChIP-chip, as mentioned in introduction. Unfortunately, routine assay methods such as Western blotting cannot be used to test whether an antibody would be suitable for ChIP-chip. To determine whether an antibody is suitable for ChIP-chip experiments requires an actual ChIP assays followed by qPCRs of several known-binding sites. To address the limited use of antibodies, epitope-tagging methods have been developed for ChIP-chip of yeast and *E. coli* without alteration in function of target proteins [4, 21]. The main advantages of epitope-tagging are the availability of a universal antibody and the ability to insert multiple copies of the epitope, which increases the immunoprecipitation yield [21].

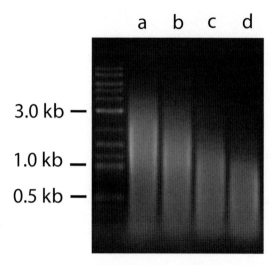

Fig. 9.2 The size distribution of the fragmented DNA by sonication: (a) one cycle, (b) two cycles, (c) three cycles, (d) four cycles

7. When epitope-tagged proteins are being used, wild-type cells can be used as a control. In this case, add the same antibody against the epitope used to both the epitope-tagged sample and the wild-type sample [21].

8. Alternatively, the antibodies can be conjugated to the Dynabeads Pan mouse IgG beads prior to the immunoprecipitation. Wash 50 μL of the magnetic beads three times, using the bead washing buffer, and resuspend the washed beads in 250 μL of bead washing buffer. Add 10 μg of antibody to the magnetic beads and incubate overnight at 4 °C on a rocker. Wash the beads three times in 1 mL of bead washing buffer and collect the beads using a MPC magnet prior to use. Then, add 0.5 mL of the sheared chromatin solution to the antibody preconjugated magnetic beads. Continue to incubate overnight at 4 °C on a rocker.

9. Add 1 mL of the washing buffer to each tube and gently resuspend beads. This can be done by removing the tubes from the MPC magnet and rotating the tubes for 1 min with rocker. Collect the magnetic beads using an MPC magnet and remove the supernatant by aspiration. Normally, to remove chromatin complexes that are nonspecifically bound to the antibody or magnetic beads, intensive washing steps are needed. Do not minimize this step.

10. Because not every protein-antibody interaction can bear the high salt and detergent conditions, we recommend washing the beads four times using only W1 buffer and once using TE buffer. If more stringent conditions are required, increase salt concentrations such as NaCl and LiCl in W2 and W3 buffer, respectively. At the final washing step using TE, the magnetic beads are collected on the tube wall loosely. Do not use aspiration but only a pipette.

11. Instead of a Qiagen PCR Purification Kit, it is possible to use the conventional purification method, consisting of a phenol:chloroform:isoamyl alcohol

Fig. 9.3 Typical qPCR results of chromatin immunoprecipitation of *Escherichia coli* RNA polymerase

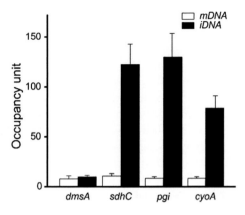

extraction and ethanol precipitation. We chose the Qiagen kit to maintain consistency among samples.

12. Examples for qPCR are shown in Fig. 9.3. We used the ChIP-chip approach to study the association of RNA polymerase across the genome of *E. coli* under aerobic conditions. The ChIP DNA samples were used as a template for qPCR with primer pairs of promoter regions from *pgi, cyoA*, and *sdhC*, whose gene expressions were up-regulated under aerobic conditions. As a control, the promoter region of *dmsA* was used, as the gene is down-regulated under aerobic conditions.

13. DNA amplification consists of two steps. Round A involves two rounds of DNA synthesis using the immunoprecipitated DNA as template, a partially degenerate primer (Rand 9-Ns primer), and T7 Sequenase. Round B consists of 25–30 cycles of PCR using a primer (Rand universal primer) that anneals to the specific region of the Rand 9-Ns primer.

14. The number of cycles should be optimized prior to the PCR amplification to prevent amplification bias. The random amplification method does not amplify DNA linearly, so the linearity depends on the number of cycles. Prepare five amplification tubes and sample each tube at the 15th, 20th, 25th, 30th, and 35th cycles. Measure the DNA concentration and perform qPCR as described in Note 12. Compare both the quantity and quality of each cycle.

15. The tubes should be protected from light during handling to prevent photobleaching of the light-sensitive Cy dyes.

16. CPK6 50-mer oligos are include in the hybridization as controls and hybridize to alignment features on NimbleGen arrays. They are required for proper extraction of array data from the scanned image.

17. This procedure describes the process for hybridization of samples prepared by chromatin immunoprecipitation and amplified by random PCR on NimbleGen custom microarrays (Design ID: 1881, *Escherichia coli* whole-genome tiling array, consisting of 371,034 oligonucleotides spaced 25 bp apart across the whole genome). The use of ORF arrays has limited power of ChIP experiments, since most transcription factor binding sites are located in the intergenic regions

and, therefore, not included on these arrays. The most robust array design for ChIP-chip has contiguous tiled DNA fragments that represent the entire genome, including the noncoding regions [1, 3, 17].

18. Prior to removing the array from the MAUI Hybridization Station, prepare two 250 mL dishes of Wash I, and one each for Wash II and Wash III. One dish for Wash I should be shallow and be wide enough to accommodate the array and mixer loaded in the MAUI assemble/disassembly jig. The lid from a 1-mL pipette tip box works well. Place the remaining three wash solutions in 300 mL Tissue-Tek slide staining dishes.

19. Peel the hybridization chamber off very slowly to prevent the slide from cracking. Do not let the surface of the slide dry out at any point during washing.

20. Various normalization methods can be used to normalize the tiling array data. Alternatively, another normalization method for the tiling array data is the use of a biweight mean [24]. Calculate the $\log_2$ of the ratio of Cy5 to Cy3 for each data point and then subtract the biweight mean of this $\log_2$ from each data point.

Acknowledgments The protocol described here was based on previous work by many other research groups in this field. The pioneers in this field are Dr. Young's group at MIT, Dr. Lieb's group at the University of North Carolina, Dr. Grunstein's group at Yale University, Dr. Ren's group at UCSD, and others. We thank anyone whose work was not referenced in here. This work is supported by NIH research grant no. GM62791.

References

1. Kim TH, Barrera LO, Zheng M, Qu C, Singer MA, Richmond TA, WuY, Green RD, Ren B (2005) A high-resolution map of active promoters in the human genome. Nature 436:876–880
2. Ren B, Robert R, Wyrick JJ, Aparicio O, Jennings EG, Simon I. Zeitlinger J, Schreiber J, Hannett,N, Kanin E, Volkert TL,Wilson CJ, Bell SP, Young RA (2000) Genome-wide resolution and function of DNA binding proteins. Science 290:2306–2309
3. Herring CD, Raffaelle M, Allen TE, Kanin EI, Landick R, Ansari AZ, Palsson BO (2005) Immobilization of *Escherichia coli* RNA polymerase and location of binding sites by use of chromatin immunoprecipitation and microarray. J Bacteriol 187:6166–6174
4. Lee TI, Rinaldi NJ, Robert F, Odom DT, Bar-Joseph Z, Gerber GK, Hannett NM, Harbison CT, Thompson CM, Simon I, Zeitlinger J, Jennings EG, Murray HL, Gordon DB, Ren B, Wyrick JJ, Tagne J, Volkert TL, Fraenkel E, Gifford DK, Young RA (2002) Transcriptional regulatory networks in *Saccharomyces cerevisiae*. Science 298:799–804
5. Wardle FC, Odom DT, Bell GW, Yuan B, Danford TW, Wiellette EL, Herbolsheimer E, Sive HL, Young RA, Smith JC (2006) Zebrafish promoter microarrays identify actively transcribed embryonic genes. Genome Biol 7:R71
6. Gilchrist M, Thorsson V, Li B, Rust AG, Korb M, Kennedy K, Hai T, Bolouri H, Aderem A (2006) Systems biology approaches identify ATF3 as a negative regulator of Toll-like receptor 4. Nature 441:173–178
7. Orian A, Steensel BV, Delrow J, Bussemaker HJ, Li L, Sawado T, Williams E, Loo LWM, Cowley SM, Yost C, Pierce S, Edgar BA, Parkhurst,SM, Eisenman RN (2003) Genomic binding by the *Drosophila* Myc, Max, Mad/Mnt transcription factor network. Genes Dev. 17, 1101–1114.

8. Chua YL, Mott E, Brown AP, MacLean D, Gray JC (2004) Microarray analysis of chromatin-immunoprecipitated DNA identifies specific regions of tobacco genes associated with acetylated histones. Plant J 37:789–800

9. Grainger DC, Hurd D, Harrison M, Holdstock J, Busby JW (2005) Studies of the distribution of *Escherichia coli* cAMP-receptor protein and RNA polymerase along the *E. coli* chromosome. Proc Natl Acad Sci USA 102:17693–17698

10. Carter NP, Vetrie D (2004) Applications of genomic microarrays to explore human chromosome structure and function. Hum Mol Genet 13:R297–R302

11. MacAlpine DM, Bell SP (2005) A genomic view of eukaryotic DNA replication. Chromosome Res 13:309–326

12. Workman CT, Mak HC, McCuine S, Tagne JB, Agarwal M, Ozier O, Begley TJ, Samson LD, Ideker T (2006) A systems approach to mapping DNA damage response pathways. Science 312:1054–1059

13. Wade JT, Roa DC, Grainger DC, Hurd D, Busby JW, Struhl K, Nudler E (2006) Extensive functional overlap between σ factors in *Escherichia coli*. Nat Struct Mol Biol. doi: 10.1038/nsmb1130

14. Hall DA, Zhu H, Zhu X, Royce T, Gerstein M, Snyder M (2004) Regulation of gene expression by a metabolic enzyme. Science 306:482–484

15. Kim TH, Ren B (2006) Genome-wide analysis of protein-DNA interactions. Annu Rev Genomics Hum Genet 7:81–102

16. Lee TI, Johnstone SE, Young RA (2006) Chromatin immunoprecipitation and microarray-based analysis of protein location. Nat Protocols 1:729–748

17. Buck MJ, Lieb JD (2004) ChIP-chip: Consideration for the design, analysis, and application of genome-wide chromatin immunoprecipitation experiments. Genomics 83:349–360

18. Negre N, Lavrov S, Hennetin J, Bellis M, Cavalli G (2006) Mapping the distribution of chromatin proteins by ChIP on Chip. Methods Enzymol 410:316–341

19. Hecht A, Grunstein M (1999) Mapping DNA interaction sites of chromosomal proteins using immunoprecipitation and polymerase chain reaction. Methods Enzymol 304:399–414

20. Ren B,d Dynlacht BD (2004) Use of chromatin immunoprecipitation assays in genome-wide location analysis of mammalian transcription factors. Methods Enzymol 376:304–315

21. Cho BK, Knight EM, Palsson BØ (2006) PCR-based tandem epitope tagging system for *Escherichia coli* genome engineering. Biotechniques 40:67–72

22. Iyer VR, Horak CE, Scafe CS, Botstein D, Snyder M, Brown PO (2001) Genomic binding sites of the yeast cell-cycle transcription factors SBF and MBF. Nature 409:533–538

23. Buck MJ, Nobel AB, Lieb JD (2005) ChIPOTle: A user-friendly tool for the analysis of ChIP-chip data. Genome Biol 6:R97

24. Bieda M, Xu X, Singer MA, Green R, Farnham PJ (2006) Unbiased location analysis of E2F1-binding sites suggests a widespread role for E2F1 in the human genome. Genome Res 16:595–605

25. Ji H, Wong WH (2005) TileMap: Create chromosomal map of tiling array hybridizations. Bioinformatics 21:3629–3636

26. Qi Y, Rolfe A, MacIsaac KD, Gerber GK, Pokholok D, Zeitlinger J, Danford T, Dowell RD, Fraenkel E, Jaakkola TS, Young RA, Gifford DK (2006) High-resolution computational models of genome binding events. Nat Biotechnol 24:963–970

27. Johnson WE, Li W, Meyer CA, Gottardo R, Carroll JS, Brown M, Liu XS (2006) Model-based analysis of tiling-arrays for ChIP-chip. Proc Natl Acad Sci USA 103:12457–12462

28. Gibbons FD, Proft M, Struhl K, Roth FP (2005) Chipper: Discovering transcription-factor targets from chromatin immunoprecipitation microarray using variance stabilization. Genome Biol 6:R96

Chapter 10
Transcriptional Profiling of Small Samples in the Central Nervous System

Stephen D. Ginsberg

Abstract RNA amplification is a series of molecular manipulations designed to amplify genetic signals from small quantities of starting materials (including single cells and homogeneous populations of individual cell types) for microarray analysis and other downstream genetic methodologies. A novel methodology named *terminal continuation* (TC) *RNA* amplification has been developed in this laboratory to amplify RNA from minute amounts of starting material. Briefly, an RNA synthesis promoter is attached to the 3′ and/or 5′ region of cDNA utilizing the TC mechanism. The orientation of amplified RNAs is "antisense" or a novel "sense" orientation. TC RNA amplification is utilized for many downstream applications, including gene expression profiling, microarray analysis, and cDNA library/subtraction library construction. Input sources of RNA can originate from a myriad of in vivo and in vitro tissue sources. Moreover, a variety of fixations can be employed, and tissues can be processed for histochemistry or immunocytochemistry prior to microdissection for TC RNA amplification, allowing for tremendous cell type and tissue specificity of downstream genetic applications.

Keywords expression profiling, functional genomics, IVT (in vitro transcription), microarray, postmortem human brain, RNA amplification

1 Introduction

Conventional molecular biology techniques allow for gene expression analysis in a wide variety of paradigms and experimental systems. Methods include in situ hybridization (RNA detection), Northern analysis (RNA detection), polymerase chain reaction (PCR; DNA detection), reverse transcription-PCR (RT-PCR; RNA detection), ribonuclease (RNase) protection assay, and Southern analysis (DNA detection), among others. Most of these methods evaluate the abundance of individual elements one at a time (or a few at a time). Advancements in high-throughput

From: Methods in Molecular Biology, Vol. 439: *Genomics Protocols: Second Edition*   147
Edited by Mike Starkey and Ramnath Elaswarapu © Humana Press Inc., Totowa, NJ

genomic methodologies now enable the assessment of dozens to hundreds to thousands of genes simultaneously in a coordinated manner [1].

RNA can originate from a variety of in vivo and in vitro sources. Fresh, frozen, and fixed tissues are useful to varying degrees, depending on the paradigm and tissue quality. Many laboratories isolate genomic DNA and total RNA from paraffin-embedded fixed tissues as well as fresh and frozen tissues [2–6]. In addition, genetic material preserved by fixation is a superior resource because fresh or frozen tissues frequently are not available, whereas archived fixed human and animal model tissues exist in many tissue banks and individual laboratories. With mRNA as starting material, it cannot be emphasized enough the importance of preservation of RNA integrity in tissues and cells. RNA species are particularly sensitive to degradation by RNases. Because they play an important role in nucleic acid metabolism, RNases are found in virtually every cell type [7]. RNases degrade RNA species through endonuclease and exonuclease activity. RNases are quite stable and retain their activity over a broad pH range [7, 8]. Thus, RNase-free precautions need to be taken prior to and during all RNA amplification procedures. All biological samples require prompt handling, either through rapid RNA extraction, flash freezing, or fixation to minimize the degradation of intracellular RNAs by RNases.

Significant discourse centers on devising optimal methods to prepare brain tissues for downstream genetic analyses. Consensus on which protocol should be utilized has not yet been achieved. RNAs have been mined from tissue samples using cross-linking fixatives, including 10% neutral buffered formalin and 4% paraformaldehyde, as well as precipitating fixatives, such as 70% ethanol buffered with 150 mM sodium chloride [3, 4, 9, 10]. A useful method for assessing RNA quality in tissue sections prior to performing expression profiling studies is acridine orange (AO) histofluorescence. AO is a fluorescent dye that intercalates selectively into nucleic acids and has been used to detect RNA and DNA in brain tissues [11–14]. AO can be employed in combination with immunocytochemistry to identify cytoplasmic RNAs and specific antigens of interest and is compatible with confocal microscopy [15, 16]. In brain tissue sections, AO-positive neurons are in contrast to the pale background of white matter tracts that lack abundant nucleic acids. Nonneuronal cells tend to have less AO histofluorescence than neurons and brain tumor cells [17]. Importantly, individual RNA species (e.g., mRNA, rRNA, and tRNA) cannot be delineated by AO histofluorescence. Rather, this method provides a diagnostic test employed on adjacent tissue sections to ensure the likelihood that an individual case has abundant RNA prior to performing expensive expression profiling studies. A more thorough examination of RNA quality can be obtained via bioanalysis (e.g., 2100 Bioanalyzer, Agilent Technologies), which utilizes capillary gel electrophoresis to detect RNA quality and quantitate relative abundance [18–20]. Bioanalysis displays the analytical assessment of RNAs in an electropherogram or digital gel format, with relatively high sensitivity.

Microarray analysis has emerged as a useful and relatively cost-effective tool to assess transcript levels in a myriad of systems and paradigms. A disadvantage to these high-throughput technologies is a requirement for significant amounts of high-quality input sources of RNA for increased sensitivity and reproducibility. Whole organism and regional studies can generate significant input amounts of RNA species without any amplification procedures [21, 22]. Unfortunately, the quantity of RNA harvested from a single cell, estimated to be approximately 0.1–1.0 picograms, is not sufficient for standard RNA extraction procedures [23–25]. As a result, molecular-based methods have been developed to increase the amount of input RNA for downstream genetic analyses, including exponential PCR-based analyses and linear RNA amplification procedures. This protocol describes one RNA amplification procedure, terminal continuation (TC) RNA amplification.

A new procedure was developed in our laboratory, TC RNA amplification (Fig. 10.1) [18, 19, 26]. TC RNA amplification of genetic signals includes synthesizing the first-strand cDNA complementary to the mRNA template, subsequently generating second-strand cDNA complementary to the first-strand cDNA, and

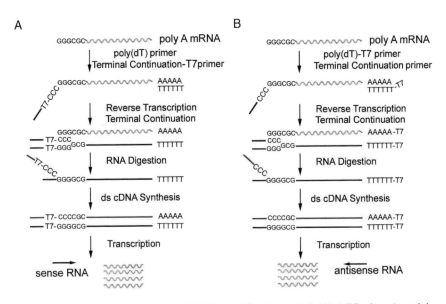

Fig. 10.1 Overview and analysis of the TC RNA amplification method. (A) A TC primer (containing a bacteriophage promoter sequence for sense orientation) and a poly d(T) primer are added to the mRNA population to be amplified. First-strand synthesis occurs as an mRNA-cDNA hybrid is formed after reverse transcription and terminal continuation of the oligonucleotide primers. Following RNase H digestion to remove the original mRNA template strand, second-strand synthesis is performed using Taq polymerase. The resultant double-stranded product is utilized as template for IVT, yielding high-fidelity, linear RNA amplification of sense orientation (rippled lines). (B) Schematic similar to A, illustrating the TC RNA amplification procedure amplifying RNA in the antisense orientation (rippled lines). Adapted from Ginsberg 2005 [26]

finally in vitro transcription (IVT) using the double-stranded cDNA as a template [18, 19, 26]. First-strand cDNA synthesis complementary to the template mRNA entails the use of two oligonucleotide primers, a first-strand poly d(T) primer and a TC primer. The poly d(T) primer is similar to conventional primers that exploit the poly(A) sequence present on most mRNAs. The TC primer contains a span of three cytidine triphosphates (CTPs) or guanosine triphosphates (GTPs) at the 3′ terminus

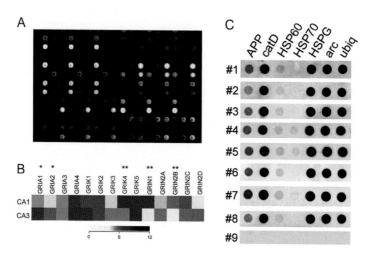

Fig. 10.2 Representative array platforms illustrating utility of RNA amplification procedures for single cell and population cell analysis. (A) Spotted cDNA array platform using RNA amplified from individual hippocampal CA1 pyramidal neurons from normal control brains and from neurofibrillary tangle (NFT) bearing CA1 neurons from Alzheimer's disease (AD) patients.

(B) Dendrogram demonstrating relative expression levels of representative genes in CA1 and CA3 pyramidal neurons microaspirated from human hippocampus. A heatmap matrix plot illustrates relative expression levels for individual glutamate receptor transcripts in CA1 and CA3 neurons. A single asterisk indicates a significant increase in expression in CA3 neurons as compared to CA1 neurons and a double asterisk denotes a significant increase in expression in CA1 neurons as compared to CA3 neurons. Key: GRIA 1, alpha-amino-3-hydroxy-5-methyl-4-isoxazolepropionate receptor 1 (AMPA1); GRIA2, AMPA2, GRIA3, AMPA3, GRIA4, AMPA4; GRIK1, kainate (KA) receptor GluR5; GRIK2, KA GluR6; GRIK3, KA GluR7; GRIK4, KA receptor KA1; GRIK5, KA receptor KA2; GRIN1, N-methyl D-aspartate receptor 1 (NR1); GRIN2A, NR2A; GRIN2B, NR2B; GRIN2C, NR2C; GRIN2D, NR2D. Adapted from Ginsberg and Che 2005 [32].

(C) Single-cell array analysis of human CA1 pyramidal neurons using custom-designed cDNA arrays and TC RNA amplification. Representative arrays illustrate a wide dynamic range of hybridization signal intensities for eight human postmortem cases (numbers 1–8). The negative control (number 9) is a single CA1 pyramidal neuron from the first case (number 1) that does not have the primers necessary for TC RNA amplification. In addition, a moderate variation of gene level expression across the eight human cases is observed, indicating the utility of using postmortem human samples for normative and neuropathological investigations. Key: APP, amyloid-β precursor protein; catD, cathepsin D; HSP60, heat shock protein 60; HSP70, heat shock protein 70; arc, activity regulated cytoskeletal-associated protein; ubiq, ubiquitin thiolesterase. Adapted from Ginsberg and Che 2002 [10]

[18]. Adenosine triphosphates (ATPs) or thymidine triphosphates (TTPs) do not perform well as constituents of the TC primer [20]. In this configuration, second-strand cDNA synthesis can be initiated by annealing a second oligonucleotide primer complementary to the attached oligonucleotide [18], and can be performed with robust DNA polymerases, such as Taq polymerase. One round of amplification is sufficient for downstream genetic analyses [18, 26]. Additionally, TC RNA transcription can be performed using a promoter sequence (e.g., T7, T3, or SP6) attached to either the 3′ or 5′ oligonucleotide primers. Therefore, transcript orientation can be in an antisense orientation (similar to conventional amplified antisense RNA methods) when the bacteriophage promoter sequence is placed on the first-strand poly d(T) primer, or in a sense orientation when the promoter sequence is attached to the TC primer, depending on the design of the experimental paradigm (Fig. 10.1) [19, 27]. Regional and single cell gene expression studies within the brains of animal models and human postmortem brain tissues have been performed via microarray analysis coupled with TC RNA amplification (Fig. 10.2) [1, 14, 27–31].

2   Materials

The methodology described here is a step-by-step protocol for TC RNA amplification as described by Che and Ginsberg [18, 19]. For clarity, the protocol begins at the point where cells are captured via laser-capture microdissection (LCM) or a microaspiration strategy and follows through IVT using biotinylated, fluorescent, or radioactive methods to label the TC RNA amplified products (*see* **Notes 1 and 2**).

2.1   Isolation of RNA

1. Trizol reagent (Invitrogen, Carlsbad, CA), stored at 4 °C.
2. 25:24:1 (v/v) Phenol:chloroform:isoamyl alcohol (Invitrogen), stored at 4 °C.
3. 5 mg/mL linear acrylamide (Ambion, Austin, TX), stored at −20 °C.

2.2   First-Strand cDNA Synthesis

1. First strand synthesis primers: 10 ng/μL poly d(T) and 10 ng/μL TC primer (Table 10.1).
2. Freshly prepared (on wet ice) reverse transcription (RT) master mix (*see* **Note 2**), comprising (for each RNA sample)

    4 μL of 5 × first-strand buffer (Invitrogen).
    1 μL of 4 × 10 mM dNTPs (Invitrogen).

Table 10.1 Representative oligonucleotide sequences utilized for the poly d(T) and TC primers for the TC RNA amplification method

Antisense RNA orientation

First-strand synthesis primer (66 bp): 3'- AAA CGA CGG CCA GTG AAT TGT AAT ACG
   ACT CAC TAT AGG CGC TTT TTT TTT TTT TTT TTT TTT TTT -5'

TC primer (17 bp): 5'- TAT CAA CGC AGA GTC CC -3'

Sense RNA orientation

First-strand synthesis primer (18 bp): 3'- TTT TTT TTT TTT TTT TTT -5'

TC-T7 primer (51 bp): 5'- AAA CGA CGG CCA GTG AAT TGT AAT ACG ACT CAC TAT
   AGG CGC GAG AGC CCC -3'

---

   1 μL of 0.1 M dithiothreitol (DTT).

   0.5 μL of 20 U/μL Superase-in RNase inhibitor (Ambion).

   5.5 μL of 18.2 MΩ RNase-free water.

3. 200 U/μL Superscript III (Invitrogen); stored at −20 °C.

2.3   Second-Strand cDNA Synthesis

1. Freshly prepared (on wet ice) second-strand master mix, comprising (for each first-strand cDNA sample)

   10 μL of 10 × PCR buffer, containing 15 mM $MgCl_2$ (PE Biosystems, Foster City, CA).

   0.5 μL of 1 U/μL RNase H (Invitrogen).

   69 μL of 18.2 MΩ RNase-free water.

2. 5 U/μL Taq polymerase (PE Biosystems), stored at −20 °C.

2.4   Double-Stranded cDNA Preparation

1. 25:24:1 (v/v/v) Phenol:chloroform:isoamyl alcohol, stored at 4 °C.
2. 50 mL conical tubes (Falcon) filled with 18.2 MΩ RNase-free water.
3. 0.025 μm membrane filters for drop dialysis (VSWP01300, Millipore, Billerica, MA).

2.5   IVT for TC RNA Amplification: Biotinylation/Fluorescent Probe Labeling

Freshly prepared (on wet ice) IVT MASTER Mix, comprising (for each double-stranded cDNA sample)

4 µL of 10 × RNA amplification buffer (Ambion; *see* **Note 3**).
4 µL of 10 × Biotin-labeled ribonucleotides (Enzo Life Sciences, Farmingdale, NY; *see* **Note 4**).
4 µL of 10 × DTT (Invitrogen).
4 µL of 10 × RNase inhibitor mix (Enzo).
12 µL of 18.2 MΩ RNase-free water.
2 µL of 1,000 U/µL T7 RNA polymerase (Epicentre, Madison, WI).

2.6 IVT for TC RNA Amplification: Radioactive Probe Labeling

1. Freshly prepared (on wet ice) IVT master mix, comprising (for each double-stranded cDNA sample)

   2 µL of 10 × RNA amplification buffer (Ambion).
   1 µL of 0.1 M DTT (Invitrogen).
   2 µL of 3 × 2.5 mM rATP, rCTP, and rGTP (Invitrogen).
   2 µL of 100 µM UTP (Invitrogen).
   0.5 µL of 20 U/µL Superase-in RNase inhibitor (Ambion).

2. 20 mCi/mL ^{33}P-UTP (GE Healthcare, Piscataway, NJ).
3. 1,000 U/µL T7 RNA polymerase (Epicentre, Madison, WI).

3 Methods

3.1 Isolation of RNA

1. Add 250 µL of Trizol reagent to the 500 µL thin-wall PCR tube that will receive the microdissected regions or cells, acquired via LCM, and keep on wet ice.
2. Invert the tube so that the Trizol reagent bathes the microdissected material, and keep on wet ice (samples can be stored at −80 °C at this juncture for future use).
3. Add 50 µL of phenol:chloroform:isoamyl alcohol, vortex vigorously for 15 s and centrifuge at 12,000 g for 15 min at 4 °C.
4. Collect the clear upper aqueous phase, containing RNA, by aspirating with a pipette.
5. Add 125 µL of 100% (v/v) 2-propanol and 5 µL of linear acrylamide to precipitate the RNA from the aqueous phase (store at −80 °C to precipitate RNA if desired).
6. Vortex vigorously for 15 s and centrifuge at 12,000 g for 15 min at 4 °C.
7. Decant the supernatant, by inverting the tube, being careful not to dislodge the pellet.

8. Add 250 μL of 75% (v/v) ethanol to wash RNA pellet. Vortex vigorously for 15 s and centrifuge at 7,500 g for 5 min at 4 °C.
9. Decant the supernatant, by inverting the tube, being careful not to dislodge the pellet.
10. Air dry the sample by inverting the tube for 5 min in a fume hood (*see* **Note 5**).
11. Resuspend the pellet in 5 μL of 18.2 MΩ RNase-free water.

3.2   First-Strand cDNA Synthesis

1. Add 1 μL of first-strand synthesis primer (Table 10.1) to a 5 μL RNA sample. Pulse centrifuge for 10 s.
2. Heat the mixture for 2 min at 75 °C. Pulse centrifuge for 10 s and place on ice.
3. Prepare the RT master mix (on wet ice; *see* **Note 2**).
4. Warm the RT master mix (without primers or Superscript III) for 2 min at 50 °C.
5. Heat denature an aliquot (1 μL for each RNA sample) of 10 ng/μL TC primer (Table 10.1) for 2 min at 70 °C, place on ice for several min and add to the RT master mix (*see* **Note 6**).
6. Add an aliquot of Superscript III (1 μL for each RNA sample) to the RT master mix (*see* **Note 7**). The RT master mix should constitute 14 μL per RT reaction.
7. Add the RT master mix to the 6 μL of sample, pipette vigorously, and centrifuge briefly. Incubate the mixture for 60 min at 50 °C.
8. Inactivate the Superscript III by heating the reaction mixture at 65 °C for 15 min.
9. Centrifuge briefly and cool immediately on wet ice. Samples can be stored at −20 °C for a short time, or at −80 °C for the long term.

3.3   Second-Strand cDNA Synthesis

1. Prepare the second-strand master mix (on wet ice).
2. Dispense 79.5 μL (for each second-strand synthesis reaction) of second-strand master mix into a 0.5-mL thin-wall PCR tube.
3. Add 20 μL of first-strand cDNA to the second-strand master mix and mix thoroughly with a pipette tip.
4. Place in a thermal cycler (*see* **Note 8**), programmed with the following second-strand cDNA synthesis protocol: 37 °C, 10 min (RNA degradation); 95 °C, 3 min (hot-start denaturation); 60 °C, 3 min (annealing); 75 °C, 30 min (elongation).
5. Start the thermocycle program. As soon as the block temperature reaches 95 °C, pause the reaction (this is the hot-start) and (using a dedicated PCR pipette) add 0.5 μL (for each reaction) of Taq polymerase.
6. Mix thoroughly with a pipette tip (*see* **Note 9**).
7. Continue the second-strand synthesis program by pushing the Continue function.
8. Once the program is complete, centrifuge briefly and store at −20 °C for the short term or at −80 °C for the long term.

3.4   Double-Stranded cDNA Preparation

1. To each sample, add 100 µL of 5 M ammonium acetate and 170 µL of phenol: chloroform:isoamyl alcohol.
2. Vortex vigorously for 30 s, centrifuge at 14,000 g for 5 min at 22 °C, and collect the upper aqueous phase in a fresh 1.7 mL microcentrifuge tube.
3. To precipitate each sample, add 1mL of 100% (v/v) ice-cold ethanol, and centrifuge at 14,000 g for 30 min at 4 °C.
4. Carefully discard the supernatant by inverting the tube and blotting on a piece of lint-free tissue. Air dry the pellet, in an inverted position, for 5 min in a fume hood.
5. Resuspend the double-stranded cDNA pellet in 20 µL of 18.2 MΩ RNase-free water.
6. Fill a 50-mL conical tube with 18.2 MΩ RNase-free water and float one 0.025-µm millipore membrane filter on the surface of the water (*see* **Note 10**). Load a cDNA sample onto the center of the floating membrane and carefully place the cap back on the conical tube. Allow it to stand undisturbed for 4 h at 22 °C.
7. Carefully remove the drop dialyzed sample into a new microfuge tube using a pipette tip.
8. Centrifuge briefly and store at −20 °C for the short term or at −80 °C for the long term.

3.5   IVT for TC RNA Amplification: Biotinylation/Fluorescent Probe Labeling

1. Prepare the IVT master mix (on wet ice).
2. Add 10 µL of double-stranded cDNA sample to the IVT master mix (30 µL for each cDNA sample).
3. Mix thoroughly (very important). Centrifuge briefly and incubate for 5 h at 37 °C.
4. The TC RNA amplified products are now ready for purification, fragmentation, and application to cDNA or oligonucleotide microarrays (*see* **Note 11**, Fig. 10.2A).

3.6   IVT for TC RNA Amplification: Radioactive Probe Labeling

1. Prepare the IVT master mix (on wet ice).
2. Centrifuge briefly and (on wet ice) add 7.5 µL of double-stranded cDNA sample to the IVT master mix (7.5 µL for each cDNA sample).
3. Add 4 µL of ^{33}P-UTP and 1 µL of T7 RNA polymerase.
4. Mix thoroughly (very important), centrifuge briefly, and incubate for 4 h at 37 °C.
5. The TC RNA amplified products now are ready to be hybridized to membrane-based arrays (*see* **Note 12**, Fig. 10.2C).

4   Notes

1. All solutions used throughout the entire protocol should be made in 18.2 MΩ RNase-free water (e.g., Nanopure Diamond, Barnstead, Dubuque, IA), which is referred to 18.2 MΩ RNase-free water in the text.
2. Prepare sufficient master mix for each reaction step using this basic formula: Tabulate the total number of samples plus two extra (i.e.. calculate the volumes of the mix components for χ number of samples + control + one to control for any volume loss or experimenter error).
3. An alternative receipe for 10 × transcription buffer (store in 1 ml aliquots) is 400 mM Tris-HCl, pH 7.5, 70 mM $MgCl_2$, 100 mM NaCl, 20 mM spermidine.
4. This protocol is based on using the BioArray RNA Transcript Labeling Kit (Enzo), although any biotinylation or fluorescent labeling protocols are suitable (minor modifications may apply).
5. Do not air dry RNA pellets longer than 5 min, as the pellets become difficult to resuspend.
6. Each RNA sample to which the RT master mix is added receives 1 μL of TC RNA primer; therefore, if there are 8 RNA samples, the RT master mix must be sufficient for ten RNA samples and 10 μL of the TC primer must be added to the RT master mix prior to adding it to the RNA samples.
7. When removing aliquots of Superscript III from the tube, be extremely careful to not contaminate the vial of enzyme with primers or RNA samples.
8. For the PCR cycling of the second-strand synthesis, it is a good idea to create a specific stepwise program using a final volume of 100 μL.
9. A crucial mistake that is commonly made is failure to mix the hot-start second-strand synthesis reaction when the Taq polymerase is added. Thorough mixing with a pipette tip ensures a suitable reaction environment.
10. This step requires a little practice. Float the membrane filter shiny side up, on the surface of the water using a pair of dedicated RNase-free tissue forceps. Take care not to sink the membrane. Do not overfill the 50-mL conical tube. Conversely, underfilling (<40 mL) the conical tube makes placement of the filter difficult.
11. There are numerous Internet sites and published protocols for the hybridization of labeled probes to microarray platforms, subsequent washing, and imaging protocols (e.g., www.affymetrix.com, www.enzo.com,. [26, 27, 33]).
12. Procedures for cDNA membrane array manufacture, prehybridization, hybridization, washing conditions, and imaging are described in detail in several published articles from this laboratory [10, 18–20, 32].

Acknowledgments I thank Shaoli Che, MD, PhD; Melissa Alldred, PhD; Irina Elarova; Shaona Fang; and Krisztina M. Kovacs for expert technical assistance Support for this project comes from the NINDS (NS43939, NS48447) and NIA (AG10668, AG14449, AG17617, AG09466) and Alzheimer's Association.

References

1. Ginsberg SD, Mirnics K (2006) Functional genomic methodologies. Prog Brain Res **158**:15–40
2. Kabbarah O et al (2003) Expression profiling of mouse endometrial cancers microdissected from ethanol-fixed, paraffin-embedded tissues. Am J Pathol 162:755–762
3. Su, JM et al (2004) Comparison of ethanol versus formalin fixation on preservation of histology and RNA in laser capture microdissected brain tissues. Brain Pathol 14:175–182
4. Tanji N et al (2001) Effect of tissue processing on the ability to recover nucleic acid from specific renal tissue compartments by laser capture microdissection. Exp Nephrol 9: 229–234
5. Fend F et al (1999) Immuno-LCM: Laser capture microdissection of immunostained frozen sections for mRNA analysis. Am J Pathol 154:61–66
6. Kinnecom K, Pachter JS (2005) Selective capture of endothelial and perivascular cells from brain microvessels using laser capture microdissection. Brain Res Protoc 16:1–9
7. Farrell RE Jr (1998) RNA methodologies, 2nd edn. Academic Press, San Diego
8. Blumberg DD (1987) Creating a ribonuclease-free environment. Methods Enzymol 152:20–24
9. Goldsworthy SM et al (1999) Effects of fixation on RNA extraction and amplification from laser capture microdissected tissue. Mol Carcinog 25:86–91
10. Ginsberg SD, Che S (2002) RNA amplification in brain tissues. Neurochem Res 27:981–992
11. Mai JK, Schmidt-Kastner R, Tefett H-B (1984) Use of acridine orange for histologic analysis of the central nervous system. J Histochem Cytochem 32:97–104
12. Vincent VA et al (2002) Analysis of neuronal gene expression with laser capture microdissection. J Neurosci Res 69:578–586
13. Ginsberg SD et al (1997) Sequestration of RNA in Alzheimer's disease neurofibrillary tangles and senile plaques. Ann Neurol 41:200–209
14. Mufson EJ, Counts SE, Ginsberg SD (2002) Single cell gene expression profiles of nucleus basalis cholinergic neurons in Alzheimer's disease. Neurochem Res 27:1035–1048
15. Ginsberg SD et al (1998) RNA sequestration to pathological lesions of neurodegenerative disorders. Acta Neuropathol 96:487–494
16. Tatton NA, Kish SJ (1997) In situ detection of apoptotic nuclei in the substantia nigra compacta of 1-methyl-4-phenyl-1,2,3,6-tetrahydropyridine-treated mice using terminal deoxynucleotidyl transferase labelling and acridine orange staining. Neuroscience 77:1037–1048
17. Sarnat HB et al (1987) Gliosis and glioma distinguished by acridine orange. Can J Neurol Sci 14:31–35
18. Che S, Ginsberg SD (2004) Amplification of transcripts using terminal continuation. Lab Invest 84:131–137
19. Che S, Ginsberg SD (2006) RNA amplification methodologies. In: McNamara PA (ed.)Trends in RNA Research. Nova Science Publishing, Hauppauge. NY, pp. 277–301
20. Ginsberg SD, Che S (2004) Combined histochemical staining, RNA amplification, regional, and single cell analysis within the hippocampus. Lab Invest. 84:952–962
21. Shaulsky G, Loomis WF (2002) Gene expression patterns in Dictyostelium using microarrays. Protist 153:93–98
22. Alter O, Brown PO, Botstein D (2003) Generalized singular value decomposition for comparative analysis of genome-scale expression data sets of two different organisms. Proc Natl Acad Sci USA 100:3351–3356
23. Kacharmina JE, Crino PB, Eberwine J (1999) Preparation of cDNA from single cells and subcellular regions. Methods Enzymol 303:3–18
24. Phillips J, Eberwine JH (1996) Antisense RNA amplification: A linear amplification method for analyzing the mRNA population from single living cells. Methods Enzymol Suppl 10:283–288

25. Sambrook J, Russell DW (2001) Molecular cloning: A laboratory manual, 3rd edn. Cold Spring Harbor Laboratory Press, Cold Spring Harbor, NY

26. Ginsberg SD (2005) RNA amplification strategies for small sample populations. Methods 37:229–237

27. Ginsberg SD et al (2006) Cell and tissue microdissection in combination with genomic and proteomic applications. In: Zaborszky L, Wouterlood FG, Lanciego JL (eds) Neuroanatomical tract tracing 3: Molecules, neurons, and systems. Springer, New York, pp. 109–141

28. Ginsberg SD et al (2006) Shift in the ratio of three-repeat tau and four-repeat tau mRNAs in individual cholinergic basal forebrain neurons in mild cognitive impairment and Alzheimer's disease. J Neurochem 96:1401–1408

29. Ginsberg SD et al (2006) Down regulation of trk but not p75 gene expression in single cholinergic basal forebrain neurons mark the progression of Alzheimer's disease. J Neurochem 97:475–487

30. Ginsberg SD et al (2006) Single cell gene expression profiling in Alzheimer's disease. NeuroRx 3:302–318.

31. Counts SE et al (2006) Galanin fiber hypertrophy within the cholinergic nucleus basalis during the progression of Alzheimer's disease. Dement Geriatr Cogn Disord 21:205–214

32. Ginsberg SD, Che S (2005) Expression profile analysis within the human hippocampus: Comparison of CA1 and CA3 pyramidal neurons. J Comp Neurol 487:107–118

33. Cheung VG et al (1999) Making and reading microarrays. Nat Genet 21:15–19

Chapter 11
Quantitative Expression Profiling of RNA from Formalin-Fixed, Paraffin-Embedded Tissues Using Randomly Assembled Bead Arrays

Marina Bibikova, Joanne M. Yeakley, Jessica Wang-Rodriguez, and Jian-Bing Fan

Abstract Formalin-fixed, paraffin-embedded (FFPE) tissues represent an invaluable resource for gene expression analysis, as they are the most widely available materials for studies of human disease. However, degradation of RNA during tissue fixation and storage makes FFPE-derived RNAs hard to work with using conventional microarray technology. Most gene expression studies done using FFPE tissues rely on quantitative RT-PCR (qPCR), but this approach has had limited success because of the short cDNA templates available in these samples.

In this chapter, we describe the DASL® (c*D*NA-mediated *a*nnealing, *s*election, extension, and *l*igation) Assay, a flexible, sensitive, and reproducible gene expression profiling system for parallel analysis of several hundred mRNA transcripts on the BeadArray™ platform. This technology is especially useful for determining cancer prognosis or therapy response, because it allows not only prospective studies but also retrospective analyses. Using the DASL Assay, gene expression analyses can be performed on routinely stored tumor specimens from patients with known outcomes.

Keywords Formalin-fixed, paraffin-embedded (FFPE) tissues; archived samples; RNA; gene expression analysis; microarray; bead array; DASL Assay; biomarker

1   Introduction

1.1   Introduction to BeadArray Technology and Randomly Assembled Bead Arrays

Illumina's BeadArray technology is based on the random self-assembly of a bead pool onto a patterned substrate. The arrays are assembled into a matrix (the Sentrix® Array Matrix, 96 samples, or Sentrix BeadChip, 16 samples), allowing efficient processing of thousands of samples [1].

From: Methods in Molecular Biology, Vol. 439: *Genomics Protocols: Second Edition*   159
Edited by Mike Starkey and Ramnath Elaswarapu © Humana Press Inc., Totowa, NJ

The manufacture of Illumina microarrays has been described elsewhere [2]. Briefly, universal BeadArrays are assembled by loading pools of glass beads (3 µm in diameter) derivatized with oligonucleotide capture probes onto the etched ends of fiber-optic bundles or the etched surface of silica slides (Fig. 11.1). Because the beads are positioned randomly, a decoding process is carried out to determine the location and identity of each bead in every array location [3]. Decoding is an automated part of array manufacturing and has the added benefit of validating the hybridization capability of each feature.

1.2   The DASL Assay for Gene Expression Profiling

There were over 300 million archived tissue samples were in the United States in 1999, with more samples accumulating at a rate of over 20 million per year [4]. These samples represent an extraordinary resource for basic research as well as for

Fig 11.1 Sentrix Array Matrix and Sentrix BeadChip. Sentrix Platform. The randomly assembled arrays are manufactured on fiber optic bundle-based substrates for high throughput (above) or on silicon wafers for large feature sets (below)

biomarker discovery and validation, yet they are largely inaccessible to standard microarray technology due to extensive degradation of the RNA [5–8].

In contrast to most array technologies that use an in vitro transcription (IVT)-mediated sample labeling procedure [9], the DASL Assay uses random priming in the cDNA synthesis and, therefore, does not depend on an intact poly-A tail for oligo-d(T) priming (Fig. 11.2). The DASL Assay interrogates specific cDNA sequences with oligos linked to unique address sequences that can hybridize to their complementary strand on the universal arrays [10]. The assay requires a relatively short target sequence of about 50 nucleotides for query oligonucleotide annealing. Thus, the DASL Assay allows analyses of partially degraded RNAs extracted from FFPE tissues [11, 12]. The DASL Assay uses one set of universal PCR primers to amplify all the targets and generates amplicons of ~100 bp. This uniformity results in a relatively unbiased amplification of the PCR template population [10]. The approach combines the advantages of array-based gene expression analysis with those of multiplexed qPCR, thereby offering much higher multiplexing capacity and throughput at lower cost. The assay multiplexes to over 1,500 sequence targets, and expression differences obtained with this assay compare well to those determined by qPCR [10]. Cell-specific expression was readily detected, and expression data allowed biologically appropriate sample classification [11].

With automation, the DASL Assay coupled with the array readout can be used for high-throughput expression profiling of hundreds of genes in hundreds

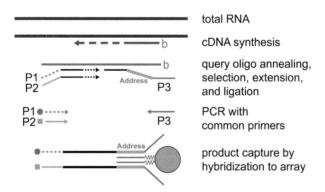

Fig. 11.2 DASL Assay scheme. In the DASL Assay, three oligos are designed to target a specific gene sequence. The total RNA first is converted to cDNA by random priming. The corresponding query oligos bind to the cDNA and become extended and ligated enzymatically. The ligated products then are amplified and fluorescently labeled during PCR and finally detected by hybridization to address sequences on the array

to thousands of FFPE samples. This opens up new avenues to large-scale discovery, validation, and clinical application for expression markers of disease.

2 Materials

The Materials and Methods sections assume that the user has access to the Illumina
BeadStation or BeadLab systems and associated reagents and equipment.

2.1 Recommended Kit for RNA Extraction from FFPE Tissues

Of the few RNA preparation kits we tested, the High Pure RNA Paraffin Kit from
Roche Applied Science (catalogue # 3 270 289) yielded the highest quality RNA
from FFPE samples for use in the DASL Assay (*see* **Notes 1 and 2**). Recently, more
RNA extraction kits have become available. If a different kit is used for RNA
extraction, the prequalification steps described in the Notes section should be used
to increase the likelihood of success in the DASL Assay (*see* **Notes 3 and 4**).

2.2 Illumina Supplied Reagents for the DASL Assay

1. MCS/MCM (master mix for cDNA synthesis for single or multiple use), reagent for cDNA synthesis.
2. DAP (DASL Assay oligo pool), mixture of oligonucleotides designed to query
   cDNA target sequences.
3. OB1 (oligo binding buffer 1), oligo annealing buffer that also contains paramagnetic particles to optimize the washing, extension, and ligation steps of the assay.
4. AM1 (add MEL buffer 1), wash buffer used to remove excess of mis- or unhy-
   bridized query oligonucleotides.
5. UB1 (universal wash buffer 1), wash buffer for several pre-PCR steps.
6. MEL (master mix for extension and ligation), optimized mixture of enzymes
   for the extension and ligation steps.
7. MMP (Master mix for PCR), PCR master mix that contains fluorescent common primers for the multiplexed ligated oligo templates.
8. IP1 (inoculate PCR buffer 1), elution buffer for inoculating the PCR reaction
   with the ligated templates.
9. MPB (magnetic particle buffer B), a suspension of paramagnetic particles used
   to bind PCR products.
10. UB2 (universal buffer 2), a wash buffer used in post-PCR processes.
11. MH1 (make Hyb buffer 1), a neutralization buffer used in making the final sam-
    ple single-stranded for hybridization to the Sentrix Array Matrix or BeadChip.

12. IS1 (image SAM buffer 1), a buffer needed to dry bundles and protect fluors before imaging.
13. WC1 (wash chip buffer 1), a buffer used to wash Sentrix BeadChips after hybridization.
14. AC1 (add coverslip buffer 1), a buffer used for protecting fluors in imaging the BeadChip.

Please refer to Table 11.1 for reagent storage conditions.

2.3 Other Reagents Required for the Assay

1. 0.1 N NaOH.
2. TE buffer: 10 mM Tris-HCl, pH 8.0, 1 mM EDTA.
3. DNA polymerase (Clontech BD, 639208 or 639209).
4. Uracil DNA glycosylase, 1 U/μL (Invitrogen, 18054-015).
5. 95% EtOH (v/v).
6. 2-butanol 99% (Sigma-Aldrich, B85919).
7. RiboGreen RNA Quantitation Kit (Molecular Probes Invitrogen, R-11490).
8. Xylenes (EMD Chemicals, XX0055-13) for FFPE sample deparaffinization.
9. Serological pipettes (10, 25, and 50 mL).
10. 96-well 0.2-mL skirted microplates (Bio-Rad, MSP-9601).
11. 0.8-mL storage plate, conical well bottom (ABgene, AB-0765).
12. 96-well, black, flat-bottom Fluotrac 200 plates (Greiner, 655076).
13. 96-well sealing mats, round cap, autoclavable (ABgene, AB-0674).

Table 11.1 Reagent storage conditions

Reagent	Storage Conditions	Shelf Life	Comments
MCS/MCM	−20 °C	6 months	Aliquot to refreeze
DAP	−20 °C	2 years	Can be stored at 4 °C up to 2 weeks
OB1	−20 °C	6 months	Does not completely freeze
AM1	4 °C	6 months	
UB1	4 °C	6 months	
MEL	−20 °C	6 months	Aliquot to refreeze
MMP	−20 °C	6 months	Aliquot to refreeze after adding DNA polymerase
IP1	−20 °C	1 year	
MPB	4 °C	6 months	Do not freeze
UB2	RT	6 months	
MH1	RT	6 months	Keep away from light
IS1	Room temperature while dry −20 °C when reconstituted	6 months	Must resuspend at least 1 d in advance
WC1	−20 °C	6 months	Must dilute
AC1	Room temperature	6 months	

14. 96-well cap mats, sealing mats, round cap, pierceable, nonautoclavable (ABgene, AB-0566).
15. Heat sealing foil sheets, thermo seal (ABgene, AB-0559).
16. Microplate clear adhesive film, 2-mil Sealplate adhesive film, nonsterile (Phenix Research Products, LMT-SEAL-EX).
17. Microseal A film, PCR plate sealing film (Bio-Rad, MSA-5001).
18. Microseal F film, aluminum adhesive film (Bio-Rad, MSF-1001).
19. 96-well V-bottom plates, Corning Costar polypropylene (Fisher Scientific 07-200-695 or VWR, 29444-102).
20. Multiscreen filter plates, 0.45-μm, clear, styrene (Millipore, MAHV-N45 10/50).
21. Nonsterile solution basins (Labcor Products Inc., 730-001 or VWR, 21007-970).
22. Cliniplate 384-well microplates (Thermo Labsystems 95040000).
23. Omni trays (Nunc, 242811 or VWR, 62409-600).

2.4   Fiber-Optic 96-Array Matrix and Other Array Formats

Illumina produces microarrays in two formats: the Sentrix Array Matrix (SAM) and the Sentrix BeadChip (see Fig. 11.1). In the array matrix, the fiber-optic bundles, each containing ~50,000 5-μm fibers, are arranged into an Array of Arrays™ format that is compatible with and can access the wells of a 96-well microtiter plate. The fiber-optic bundles in the array matrix assembly are polished flat on both ends. On one end, the cores of each fiber are etched to form nanowells that will accept 3-μm silica beads. Each of the beads has been derivatized with several hundred thousand oligonucleotides of a particular sequence. In the BeadChip format, several microarrays are arranged on silicon slides that have been processed by microelectromechanical systems (MEMS) technology to also have nanowells that support self-assembly of beads. In both formats, an average of over 30 beads of each type is maintained, a strategy that provides the accuracy of multiple measurements and statistical power. Beads are self-assembled in the wells during the manufacturing process, and each array is decoded to determine the map of bead type positions. Combinatorial series of brief hybridizations and rinses are used for decoding that result in a level of accuracy well beyond the requirements of any application [3]. This decoding process also provides a quality control measure of the function of each bead that is incorporated into the final bead map. BeadArray technology is the only microarray technology with functional quality control of every element in the array.

2.5   Bead Array Imaging

The Illumina BeadArray Reader is required for array imaging. Each assay results in a fluorescent signal associated with individual bead types on the array. To read out these signals, we developed the BeadArray Reader, a two-channel, 0.8-μm resolution confocal laser scanner that can simultaneously (via menu-driven software) scan a BeadChip or SAM at two wavelengths and create an image file for each channel. For Illumina BeadArray Reader instructions, see the *Illumina BeadArray Reader User Guide*.

3   Methods

3.1   RNA Extraction from FFPE Tissues

RNA extraction from formalin-fixed, paraffin-embedded samples is a critical step for success in the DASL Assay. We recommend the High Pure RNA Paraffin Kit (Roche) for RNA purification, using five 5-μm sections of FFPE tissues. All buffers and reagents are provided with the kit. The RNA extraction is carried out following manufacturers' recommendations (*see* **Notes 1 and 2** for details and Table 11.2 for RNA yields from different tissue sources). RNA integrity can be examined with the Agilent 2100 Bioanalyzer using RNA 6000 Pico Assay Kit.

3.2   DASL Assay Protocols

These protocols are not intended to replace the DASL Assay manual supplied with Illumina systems but rather give a detailed overview of the process.

Table 11.2   RNA yield from five 5-μm sections of FFPE tissue

Tissue	Diagnosis	Yield (μg)
Prostate	Normal	3.81
Prostate	Neoplasia	2.09
Prostate	Adenocarcinoma	1.31
Lung	Normal	0.34
Lung	Adenocarcinoma	11.45
Lung	Bronchio-alveolar adenocarcinoma	7.80
Lung	Small cell adenocarcinoma	4.73
Lung	Squamous cell adenocarcinoma	13.22
Breast	Normal	0.65
Breast	Fibroadenoma	4.49
Breast	Invasive ductal adenocarcinoma	17.14
Breast	Lobular adenocarcinoma	5.56
Colon	Normal	10.07
Colon	Adenocarcinoma mucoid	2.62
Colon	Adenocarcinoma sygmoid	6.40
Colon	Adenocarcinoma ulcer	14.46

3.2.1 Assay Probe Design

The DASL Assay is based on the sequence-specific extension and ligation of correctly hybridized query oligos (*see* **Note 5**), which are distinguished by their shared primer landing sites (for an overview of the DASL Assay workflow, see http://www.illumina.com/downloads/DASLTECHBULLETIN.pdf). Query oligos are designed to target nonpolymorphic sites in cDNAs, and the assay uses two upstream oligos (ASOs) for each downstream oligo (LSO) (Fig. 11.2). The result is two separate measurements of the population of a single transcript, one in each fluorescent channel. Used only for expression profiling, the DASL Assay need not be performed with two fluors. However, its current configuration allows the assay to be compatible with the reagents used in the Illumina GoldenGate® Genotyping Assay. This arrangement gives the researcher the flexibility to investigate the allele-specific expression by "genotyping" cDNA as well as genomic DNA.

For array analysis, three query oligonucleotides are designed to interrogate each target site on the cDNA, with up to ten target sites per gene. Query oligos are designed to target intraexonic sequences, using bioinformatic and biochemical parameters optimized for the assay biochemistry (*see* **Notes 6 and 7**). The ASO oligos consist of two parts: the gene-specific sequence and the universal PCR primer sequences, with P1 and P2 at the 5′-end (Fig. 11.2). The downstream oligo consists of three parts: the gene-specific sequence, a unique address sequence that is complementary to 1 of 1,536 capture sequences on the array, and a universal PCR primer sequence (P3) at the 3′-end. A single address sequence is uniquely associated with a single target site. This address sequence allows the PCR-amplified products to hybridize to a universal microarray bearing the complementary probe sequences.

3.2.2 DASL Assay Protocol

3.2.2.1 The Make SUR (Single Use RNA) Process for cDNA Synthesis

1. Preheat the heat block to 42 °C and allow temperature to stabilize. Thaw the cDNA synthesis reagent (MCS, Illumina) to room temperature and vortex to fully mix tube contents.
2. Normalize the RNA samples to 40–100 ng/μL with DEPC-treated H_2O. We recommend using 200 ng of RNA extracted from FFPE samples for one assay (*see* **Notes 8 and 9**).
3. Add 5 μL of MCS to each well of microtiter plate labeled SUR. Transfer 5 μL of normalized RNA sample to each well of the SUR plate, heat seal with foil, and vortex at 2,300 rpm for 20 s. Pulse-centrifuge the sealed plate to 250 g to prevent the wells from evaporating during the incubation.
4. Incubate the SUR plate at room temperature at least 10 min, 1then continue incubation at 42 °C for 60 min in the preheated heat block.
5. When incubation is complete, remove the plate from the heat block and pulse-centrifuge to 250 g to collect any condensation.

6. If the experiment will be continued the same day, set the heat block to 70 °C. If not proceeding with the experiment immediately, the SUR plate can be stored up to 4 h at 4 °C or overnight at −20 °C.

3.2.2.2 The Make ASE (Assay-Specific Extension) Process for Annealing Query Oligonucleotides to cDNA

1. Preheat heat block to 70 °C and allow temperature to stabilize.
2. Remove DASL Assay oligonucleotides (DAP) tube from refrigerator (if frozen, thaw, vortex, then centrifuge). Thaw the oligonucleotide annealing reagent (OB1) to room temperature and vortex. Do not centrifuge the OB1 tube.
3. Dispense 10 μL of DAP to each well of a new, 96-well, 0.2-mL skirted microplate labeled ASE. Add 30 μL of well-resuspended OB1 to each well of that plate.
4. Centrifuge the SUR plate to 250 g to collect samples at the bottom of the wells. Transfer 10 μL of the biotinylated cDNA to the ASE plate containing the DAP and OB1 to bring the final volume to 50 μL. Heat seal the plate and vortex briefly at 1,600 rpm to mix the content of the wells. Place the ASE plate in the 70 °C heat block and immediately reduce the temperature setting to 30 °C. This will carry out assay oligonucleotide annealing to the cDNA target by ramping the temperature over approximately 2 h then holding at 30 °C until the next processing step.

3.2.2.3 The Add MEL Process for Assay Oligonucleotide Extension and Ligation

1. Remove the ASE plate from the heat block, reset it to 45 °C and allow the temperature to stabilize. Thaw the extension and ligation reagent (MEL) to room temperature.
2. Place the ASE plate with assay oligonucleotides annealed to the cDNA template on the Illumina-supplied magnetic plate for at least 2 min, or until the beads are completely captured. Washing the beads removes excess and mis-hybridized oligonucleotides.
3. After the paramagnetic particles are captured, remove the heat seal from the plate and remove and discard all liquid (~50 μL) from the wells, retaining the beads. Add 50 μL Add MEL buffer (AM1) to each well of the assay plate. Seal the plate with adhesive film and vortex at 1,600 rpm for 20 s or until all beads are resuspended.
4. Place the ASE plate on the magnet for at least 2 min, or until the beads are completely captured. Remove all the AM1 from each well, leaving the beads in the wells. Repeat the addition of 50 μL AM1, vortexing, and removal of buffer.
5. Remove the ASE plate from the magnet and add 50 μL of universal wash buffer (UB1) to each well.
6. Place the ASE plate onto the magnet for at least 2 min, or until the beads are completely captured. Remove all UB1 from each well. Repeat the addition of 50 μL UB1 and removal of buffer.

7. Add 37 μL of extension and ligation master mix (MEL; Illumina) to each well of the ASE plate. Seal plate with adhesive film and vortex at 1,600 rpm for 1 min.
8. Incubate the ASE plate on the preheated 45 °C heat block for 15 min.

3.2.2.4 The Make PCR and Inoc PCR Processes for Preparing the PCR Mix and Setting up the PCR Reaction

1. Prepare PCR master mix by adding 64 μL of DNA polymerase and 50 μL UDG to the tube of PCR reagent (MMP). Invert the MMP tube several times to mix contents and aliquot 30 μL of mixture into each well of a new 96-well 0.2-mL microplate (the PCR plate).
2. Remove the ASE plate from the heat block after the extension and ligation step and reset the heat block to 95 °C.
3. Place the ASE plate on the magnet for at least 2 min, or until the beads are captured. Remove the clear adhesive film from the assay plate, and remove and discard supernatant (~50 μL) from all wells of the ASE plate, leaving the beads in wells. Leave the ASE plate on the magnet and add 50 μL of the universal wash buffer UB1 to each well of the plate.
4. Allow the ASE plate to rest on magnetic plate for at least 2 min to collect the paramagnetic particles. Remove and discard all supernatant (~50 μL) from all wells of the ASE plate, leaving beads in wells.
5. Add 35 μL of the elution buffer IP1 to each well of the assay plate and seal it with adhesive film. Vortex the plate at 1,800 rpm for 1 min, or until all the beads are resuspended. Place the plate on the 95 °C heat block for 1 min.
6. Remove the ASE plate from heat block and place it onto the magnet for at least 2 min, or until the beads have been completely captured. Transfer 30 μL of supernatant from the first column of the ASE plate into the first column of the PCR plate. Repeat the transfer for the remaining columns, using new pipette tips for each column.
7. Seal the PCR plate with Microseal A PCR plate sealing film. Immediately transfer the plate to the thermal cycler and run the following cycling program: 10 min at 37 °C; 34 cycles (35 s at 95°, 35 s at 56°, 2 min at 72°); 10 min at 72°; and 4 °C for 5 min.
8. Proceed immediately to the preparation of the PCR product for array hybridization or seal and store the PCR plate at −20 °C.

3.2.2.5 The Make HYB Process to Prepare Samples for Array Hybridization

1. Vortex the tube with the suspension of paramagnetic particles (MPB) until the beads are completely resuspended.
2. Dispense 20 μL of resuspended MPB into each well of the PCR plate. Mix the beads with PCR product by pipetting up and down, then transfer the mixed solution to the filter plate. Cover the filter plate with its cover and store at room temperature, protected from light, for 60 min.

3. Place the filter plate containing the bound PCR products onto a new 96-well V-bottom waste plate using a filter plate adapter. Centrifuge at $1,000\,g$ for 5 min at 25 °C.

4. Remove the filter plate lid. Add 50 µL of universal wash buffer 2 (UB2) to each well of the filter plate. Dispense slowly, so as not to disturb the beads. Relid the filter plate and centrifuge at $1,000\,g$ for 5 min at 25 °C.

5. Prepare a new 96-well V-bottom plate and dispense 30 µL of hybridization buffer (MH1) to all wells of the new intermediate plate. Place the filter plate onto the intermediate plate such that column A1 of the filter plate matches column A1 of the intermediate plate.

6. Dispense 30 µL of $0.1\,N$ NaOH to all the wells of the filter plate. Relid the filter plate and centrifuge immediately at $1,000\,g$ for 5 min at 25 °C. Gently mix the contents of the intermediate plate by moving it from side to side, without splashing.

7. Prepare a HYB Cliniplate 384-well microplate. First, dispense 30 µL of UB2 buffer in every other well (B2, B4, D2, D4, etc.) of the hybridization plate for humidity control. Then transfer 50 µL of neutralized hybridization solution from the intermediate plate into appropriate wells (A1, A3, C1, C3, etc.) of hybridization plate. It is helpful to use separate templates for humidity control wells and hybridization wells to assist in sample dispensing.

8. Seal the plate with clear adhesive film and centrifuge at $3,000\,g$ for 4 min at 25 °C. If the hybridization is not performed immediately, the HYB plate can be stored at −20 °C.

3.3 Hybridization to Bead Arrays

1. Preheat the oven to 60 °C and allow the temperature to equilibrate. If the HYB plate has been frozen, thaw it completely at room temperature in a light-protected drawer. Centrifuge the plate at $3,000\,g$ for 4 min.

2. Prepare two OmniTrays, one with 70 mL of UB2 and another with 60 mL of 0.1N NaOH. Carefully place the Sentrix Array Matrix (SAM), with the barcode facing up and the fiber bundles facing down, into the UB2 tray. Agitate the SAM gently (10 s only) to remove bubbles from the bottoms of the arrays. After 3 min, move the SAM into a NaOH tray and incubate 30 s, then move the SAM back into the UB2 conditioning tray. Allow the SAM to sit in the UB2 conditioning tray for at least 30 s to neutralize the NaOH.

3. Remove the clear adhesive film from the HYB plate. Insert the SAM fiber-optic bundles into appropriate wells of the HYB plate using the SAM alignment fixture or a SAM hybridization cartridge. Place the HYB plate/SAM pair into the hybridization oven preheated to 60 °C and incubate for 30 min.

4. Hybridization then is conducted under a temperature gradient program from 60 to 45 °C over approximately 12 h.

3.4   Array Imaging

3.4.1   Wash and Dry Array Matrix

1. Reconstitute the array drying reagent (IS1; Illumina) by adding 94 mL of an equal mixture of 95% ethanol and 2-butanol to 6 mL of IS1 to prepare enough reagent to image one array.
2. For each SAM to be imaged, prepare two OmniTrays with 70 mL of UB2 and one with 70 mL of diluted IS1.
3. Carefully separate the SAM from the HYB plate. Place the SAM into the first UB2 tray and agitate 1 min at room temperature. Transfer the SAM to the second UB2 tray and repeat the agitation for 1 min at room temperature. Ensure no bubbles are on the bottom of the array fiber bundles. These bubbles may prevent certain areas of the fiber bundle from contacting the liquid.
4. Dip the SAM into the IS1 tray. It is important to dip the SAM several times to ensure that the UB2 buffer is completely exchanged. After 5 min, remove the SAM from the IS1 and place it on an empty OmniTray to air dry 20 min, fiber bundles up. The dry array is ready for imaging.

3.4.2   Imaging the BeadArray

The arrays are imaged using the BeadArray Reader (Illumina). Image processing and intensity data extraction is performed by the BeadScan software included with the BeadArray Reader.

3.5   Data Collection and Analysis

The BeadStudio software package is included with the Illumina DASL Assay product and is used as a tool for analyzing gene expression data from scanned microarray images collected from the Illumina BeadArray Reader. Alternatively, BeadStudio can be used to export the array intensity data for processing by most standard gene expression analysis programs.

Specifically, BeadStudio executes two types of data analysis:

1. Gene analysis, determining gene expression levels and whether gene expression is detected.
2. Differential analysis, determining if gene expression levels are different between any experimental groups.

Analysis can be performed on individual arrays or groups of arrays treated as replicates. BeadStudio reports experiment performance based on built-in controls that accompany each experiment (*see* **Notes 10–13**). In addition, BeadStudio provides scatterplotting and dendrogram tools, facilitating quick, visual means for exploratory analysis (*see* **Note 14**).

3.6   Examples of DASL Assay Applications

Our data showed that the DASL Assay was capable of validating tissue and cancer markers previously identified using RNA prepared from fresh-frozen tissues [11]. We also demonstrated that, using the DASL Assay system, highly reproducible tissue- and cancer-specific gene expression profiles could be obtained with RNA isolated from formalin-fixed tissues that had been stored from 1 to over 10 years [11]. Furthermore, the marker sensitivity and specificity measured by the DASL analysis were comparable to those determined for qPCR. Overall, the DASL Assay analysis outperformed qPCR for these samples (*see* **Note 15**).

In addition, we constructed a splicing array to interrogate >1,500 mRNA isoforms from a panel of genes previously implicated in prostate cancer. We derived a highly specific panel of mRNA isoforms for prostate cancer using independent sets of samples for marker selection and validation [13, 14].

4   Notes

1. RNA from FFPE specimens can be difficult to extract, since the RNA becomes cross-linked and degraded during the fixation and storage process; in addition, the amount of tissue in the FFPE specimen can be very small (for example, normal lung samples consist mostly of extracellular air space). Therefore, it is essential to have a robust method to retrieve high-quality RNA from FFPE tissue efficiently. Various RNA extraction kits for this application are commercially available, but we found the Roche High Pure RNA Paraffin Kit worked the best in our experiments. However, this protocol is tedious and time consuming, which may become a limiting step for high-throughput.

2. We recommend using several [5, 6] 5-µm sections of FFPE tissue for RNA extraction. The average tissue sections from paraffin blocks were approximately 1 cM^2 with the solid tissue in the middle. Our results suggest that 5-µm tissue sections gave better RNA yield than 20-µm sections, probably due to more efficient deparaffinization and cell lysis. We also isolated RNA from tissue sections mounted on slides.

3. RNA quantitation is an important step to ensure that sufficient material is used in the DASL Assay to generate high-quality data. We recommend the Molecular Probes RiboGreen Assay Kit for quantitation of RNA samples. The RiboGreen assay can quantitate small RNA volumes and measure RNA directly. Other techniques may pick up contamination, such as small molecules and proteins. We recommend using a fluorometer, as fluorometry provides RNA-specific

quantitation, whereas spectrophotometry may be affected by DNA contamination, leading to artificially inflated amounts.

4. To prequalify RNA samples prior to array analysis, we use a real-time PCR-based method to assess the intactness of the RNA samples. We find this approach more effective than using a combination of RNA quantitation and a gel-based size analysis.

There are various qPCR methods, many of which could be used for prequalification of degraded RNA samples. We developed the following protocol based on amplification of a fragment of the highly expressed RPL13a ribosomal protein gene (GenBank accession # NM_012423.2), with SYBR Green detection.

The RPL13a primers (forward primer 5′ GTACGCTGTGAAGGCATCAA 3′ and reverse primer 5′ GTTGGTGTTCATCCGCTTG 3′) amplify a 90-base-pair fragment and are designed to span an intron, so they should produce a correctly amplified product only from cDNA but not genomic DNA.

Using a reaction containing 5 μL of SYBR Green PCR Master Mix, 1 μL of cDNA product diluted 1:10, and 250 nM each forward and reserve primers in a total reaction volume of 10 μL, and a PCR profile of 95 °C for 12 min, followed by 40 cycles of 95 °C for 20 s, 54 °C for 2 s, 72 °C for 30 s, we routinely measure a crossover threshold (Ct) value of about 17 cycles for intact human reference total RNA (BD Biosciences Clontech, catalog # 636538) using the RPL13a primers and the reagents listed previously (the detection threshold set at 0.2 for the ABI Prism).

Based on analyses of more than 300 RNA samples derived from formalin-fixed, paraffin-embedded tissues, we found that samples should not exhibit a Ct of more than 28 cycles under these conditions to assure a reasonable expectation of reliable data in the DASL Assay (see Fig. 11.3). It is important to recognize that the Ct values we observed are entirely dependent on the equipment and reagents we use.

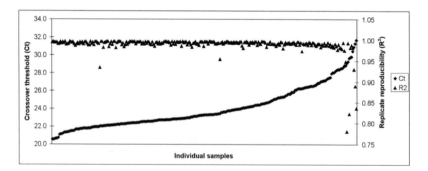

Fig. 11.3 Correlation between RNA sample quality control and array data quality. The DASL Assay can effectively generate high-quality gene expression data with samples that have up to eight cycle number differences in qPCR for the highly expressed housekeeping gene, RPL13A (~170-fold difference in "workable" RNA input). Under our qPCR conditions, any RNAs with a Ct number exceeding 28 in the prequalification assay may compromise array data quality

If other qPCR systems are used, different values may be expected. But, in general, the difference in cycle number at detection between cryo-preserved and paraffin-derived RNAs should not exceed 8 cycles for good performance in the DASL Assay.

5. DASL Assay query oligos are designed to target intraexonic sequences by default. But the oligos can be designed to span exon junctions for alternative splicing monitoring, for example. As a matter of fact, since DASL uses random priming in the cDNA synthesis, the probes can be designed to target any unique regions of the transcripts. There is no need to limit the selection of optimal probes to the 3′-end of the transcripts. The DAP oligo design program can take gene names, RefSeq accession numbers (NM or GI numbers), or sequences.

   With regard to the number of probes per gene, our analysis has shown that three optimally designed probes performed comparably to four or more probes as to their ability to detect expressed genes as well as differential expression in RNA samples. Further lowering the probe number negatively affected assay reproducibility slightly, but people are free to do so, especially if expression from more genes is required.

   A common question in designing a DAP concerns the distribution of query oligos among genes with widely differing expression levels. While it is theoretically possible to swamp out detection of a very low expressor by including query oligos for very high expressors, our experience is that these effects are less dramatic. For DAPs targeting more than 500 genes, expression levels cover a wide range across biological samples, and addition of query oligos for highly expressed genes, such as housekeeping genes, have little effect. In one case, we observed that adding query oligos for high expressors into a low-plexity DAP (~30 genes) also had little impact on detection of low expressors. We presume that compression at the high end of the dynamic range in the DASL Assay is due to competition for priming late in the PCR from previously amplified products.

6. Careful experimental design may help maximize the utility of the DASL Assay. Because DAP oligo designs target intraexonic sequences, genomic DNA samples can be used to monitor assay performance and troubleshoot questionable DASL Assay results for FFPE-derived RNA samples. As an option, one can use the Make SUD (make single-use DNA) process to prepare activated gDNA to include as a positive control sample. The data derived from the gDNA samples can then be used to qualify individual target assays during data analysis in the BeadStudio application.

   A well-designed DASL Assay experiment includes replicates and both gDNA and intact reference RNA positive control samples. We find that the gDNA samples are useful to monitor assay performance in the steps following cDNA synthesis and the reference RNA sample can be used as a positive control relative to the degraded FFPE-derived RNA. Replicates can identify those samples that are less robust and less trustworthy by lower correlations.

7. As mentioned previously, a gDNA sample can be used to qualify the performance of the assay as well as the arrays. Ideally, intensities for all probes should

exceed background for a gDNA sample. Exceptions occur when individual probes are not uniquely mapped to the genome in RefSeq, causing a probe to cross an exon junction. Currently, this occurs for about 1% of probes, assuming a ~5% rate of genes not mapped in RefSeq coupled with a ~15% rate of randomly targeting an exon junction. In addition, there is the chance of low query oligo concentration, leading to low intensities for a few more percent of probes. In our experience, at least 95% of probes should show intensities significantly above background for gDNA samples.

8. The data quality for samples with too little input is quite similar to those with too much RNA degradation. We observed that poor samples are characterized by a decrease in overall signal across probes, a shift in the distribution of signals to lower intensities (that is, fewer genes with high signals), an increase in the intensity of the negative controls, and an increase in the variation among sample replicates at low intensity. To estimate the impact of input RNA quantity on assay performance, we tested various amounts of total FFPE-derived RNA (400 ng, 200 ng, 100 ng, and 50 ng) in the DASL Assay. Each cDNA sample was processed in duplicate for independent technical replicates. Highly reproducible results were obtained with as little as 50 ng of total RNA ($R^2 = 0.97$). However, the larger the RNA input, the more closely the samples correlated, until 200-ng samples were indistinguishable from 400-ng samples (Fig. 11.4).

9. We previously compared the number of genes detectable by the DASL Assay in RNA samples extracted from paired fresh-frozen and FFPE colon and breast tissues, both cancerous and normal. More than 90% of the genes that were detected in the fresh-frozen samples also were detected in their matching FFPE samples,

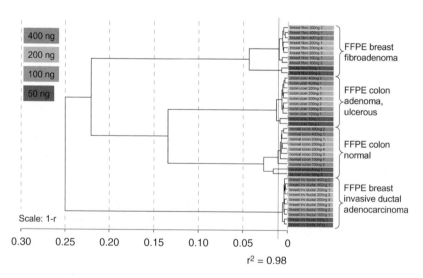

Fig. 11.4 The effect of RNA input on assay reproducibility. Lower RNA input increases scatter among technical replicates of FFPE-derived RNAs

when 200 ng of total RNA was assayed. Lists of differentially expressed genes generated from fresh-frozen and FFPE samples had highly significant overlap (with the FFPE list containing ~50% fewer genes), suggesting that sets of differentially expressed genes identified in FFPE samples resemble those identified from fresh-frozen samples.

The correlation between the gene expression profiles of paired fresh-frozen and FFPE tissues was 0.69 on average, but between each pair, this correlation ranged from 0.4 to 0.9. This observation is entirely consistent with other measurements of expression results in FFPE vs. fresh-frozen tissues, including qPCR data. The result suggests that sequence (or gene)-dependent effects may be derived largely from variations in sample processing and handling.

It is well known that the quality of RNA extracted from FFPE samples depends largely on multiple parameters: initial tissue fixation, sample processing, storage, and RNA extraction protocol. We believe that the major source of variability in expression profiles using RNA extracted from FFPE tissues is sample quality. As mentioned, we implemented a qPCR-based sample prequalification procedure to address this issue. Using this approach, the FFPE samples that showed the lowest correlations with matching frozen samples also had the poorest overall quality. In addition, RNA secondary structures may change after degradation, which may lead to changes in random primer accessibility and annealing efficiency in the cDNA synthesis. Taken together, these observations suggest that sequence-dependent effects may be responsible for part of the variation in expression data between fresh-frozen and FFPE samples.

While we realize that more experiments comparing fresh-frozen and matching FFPE samples are needed to study all possible sources and mechanisms of variation in gene expression between frozen and archived tissues, we strongly believe that FFPE tissues are an important source for gene expression profiling studies; and they can be used for biomarker discovery and validation, particularly if comparisons are made between samples treated equivalently.

10. The hybridization controls monitor the overall intensity of the arrays. In our experience, the hybridization controls should exceed ~10,000 counts in the appropriate channels. If there is an outlier in hybridization control intensities, then normalization may be attempted. If the hybridization controls are near background for any individual array, then a problem with that array or sample is likely. We observed this behavior when UB2 from the humidity control wells splashed into a sample well.

11. The negative controls should be below ~500 counts, and the standard deviation of the negative controls should be lower than their intensity. Any outlier samples that show high counts or large standard deviations on the negative controls should be considered with caution. We find that high negative controls can be associated with RNAs that are overly degraded and, therefore, unlikely to provide useful expression information. The exception to this rule is in the analysis of gDNA control samples, where we have seen high negative controls occur.

This phenomenon depends on which genes are targeted in the DAP, presumably due to the increased complexity of gDNA sequences relative to the transcriptome. When high negative controls were observed on gDNA samples, we observed that RNA samples are unaffected.

12. The control summary graph titled "Genes" reports the average intensity and standard deviation of all genes monitored. This value is expected to be lower in RNA samples than gDNA samples and lower in FFPE-derived RNAs than in intact RNAs. Outlier samples with low intensities may reflect overly degraded RNA or very low RNA input. Because the assay monitors the number of amplifiable targets, low RNA input levels can resemble overly degraded RNAs in DASL Assay performance.

13. The gap, stringency, and contamination control query oligos all target the glutaminyl amino tRNA synthase (QARS) transcript. Therefore, the intensities of these controls should reflect the level of expression of the QARS gene in the samples tested. This transcript was chosen because it is a housekeeping gene with medium expression levels in most tissues. The intensities for these controls should show general concordance across samples.

The intensities of these controls should follow a general pattern: (1) the gap control should approximate the average intensity of all genes, (2) the intensity in the red channel should dominate the intensity in the green channel for the low stringency control and the opposite should be true for the high stringency control, and (3) only one of the contamination control intensities should exceed background. The way the contamination controls work is that only one set of query oligos targeting one of four sequences in the QARS cDNA is added to any one tube of DAP. Therefore, only one of the first four columns in the Contamination control graph should show intensity that parallels the Stringency and Gap signals across samples. The last four columns indicate intensity on bead types that are targeted only in the GoldenGate Assay for SNP genotyping. If there is a contamination with amplicons in the lab, the data should show a large, constant signal across all samples, inconsistent with sample-dependent differences in QARS expression.

14. Another valuable approach to assessing data quality is to cluster the samples using the dendrogram tool in BeadStudio. If there are replicate samples or known sample relationships, outlier samples can be readily detected by their failure to associate with each other in clustering.

15. The DASL Assay combines the advantages of array-based gene expression analysis with those of multiplexed qPCR, thereby offering much higher multiplexing capacity, at high throughput and low cost. It can use as little as 50 ng of total RNA to analyze over 500 genes in FFPE samples, ~100-fold less than that required by qPCR, which usually uses 20–50 ng per reaction (per gene). Since many genes are measured simultaneously in one DASL Assay, it provides an excellent opportunity for internal data normalization, thus solving a major problem encountered by qPCR. This becomes very important when cross-sample comparisons are needed, especially when the samples under study have different degrees of RNA degradation.

Acknowledgments We thank Amir Arsanjani, Monica Kostelec, and Joshua Modder for acquisition and pathological evaluation of FFPE tissue samples; Eliza Wickham Garcia and Jing Chen for technical assistance; Lixin Zhou for help with the assay design; Eugene Chudin for data analysis; Dimitri Talantov for participation in the project; and Yixin Wang, Tracy Down, and David Barker for helpful discussions and advice.

References

1. Fan JB, Gunderson KL, Bibikova M, Yeakley JM, Chen J, Wickham Garcia E, Lebruska LL, Laurent M, Shen R, Barker D (2006) Illumina universal bead arrays. Methods Enzymol. 410:57–73
2. Barker DL, Theriault G, Che D, Dickinson T, Shen R, Kain R (2003) Self-assembled random arrays: High-performance imaging and genomics applications on a high-density microarray platform. Proceedings of SPIE 4966:1–11
3. Gunderson KL, Kruglyak S, Graige MS, Garcia F, Kermani BG, Zhao C, Che D, Dickinson T, Wickham E, Bierle J et al (2004) Decoding randomly ordered DNA arrays. Genome Res 14:870–877
4. Bouchie A (2004) Coming soon: A global grid for cancer research. Nat Biotechnol 22:1071–073
5. Cronin M, Pho M, Dutta D, Stephans JC, Shak S, Kiefer MC, Esteban JM, Baker JB (2004) Measurement of gene expression in archival paraffin-embedded tissues: Development and performance of a 92-gene reverse transcriptase-polymerase chain reaction assay. Am J Pathol 164:35–42
6. Antonov J, Goldstein DR, Oberli A, Baltzer A, Pirotta M, Fleischmann A, Altermatt HJ, Jaggi R (2005) Reliable gene expression measurements from degraded RNA by quantitative real-time PCR depend on short amplicons and a proper normalization. Lab Invest 85:1040–1050
7. Tachibana M, Shinagawa Y, Kawamata H, Omotehara F, Horiuchi H, Ohkura Y, Kubota K, Imai Y, Fujibayashi T, Fujimori T (2003) RT-PCR amplification of RNA extracted from formalin-fixed, paraffin-embedded oral cancer sections: Analysis of p53 pathway. Anticancer Res 23:2891–2896
8. Godfrey TE, Kim SH, Chavira M, Ruff DW, Warren RS, Gray JW, Jensen RH (2000) Quantitative mRNA expression analysis from formalin-fixed, paraffin-embedded tissues using 5′ nuclease quantitative reverse transcription-polymerase chain reaction. J Mol Diagn 2:84–91
9. Van Gelder RN, von Zastrow ME, Yool A, Dement WC, Barchas JD, Eberwine JH (1990) Amplified RNA synthesized from limited quantities of heterogeneous cDNA. Proc Natl Acad Sci USA 87:1663–1667
10. Fan JB, Yeakley JM, Bibikova M, Chudin E, Wickham E, Chen J, Doucet D, Rigault P, Zhang B, Shen R et al (2004) A versatile assay for high-throughput gene expression profiling on universal array matrices. Genome Res 14:878–885
11. Bibikova M, Talantov D, Chudin E, Yeakley JM, Chen J, Doucet D, Wickham E, Atkins D, Barker D, Chee M et al (2004) Quantitative gene expression profiling in formalin-fixed, paraffin-embedded tissues using universal bead arrays. Am J Pathol 165:1799–1807
12. Bibikova M, Yeakley JM, Chudin E, Chen J, Wickham E, Wang-Rodriguez J, Fan JB (2004) Gene expression profiles in formalin-fixed, paraffin-embedded tissues obtained with a novel assay for microarray analysis. Clin Chem 50:2384–2386
13. Zhang C, Li HR, Fan JB, Wang-Rodriguez J, Downs T, Fu XD, Zhang MQ (2006) Profiling alternatively spliced mRNA isoforms for prostate cancer classification. BMC Bioinformatics 7:202
14. Li HR, Wang-Rodriguez J, Nair TM, Yeakley JM, Kwon YS, Bibikova M, Zheng C, Zhou L, Zhang K, Downs T et al (2006) Two-dimensional transcriptome profiling: Identification of messenger RNA isoform signatures in prostate cancer from archived paraffin-embedded cancer specimens. Cancer Res 66:4079–4088

Chapter 12
Expression Profiling of microRNAs in Cancer Cells: Technical Considerations

Mouldy Sioud and Lina Cekaite

Abstract MicroRNAs (miRNAs) represent a class of small, noncoding RNAs. These small RNAs are involved in diverse biological processes, and it has been predicted that about one third of human messenger RNAs (mRNAs) appear to be miRNA targets, underlying the major influence of miRNAs on almost all cellular pathways. Deviation from the normal pattern of miRNA expression has been implicated in several diseases. Among human diseases, it has been shown that changes in miRNA expression correlate with various human cancers. Thus, miRNA profiling could contribute to more precise tumor classification and better prediction of the therapeutic outcome. This chapter summarizes these recent findings and highlights the technical advances in miRNA probe preparation, miRNA expression profiling and target identification.

Keywords miRNAs, siRNAs, RNA interference, tumor cells

1   Introduction

MicroRNAs (miRNAs) are noncoding, single-stranded RNAs of 21–25 nucleotides that negatively regulate the gene expression. They down-regulate protein expression by either blocking translation of proteins from mRNA or mediating the degradation of target mRNAs [1]. These tiny RNAs are encoded by hundreds of genes as long primary transcripts termed "pri-miRNAs," with hairpin structures that are processed in the nucleus by the double-strand-specific ribonuclease III Drosha into miRNA precursors of 60–80 nucleotides. These precursors are transported to the cytoplasm by Exportin 5 and further processed by the enzyme Dicer into partially double-stranded RNAs in which one strand is the mature miRNA. Subsequently, mature miRNAs are incorporated into the RNA-induced silencing complex (RISC), leading to RNA interference. In mammalian cells, the bound mRNA typically remains intact but is not translated, resulting in reduced expression of the corresponding gene.

From: Methods in Molecular Biology, Vol. 439: *Genomics Protocols: Second Edition*
Edited by Mike Starkey and Ramnath Elaswarapu © Humana Press Inc., Totowa, NJ

Presently, 474 human miRNAs have been experimentally verified. (miR-Base, version 9.1, http://microrna.sanger.ac.uk). In addition, computational predictions have estimated the existence of at least 1,000 human miRNAs [2, 3]. Current estimates suggest that about one third of human mRNAs appear to be miRNA targets [4]. This makes the miRNA genes one of the most abundant classes of regulatory genes in mammals. The picture gets even more complicated when one miRNA may have several mRNA targets and several miRNAs may regulate a single target mRNA [5]. Despite the success in identification of genes encoding miRNAs, knowledge about their function and specific mRNA targets is limited, and only a handful miRNAs have been fully characterized. In mammals functions for specific miRNAs have been described in the processes of B-cell differentiation [6], adipocyte differentiation [7], insulin secretion [8], embryonic brain development [9], and embryonic cardiogenesis [10]. Therefore, one major function is to control processes that underlie cell proliferation and differentiation in diverse organisms during normal development (reviewed in [11]). The essential roles of miRNAs in vertebrate development have been strongly demonstrated by removing the Dicer enzyme in zebra fish [9, 12] and mice [13]. In both cases, removal of Dicer resulted in severe developmental defects and embryonic death. Further, it has been demonstrated that an individual miRNA can trigger large-scale changes in development [14]. Moreover, some miRNAs are differentially expressed in stem cells [15, 16], and recently it was found that miRNAs are required for stem cell division [17].

It is speculated that abnormalities in miRNA expression might similarly contribute to the generation or maintenance of "cancer stem cells," tumor initiating cells [18]. One of the arguments connecting several miRNAs to cancer was the finding that more than half the mapped miRNA genes are localized to fragile sites and genomic regions involved in cancer [19]. And, the first evidence for direct involvement of miRNAs in cancer was the finding that miR-15 and miR-16 were deleted or underexpressed in most cases of chronic lymphocytic leukemia [20]. Reduced accumulation of specific miRNAs has further been reported for let-7 in lung cancer [21, 22]; miR26a and miR-99a in lung cancer cell lines [19]; miR-143 and miR-145 in colon cancer [23]; and miR-125b, miR-145, miR-21, and miR-155 in human breast cancer [24]. Recently, a general down-regulation of miRNAs in tumors compared with normal tissues was observed [17]. An accumulation of specific miRNAs in cancer has also been observed: miR-155 in lymphomas [25] and overexpression of seven miRNAs encoded by the gene c13orf25 amplified in certain lymphomas nominates this miRNA cluster the first noncoding oncogene [26]. Another example of a miRNA-oncogene overexpressed in brain tumor is the antiapoptotic factor miR-21 [26]. On the other hand, down-regulation of the RAS oncogene by let-7 implicates let-7 as a tumor suppressor in lung tissue [22]. Thus, depending on what genes the miRNAs regulate, the miRNAs may act either as oncogenes or tumor suppressors.

Only recently, the analysis of the entire miRNAome has become possible by the development of microarrays containing all known human miRNAs [26, 27]. One important finding of such analysis is that specific miRNAs have altered expression in tumors compared to normal cells. Furthermore, miRNA revealed distinct signatures in specific cancer forms [28], and enabled successful classification of poorly differentiated tumors that were inaccurately classified using mRNA profiles [18]. Recognition of differentially expressed miRNAs may help to identify miRNAs involved in specific processes, such as normal differentiation or carcinogenesis, and establish their roles. Given the role of miRNAs in regulating gene expression, we have developed several strategies for profiling miRNA expression in human cells [29, 30].

2   Profiling miRNA Gene Expression

Microarray technology is based on basic principles of nucleic acid hybridization [31], whereby fragmented DNA is attached to a substrate then hybridized with a known gene or fragment. Typically, spotted oligos or cDNA microarrays are hybridized with labeled cDNA probes prepared from polyA+ RNA using standard methods for making cDNA libraries. Subsequent to hybridization, the captured images are analyzed using software that quantifies the signal of each spot corresponding to individual capture probes. The intensity of each spot is expected to be proportional to the amount of mRNA present within the analyzed samples. Since the first expression microarray report was published in 1995 [32], the microarray technology has seen a number of emerging approaches. The analysis of miRNAs expression by oligonucleotides macro- and microarrays prove to be feasible [26, 29, 30, 33]. However, because of the small size of miRNA and high sequence similarities between miRNA families, the miRNA microarray approaches remain challenging. Here, we review different arrays and labeling techniques used for detecting miRNAs.

2.1   Macro- and Microarrays

Macroarrays on nitrocellulose or Nylon dot blots can be used to analyze miRNA expression in mammalian cells. To assess the high-throughput feasibility miRNAs were gel purified from total RNAs, internally Cy5-labeled during reverse transcription then hybridized to a glass slide on which specifically designed multiple miRNA-probes had been arrayed in-house. The fabricated microarrays were capable of determining expression patterns of at least 250 human miRNAs. We applied the method to three lymphoma cell lines (Daudi, BJAB, and Ramos). The breast cancer cell line MCF-7 was also included. Figure 12.1 shows the miRNA cluster of the four cells lines. With little exception, Ramos and BJAB cell lines showed a comparable expression profiles. The data suggest that certain miRNAs

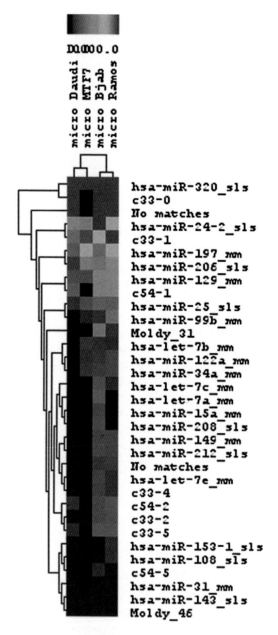

Fig. 12.1 Expression profiles of miRNAs in cancer cell lines. Purified miRNA was internally labeled during reverse transcription with Cy5 and then hybridized onto in-house fabricated microarrays containing around 250 human miRNA sequences

are overexpressed, whereas others are down-regulated. Although further work is needed, it is interesting that miRNA-260 is up-regulated in MCF-7 cells but is down-regulated in lymphoma cell lines. Notably, impaired miRNA expression has been implicated in tumorigenesis [34]. Therefore, further research into miRNA functions might lead to an advanced understanding of the mechanisms that lead to cancer. For example, a recent study showed that miR15a and miR16-1 negatively regulate the antiapoptotic gene BCL2, which often is overexpressed in many types of human cancers [35]. Several miRNA oligo sets or preprinted microarrays and labeling protocols are commercially available. For example, mirVana miRNA Bioarrays miRCURY LNA Array, based on locked nucleic acid oliconucleotides from Exiqon.

2.2   Probe Preparation

Generally, the small size of miRNA may not be adequate for microarray hybridization using direct labeling protocols. To take full advantage of the DNA microarrays, one would like to be able to prepare sensitive probes from miRNAs. Using standard procedures to prepare internally labeled cDNA probes from RNAs, we recently described a technique that allows the preparation of cDNA probes from gel-purified miRNAs. In this procedure, miRNA recovered from 15% denaturing acrylamide gel slices were reversed transcribed using hexamer random primers to produce a labeled probe for analyzing miRNA expression between two cell populations [30].

Notably, during cDNA synthesis miRNA can be tagged with fluorescently labeled molecules for miRNA expression profiling. In contrast to end labeling using T4 polynucleotide kinase, cDNA labeling incorporates multiple radioactive or fluorescent bases, and therefore the specific activity of such probes is expected to be high (see Fig. 12.2). Since most miRNAs constitute less than 0.001% of a total RNA sample, this type of labeling is very important for obtaining sensitive and accurate miRNA microarray data.

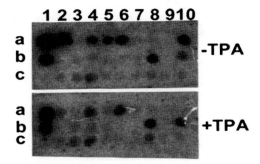

Fig. 12.2 Oligonucleotides representing 30 arbitrary miRNA sequences were spotted onto Hybond-N+ nylon membranes and then hybridized with cDNA probes made from small miRNAs prepared from either untreated or TPA-treated HL60 cells. A differential expression profile between samples was found

As we have previously indicated, purified miRNAs can be 3′ end-labeled by polyA polymerase, which adds several radioactive- or fluorescence-labeled ATP molecules [29]. This simple method also provides high sensitivity comparable to that obtained with the cDNA-labeling methods. In the case of the cDNA labeling, the microarrays should contain the processed miRNA sequences, whereas in the case of the 3′ end labeling the microarrays should contain the miRNA complementary strands. To reduce the nonspecific cross-hybridization of longer mRNA or ribosomal RNA, we purified miRNAs [30]. However, Liu and colleagues demonstrated that total RNA can be used directly with no need for miRNA purification [26]. In this method, total RNA was reverse transcribed with biotinylated random primers and hybridized to oligonucleotide spotted arrays. Subsequently, miRNA expression was detected using streptavidin-bound fluorophores. This method of labeling is crucial for small biological materials.

To further asses the feasibility to total RNA labeling, we have investigated the expression of miRNAs in biopsies from patients with B cell lymphomas. In these experiments 5 μg of total RNA from each sample were internally Cy5-labeled during reserve transcription and then hybridized to the array. Figure 12.3A shows an image of a microarray hybridized with Cy-5 cDNA probe prepared from 5 μg of lymph node total RNA. As can be seen, several miRNAs were differentially expressed in this tumor sample. Certain miRNAs are overexpressed (e.g., miRNA let7, miRNA449, miRNA130a, miRNA423), while others are not (e.g., miRNA226, miRNA107) (Fig. 12.3B). Therefore, it will be interesting to investigate the biological meaning of the expression data. However, when purified miRNAs from the same sample were used as template for probe preparations, only four miRNA sequences were detected. Similarly, a significant discrepancy was seen with other samples and total RNA prepared from cancer cell lines. In all cases, the number of positive signals obtained with probes prepared from total RNA was significantly higher than those obtained with purified miRNAs. These observations indicate that the data obtained with total RNA must be interpreted with caution and validated with additional strategies such as Northern and PCR techniques.

Some of the technical advances that we described for the first time, such as polyA tailing of miRNAs [29], are similar to those used by Ambion to label samples for miRNA array hybridization. Also, a polyA tailing protocol was implemented by several other investigators [36]. For example, polyadenylated miRNAs were tagged with unique capture sequences labeled with either Cy3 or Cy5. After ligation, tagged miRNAs were purified and then hybridized to the miRNA microarrays [36]. Also, the ULS labeling technology was used to label miRNAs [37]. ULS labeling is based on an ability to label directly guanine nucleotides with no size discrimination and no need of enzymes.

2.3   Probe Amplification

Generally 50–200 μg of total RNAs is required to prepare miRNA probe by gel-purification for each microarray hybridization. This requirement for large amounts

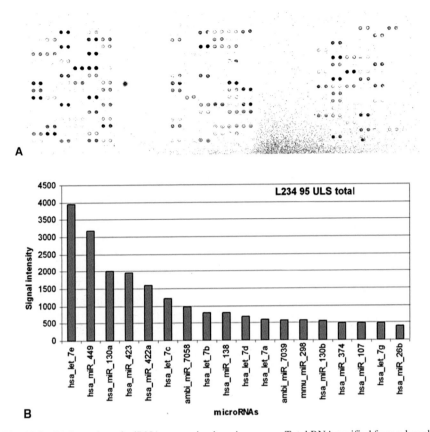

Fig. 12.3 (A) Detection of miRNA expression by microarrays. Total RNA purified from a lymph node biopsy from a lymphoma patient was ULS labeled with Cy5 and then hybridized to mir-Vana™ miRNA (Ambion oligo set) microarray printed in-house. The array contains 250 human miRNA sequences. (B) Quantification of the signals shown in A

of starting materials severely limits the use of microarrays to profile miRNA gene expression in small specimens of human tissue, such as obtained by core needle or fine needle aspiration biospsies. To overcome this problem, we developed a PCR-based amplification method that allows the analysis of miRNAs gene expression in small biopsies [29]. The amplification method we used is diagrammed in Fig 12.4. In this procedure, purified miRNA were polyA tailed using *E. coli* polyA polymerase and the RNA primed by oligo-dT carrying on its 5′ end a T7 RNA polymerase promoter sequence. The use of the reverse transcriptase PowerScript during cDNA synthesis allows the addition of few additional nucleotides, primarily deoxycytidines, to the 3′ end of the cDNA. The 3′-tailed first-strand cDNA then is primed with an SP6 RNA polymerase promoter sequence carrying GGG at its 3′ end, leading to the generation of double-stranded cDNA. The resulting double-stranded DNA can be either PCR amplified for a desired number of cycles or in-vitro

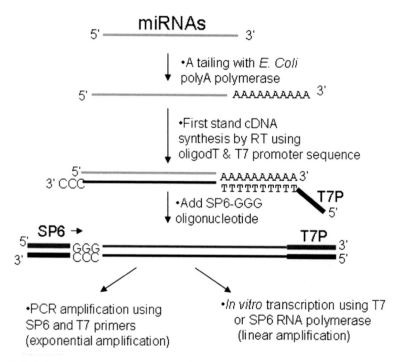

Fig. 12.4 Schematic illustration of miRNA amplification (for details see text)

transcribed using T7 (or SP6) RNA polymerase, resulting in linear amplification of miRNAs. The protocol is expected to generate large amounts of miRNA starting from few ng of miRNAs. Figure 12.5 shows the hybridization signals using an internally labeled 32p PCR probe, which was prepared from miRNA that had undergone 25 rounds of amplification. In these experiments, the probe was hybridized to dot-blots containing 160 human miRNA sequences. We obtained reproducible results between different samples and no indication of a bias due to the amplification step. Moreover, a hybridization of fluorescently labeled PCR products to a glass slide on which specifically designed miRNA-specific probes have been arrayed in-house revealed satisfying results, thus enabling the use of microarray technology for qualitative and quantitative studies of miRNAs expression in small number of cells.

3 Identification of miRNA Target Genes

Several groups developed computational algorithms designed to predict the target genes of miRNAs. In general, the basis for these prediction programs is the degree of sequence complementarity between a miRNA and a target mRNA gene. In animal cells, single-stranded miRNAs bind to specific target mRNAs through partly complementary sequences that are predominantly in the 3′ untranslated region

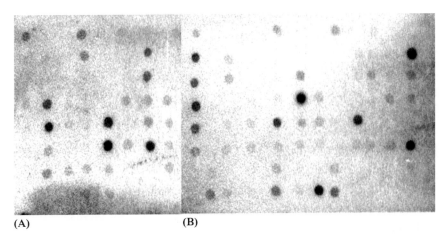

(A)   (B)

Fig. 12.5 Amplified miRNAs produce significant hybridization signals. Oligonucleotides representing 160 miRNA sequences were spotted onto Hybond-N+ nylon membranes then hybridized with a cDNA probe prepared from PCR-amplified miRNAs prepared from the breast cancer cell line MCF-7. The result is reproducible for different miRNA preparations

(3′UTR). Computational approaches to find target transcripts have been hindered by this imperfect complementarity. In addition, many known target genes of miR-NAs contain several miRNA binding sites and several genes are predicted to be regulated by a single miRNAs [38, 39]. These observations further increase the complexity of target prediction using bioinformatics. Therefore, experimental validation is required for miRNA target genes.

To date, different strategies have been used to validate miRNA target genes. Most commonly used are luciferase or GFP fusion constructs containing the miRNA target site. In cotransfection experiments, active miRNAs are expected to reduce luciferase activity or GFP expression. Notably, the same strategy was used to identify effective siRNAs in human cells. Also, specific inhibition of miRNA with antisense 2′-O-methyl-modified oligonucleotides was used. In this assay, the inhibition of miRNA is expected to lead to the up-regulation of its target gene. In addition to translation arrest, a recent study demonstrated that miRNAs can reduce the transcript levels [14]. These findings suggest that target mRNAs that show significant complementarity to miRNAs are expected to be suppressed in cell types where the miRNAs are up-regulated. Northern blot analysis or RT-PCR could be used to measure the expression levels of candidate mRNAs. In addition, the use of specific antibodies to measure the protein levels of candidate genes in cells with high and low expression of the miRNA is recommended. By transfecting cells with candidate miRNAs, it is possible to see whether the overexpression of specific miRNAs affect the expression of the identified target genes in in vitro and in vivo. However, a more global approach to find target mRNAs is to use proteomics to uncover target proteins that are affected as a consequence of miRNA expression in target cells. This strategy works only when the cell populations compared are identical except for the artificial increase of miRNA in one. Proteins that are altered as a probable consequence of higher miRNA expression may be purified from gels and subsequently sequenced.

Acknowledgments We thank Dr. Anne Dybwad for her critical reading of the manuscript and Dr. Øystein Røsok for technical help. This work is supported in part by Helse Sør. The miRNA arrays were provided by the Norwegian Microarray Consortium at the Norwegian Radium Hospital.

References

1. Bartel DP (2004) MicroRNAs: Genomics, biogenesis, mechanism, and function. Cell 116:281–297
2. Bentwich I, Avniel A, Karov Y, Aharonov R, Gilad S, Barad O, Barzilai A, Einat P, Einav U, Meiri E et al (2005) Identification of hundreds of conserved and nonconserved human microRNAs. Nat Genet 37:766–770
3. Berezikov E, Guryev V, van de Belt J, Wienholds E, Plasterk RH, Cuppen E (2005) Phylogenetic shadowing and computational identification of human microRNA genes. Cell 120:21–24
4. Lewis BP, Burge CB, Bartel DP (2005) Conserved seed pairing, often flanked by adenosines, indicates that thousands of human genes are microRNA targets. Cell 120:15–20
5. Krek A, Grun D, Poy MN, Wolf R, Rosenberg L, Epstein EJ, MacMenamin P, da Piedade I, Gunsalus KC, Stoffel,M et al (2005) Combinatorial microRNA target predictions. Nat. Genet. 37:495–500
6. Chen, CZ, Li L, Lodish HF, Bartel DP (2004) MicroRNAs modulate hematopoietic lineage differentiation. Science 303:83–86
7. Esau C, Kang X, Peralta E, Hanson E, Marcusson EG, Ravichandran LV, Sun Y, Koo S, Perera RJ, Jain R et al (2004) MicroRNA-143 regulates adipocyte differentiation. J Biol Chem 279:52361–52365
8. Poy MN, Eliasson L, Krutzfeldt J, Kuwajima S, Ma X, Macdonald PE, Pfeffer S, Tuschl T, Rajewsky N, Rorsman P et al (2004) A pancreatic islet-specific microRNA regulates insulin secretion. Nature 432:226–230
9. Giraldez AJ, Cinalli RM, Glasner ME, Enright AJ, Thomson JM, Baskerville S, Hammond SM, Bartel DP, Schier AF (2005) MicroRNAs regulate brain morphogenesis in zebrafish. Science 308: 833–838
10. Zhao Y, Samal E, Srivastava D (2005) Serum response factor regulates a muscle-specific microRNA that targets Hand2 during cardiogenesis. Nature 436:214–220
11. Ambros V (2004) The functions of animal microRNAs. Nature 431:350–355
12. Wienholds E, Koudijs MJ, van Eeden FJ, Cuppen E, Plasterk RH (2003) The microRNA-producing enzyme Dicer1 is essential for zebrafish development. Nat Genet 35:217–218
13. Bernstein E, Kim SY, Carmell MA, Murchison EP, Alcorn H, Li MZ, Mills AA, Elledge SJ, Anderson KV, Hannon GJ (2003) Dicer is essential for mouse development. Nat Genet 35:215–217
14. Lim LP, Lau NC, Garrett-Engele P, Grimson A, Schelter JM, Castle J, Bartel DP, Linsley PS, Johnson JM (2005) Microarray analysis shows that some microRNAs downregulate large numbers of target mRNAs. Nature 433: 769–773
15. Suh MR, Lee Y, Kim JY, Kim SK, Moon SH, Lee JY, Cha KY, Chung HM, Yoon HS, Moon SY et al (2004) Human embryonic stem cells express a unique set of microRNAs. Dev Biol 270"488–498
16. Houbaviy HB, Murray MF, Sharp PA (2003) Embryonic stem cell-specific MicroRNAs. Dev Cell 5:351–358
17. Hatfield SD, Shcherbata HR, Fischer KA, Nakahara K, Carthew RW, Ruohola-Baker H (2005) Stem cell division is regulated by the microRNA pathway. Nature 435:974–978

18. Lu J, Getz G, Miska EA, Alvarez-Saavedra E, Lamb J, Peck D, Sweet-Cordero A, Ebert BL, Mak RH, Ferrando AA et al (2005) MicroRNA expression profiles classify human cancers. Nature 435:834–838

19. Calin GA, Liu CG, Sevignani C, Ferracin M, Felli N, Dumitru CD, Shimizu M, Cimmino A, Zupo S, Dono M et al (2004) MicroRNA profiling reveals distinct signatures in B cell chronic lymphocytic leukemias. Proc Natl Acad.Sci USA 101:11755–11760

20. Calin GA, Dumitru CD, Shimizu M, Bichi R, Zupo S, Noch E, Aldler H, Rattan S, Keating M, Rai K et al (2002) Frequent deletions and down-regulation of micro-RNA genes miR15 and miR16 at 13q14 in chronic lymphocytic leukemia. Proc Natl Acad Sci USA 99:15524–15529

21. Takamizawa J, Konishi H, Yanagisawa K, Tomida S, Osada H, Endoh H, Harano T, Yatabe Y, Nagino M, Nimura Y et al (2004) Reduced expression of the let-7 microRNAs in human lung cancers in association with shortened postoperative survival. Cancer Res 64:3753–3756

22. Johnson SM, Grosshans H, Shingara J, Byrom M, Jarvis R, Cheng,A, Labourier E, Reinert KL, Brown D, Slack FJ (2005) RAS is regulated by the let-7 microRNA family. Cell 120:635–647

23. Michael MZ, O'Connor SM, van Holst Pellekaan NG, Young GP, James RJ (2003) Reduced accumulation of specific microRNAs in colorectal neoplasia. Mol. Cancer Res 1:882–891

24. Iorio MV, Ferracin M, Liu CG, Veronese A, Spizzo R, Sabbioni S, Magri E, Pedriali M, Fabbri M, Campiglio M et al (2005) MicroRNA gene expression deregulation in human breast cancer. Cancer Res 65:7065–7070

25. Eis PS, Tam W, Sun L, Chadburn A, Li Z, Gomez MF, Lund E, Dahlberg JE (2005) Accumulation of miR-155 and BIC RNA in human B cell lymphomas. Proc Natl Acad Sci USA 102:3627–3632

26. Liu CG, Calin GA, Meloon B, Gamliel N, Sevignani C, Ferracin M, Dumitru CD, Shimizu M, Zupo S, Dono M et al (2004) An oligonucleotide microchip for genome-wide microRNA profiling in human and mouse tissues. Proc Natl Acad Sci USA 101:9740–9744

27. Nelson PT, Baldwin DA, Scearce LM, Oberholtzer JC, Tobias JW, Mourelatos Z (2004) Microarray-based, high-throughput gene expression profiling of microRNAs. Nat Methods 1:155–161

28. Calin GA, Sevignani C, Dumitru CD, Hyslop T, Noch E, Yendamuri S, Shimizu M, Rattan S, Bullrich F, Negrini M et al (2004) Human microRNA genes are frequently located at fragile sites and genomic regions involved in cancers. Proc Natl Acad Sci USA 101:2999–3004

29. Sioud M, Rosok O (2005) High-throughput analysis of microRNA gene expression using sensitive probes. Methods Mol Biol 309:311–320

30. Sioud M, Rosok O (2004) Profiling microRNA expression using sensitive cDNA probes and filter arrays. Biotechniques 37:574–576, 578–580

31. Southern EM (1975) Detection of specific sequences among DNA fragments separated by gel electrophoresis. J Molec Biol 98:503–517

32. Schena M, Shalon D, Davis RW, Brown, PO (1995) Quantitative monitoring of gene expression patterns with a complementary DNA microarray. Science 270:467–470

33. Babak T, Zhang W, Morris, Q, Blencowe BJ, Hughes TR (2004) Probing microRNAs with microarrays: Tissue specificity and functional inference. RNA 10:1813–1819

34. Esquela-Kerscher A, Slack FJ (2006) Oncomirs—micrornas with a role in cancer. Nat Rev Cancer 6:259–269

35. Cimmino A, Calin GA, Fabbri M, Iorio MV, Ferracin M, Shimizu M, Wojcik SE, Aqeilan RI, Zupo S, Dono M et al (2005) miR-15 and miR-16 induce apoptosis by targeting BCL2. Proc Natl Acad Sci USA 102:13944–13949

36. Goff LA, Yang M, Bowers J, Getts RC, Padgett RW, Hart RP (2005) Rational probe optimization and enhanced detection strategy for microRNAs using microarrays. RNA Biol. 2

37. Wiegant JC, van Gijlswijk RP, Heetebrij RJ, Bezrookove V, Raap AK, Tanke HJ (1999) ULS: A versatile method of labeling nucleic acids for FISH based on a monofunctional

reaction of cisplatin derivatives with guanine moieties. Cytogenetics and Cell Genetics 87:47–52

38. Kiriakidou M, Nelson PT, Kouranov A, Fitziev P, Bouyioukos C, Mourelatos Z, Hatzigeorgiou A (2004) A combined computational-experimental approach predicts human microRNA targets. Genes Dev 18:1165–1178

39. Lewis BP, Shih IH, Jones-Rhoades MW, Bartel DP, Burge CB (2003) Prediction of mammalian microRNA targets. Cell 115:787–798

Chapter 13

Identification of Disease Biomarkers by Profiling of Serum Proteins Using SELDI-TOF Mass Spectrometry

Sigrun Langbein

Abstract Proteins are the main actors in all physiological and pathological processes. Since the final structure of the protein does not depend on the DNA sequence or even the mRNA sequence alone, the search for direct approaches on the proteome has gained great interest. The most complex and probably the largest proteome is serum, making it clinically the most important.

ProteinChip technology, in combination with modern mass spectrometry, allows the complex search for biomarkers, molecular interactions, signaling pathways, and the identification of novel therapeutic compounds. Here we describe the surface-enhanced laser desorption-ionization (SELDI) in combination with the time-of-flight (TOF) mass spectrometry for analyzing serum samples (SELDI was a patented technique from Ciphergen, Fremont, CA). Aluminum-based arrays contain chemical or biological surfaces allowing the capture of proteins, which interact with the surface. The bound proteins are laser desorbed and ionized for mass spectroscopy analysis. The differential mass spectral patterns reflect the protein expression bound on the chip surface and allow the comparison between various samples. Proteins of interest can be identified using peptide mass fingerprinting (PMF).

Keywords Surface-enhanced laser desorption-ionization (SELDI), time of flight (TOF), mass spectrometry (MS), serum, ProteinChip, fractionated serum, biomarker, peptide mass fingerprinting (PMF)

1 Introduction

The direct identification of biological end points is a major advantage of proteomic research. The primary sequence of proteins does not depend on the DNA or the mRNA sequence alone, various posttranslational modifications, like acetylation, glycosylation, phosphorylation, or other types of covalent alterations, can diversify proteins [1, 2]. Protein biomarkers are among the important tools for detection, diagnosis, treatment, and monitoring of diseases

From: Methods in Molecular Biology, Vol. 439: *Genomics Protocols: Second Edition*   191
Edited by Mike Starkey and Ramnath Elaswarapu © Humana Press Inc., Totowa, NJ

[3–9]. The estimated number of different human proteins is about 500,000, and it will be challenging to identify, catalogue, and functionally assign these molecules. Serum is one of the preferred mediums when searching for markers, because it is obtained with low levels of invasiveness, it is available in sufficient quantity, and it is a source with great information content [5, 6, 10–12]. Serum is probably the most complex proteome, containing other tissue proteomes as subsets [13]. Profiling human serum is ambitious, because the serum contains large proportions of albumin (55%) and glycoproteins with an enormous heterogeneity. Twenty-two protein groups, such as albumin, immunoglobulins, haptoglobin, transferins, and lipoproteins, account for 99% of the protein mass in serum. As many as 10,000 unique proteins are estimated to be present within the human serum proteome that span a dynamic range of concentrations approximately of $>10^9$ [13, 14]. Therefore, the background matrix represents a very complex milieu, and proteins observed in general analyses, in general, are of high abundance.

The ProteinChip® System was introduced with the intention of an easy to use, time- and resource-saving method by Ciphergen Biosystems (Fremont, CA, www.ciphergen.com). In the meantime the proteomics instrument business was sold to Bio-Rad Laboratories (www.bio-rad.com). Based on the surface-enhanced laser desorption-ionization process (SELDI), it combines two methods: solid phase chromatography and time-of-flight mass spectrometry (TOF-MS) [10, 15–16]. This technology integrates the complex protein expression on one integrated platform and enables the study of complex serum protein profiles [17]. Chip arrays capture individual proteins out of a probe, which are subsequently resolved by mass spectrometry (MS). Varying coatings on the chips allow the binding of different protein classes. ProteinChip arrays are available with different chromatographic properties, including hydrophobic, hydrophilic, anion exchange, cation exchange, and immobilized-metal affinity surfaces; or they are preactivated for the coupling of capture molecules. Thus, each analysis is a sort of an on-spot fractionation step, which reduces the complexity of the serum sample. This kind of fractionating is not be effective enough in selected cases, and the dynamic range of proteins and protein concentrations in human serum sometimes necessitates effective fractionating methods capable of separating high abundance from low abundance proteins. The sample analysis takes place in SELDI-TOF MS, and data are processed by complex bioinformatic software. Further detailed information and images about the SELDI-TOF MS technique can be found on the homepage of BIO-RAD (www.bio-rad.com) and in the literature [5, 15, 18–19].

2 Material

For beginners to the SELDI-TOF technique, there is a starter kit (ProteinChip SELDI System tarter Kit) available. This includes a selection of ProteinChip arrays, calibration standards, energy absorbing molecules (EAM), and a bioprocessor.

2.1 Probes

Serum samples are obtained in routinely used serum monovettes.

2.2 Fractionating the Serum

Serum may be used unfractionated or fractionated. Any method to fractionate the serum probe is fine. Expression Difference Mapping Kits (Bio-Rad Laboratories, Hercules, CA, USA) are especially designed for serum fractionating in combination with SELDI TOF analyses (procedures are described in detail in the Expression Difference Mapping™ Kit—Serum Fractionating, www.bio-rad.com), but any method to fractionate the serum is suitable.

2.3 Protein Chip Arrays

1. IMAC30 is an immobilized metal affinity capture array with a nitriloacetic acid (NTA) surface, with an updated hydrophobic barrier coating.
2. CM10 is a weak cation exchange array with carboxylate functionality, with an updated hydrophobic barrier coating.
3. Q10 is a strong anion exchange array with quaternary amine functionality, with an updated hydrophobic barrier coating.
4. H50 binds proteins through reversed phase or hydrophobic interaction chromatography with an updated hydrophobic barrier coating.
5. H4 mimics reversed phase chromatography with C16 functionality.
6. NP20 mimics normal phase chromatography with silicate functionality.
7. Au are gold chips to be used directly for MALDI-based experiments.
8. Mainly used ProteinChip surfaces for serum diagnostic: Metal Affinity Capture (IMAC3/IMAC30), Weak Cation Exchange (CM10/WCX2) reversed Phase (H50).

2.4 Binding Buffers for the Different Chip Arrays

1. PBS buffer: $2.7\,mM$ potassium chloride, $120\,mM$ sodium chloride, and $10\,mM$ phosphate buffer maintained at a pH of 7.0.
2. $100\,mM$ ammonium acetate, pH 4.0 (with or without 0.1% Triton X-100) for CM10.
3. $100\,mM$ Tris-HCl, pH 9.0 (with or without 0.1% Triton X-100) for Q10.
4. $100\,mM$ sodium phosphate, $500\,mM$ NaCl pH 7.0 (activated with either $100\,mM$ copper sulphate or $100\,mM$ nickel sulphate hexahydrate) for IMAC30.
5. 10% acetonitrile (AcN), 0.1% trifluoroacetic acid (TFA) or phosphate-buffered saline (PBS) for H50.

2.5   Energy Absorbing Molecules

Matrix solution: 50% saturated sinapinic acid, in 50% acetonnitrile and 0.5% trifluoroacetic acid, or 50% saturated α-Cyano-4-hydoxycinnamic acid, in 50% acetonitrile and 0.5% trifluoroacetic acid (C30-00004 Kit, or ProteinChip SELDI Starter Kit, Bio-Rad).

2.6   Mass Spectrometry

1. SELDI-TOF MS: PBS-IIc ProteinChip Reader or ProteinChip System Series 4000 (Bio-Rad).
2. SELDI-QTOF: QStar Pulsar or QStarXL instrument (Applied Biosystems) equipped with a PCI 1000 ProteinChip Interface (Bio-Rad).

2.7   Software

Analytical software package: ProteinChip software version 3.1 or later versions (Bio-Rad Laboratories, Hercules, CA, USA).

3   Methods

SELDI ProteinChip technology binds distinct groups of serum proteins on the chip surface after adding EAM. The bound proteins finally are read in a SELDI-TOF-MS. Generally, the serum is processed according to the technique of each institution, and many different procedures have been described. Regardless of the method chosen, standardization, with careful attention to detail, is imperative in protein research.

3.1   Sample Preparation for Storing

1. Standardize blood collection in a serum separator tube and allow it to clot for 30 min at room temperature.
2. Centrifuge the blood for 10 min at 1,000 g.
3. Aliquot the supernatant serum before flash freezing at −80 °C (see Note 1). It also is possible to aliquot the samples later (for example, first thaw) but the quality of the probes deteriorates with each freeze-thaw cycle.

3.2   Using Unfractionated Serum

1. After thawing frozen samples on ice, spin aliquots at 20,000 g for 10 min at 4 °C (*see* **Note 2**).
2. Mix 20 μL of the serum sample with 30 μL of U9 buffer and vortex samples for 20 min.
3. Dilute this sample with 50 μL of PBS buffer.
4. If necessary dilute further 1:8 to 1:10 into the respective binding buffer.

The probes now are ready to be analyzed. Continue with step 3.4.1.

3.3   Using Fractionated Serum Probes

The procedure of fractionating the serum results in aliquots, which can be used on the chip. Dilute the fractionated serum samples 1:5–1:10 into the appropriate binding buffers.

3.4   Sample Application on the Chip

For using the bioprocessor, the minimum sample requirement is 1–10 μg of the total protein per spot and a sample volume of 50–200 μL. The serum samples now can be profiled with different array surfaces. The most commonly used and widely tested surfaces with serum are Metal Affinity Capture (IMAC3/IMAC30), Weak Cation Exchange (CM10/WCX2), and Reversed Phase (H50). To find the chip that suits best, different surfaces have to be preevaluated by testing or studying the literature.

1. Add 50–150 μL to the bioprocessor and incubate for 30 min with shaking on moderate speed.
2. Wash three times with 150 μL binding buffer for 5 min with shaking.
3. Wash twice with 150 μL distilled water (1 min each wash) to remove buffer salts.
4. Add twice, 1 μL of 50% saturated matrix solution to the spot surface and allow to air dry in between.

The probes now are ready to be analyzed. If necessary, store the chips in a dark room at room temperature until processing.

3.5   Mass Analyzer

The ProteinChip arrays are placed in the SELDI-TOF MS (PBS IIc ProteinChip reader or ProteinChip System 4000) and irradiated with a pulsed UV nitrogen laser. Here, protons are transferred onto the peptides and proteins, which subsequently are accelerated by electromagnetic fields through a flight tunnel. The time of flight corresponds inversely to the molecular mass. The resulting signals are converted to mass-to-charge (m/z) ratios based on the time each species takes to pass through the TOF mass analyzer.

1. Set the mass analysis range you want to evaluate.
2. Adjust the laser intensity and some other parameters before measurements.
3. Calibrate the instrument using molecular markers.

3.5.1   Analysis of TOF MS Spectra

Complex software packages, based on genetic algorithms, decision trees, and a unified maximum separability algorithm are used for data analyses. The concept of an independent procedure, a supervised analysis with a training data set to differentiate between sample classes, is used. The results are visualized in a graph with the mass-to-charge ratio of the sample components on the x-axis and the corresponding signal intensities on the y-axis [20]. Modern software allows a wide spectrum of different views and spectra for the best visualization of the data. Widely used is the ProteinChip Biomarker Wizard™, the EDM package, and the Biomarker Patterns Software (Bio-Rad). More detailed information can be found in the literature [20] or on the homepage of Ciphergen (www.bio-rad.com).

3.5.2   Protein Identification (Peptide Mass Fingerprinting)

Using the SELDI-based technique, identification of the target proteins or suspected biomarkers is the last step. The detailed steps of protein purification and identification are beyond the scope of this chapter. Briefly, each protein of interest needs to be purified or enriched for subsequent analysis. Different methods (mini-spin columns, microplate formatted chromatographic devices, preparative scale columns, and antibodies in combination with preactivated arrays) can be used. Finally, the enriched protein is proteolytically cleaved and the masses of the resulting peptide fragments are determined using MS. The obtained data can be submitted directly to protein databases, matching the measured masses with peptide fragments of known sequences (for an overview of protein databases, see http://userpage.chemie.fu-berlin.de/biochemie/protprak.html).

4   Notes

1. Proteins are easily altered or destroyed, when thawed or frozen. Therefore, serum samples should not undergo more than one or two freeze-thaw cycles prior to the assay. The handling, including the freeze-thaw cycles, should be well documented. This helps in explaining the differences of the later results as artifacts due to sample handling.
2. If serum is likely to be stored for certain period, prepare serum samples as soon as possible soon after collection.

Acknowledgments I thank Dr. R. Bogumil for his helpful advice and constructive criticism.

References

1. Seet BT, Dikic I, Zho, MM, Pawson T (2006) Reading protein modifications with interaction domains. Nat Rev Mol Cell Biol 7:473–483
2. Hanahan D, Weinberg RA (2000) The hallmarks of cancer. Cell 100:57–70
3. Gretzer MB, Partin AW, Chan DW, Veltri RW (2003) Modern tumor marker discovery in urology: Surface enhanced laser desorption and ionization (SELDI). Rev Urol 5:81–89
4. Ciordia S, de Los Rios V, Albar JP (2006) Contributions of advanced proteomics technologies to cancer diagnosis. Clin Transl Oncol 8:566–580
5. Wiesner A (2004) Detection of tumor markers with ProteinChip technology Curr Pharm Biotechnol 5:45–67
6. Xiao Z, Prieto D, Conrads TP, Veenstra TD, Issaq HJ (2005) Proteomic patterns: Their potential for disease diagnosis, Mol Cell Endocrinol 230:95–106
7. Zhang, H, Kong B, Qu X, Jia L, Deng B, Yang Q (2006) Biomarker discovery for ovarian cancer using SELDI-TOF-MS. Gynecol Oncol 102:61–66
8. Pusztai L, Gregory BW, Baggerly KA, Peng B, Koomen J, Kuerer HM, Esteva FJ, Symmans WF, Wagner P, Hortobagyi GN, Laronga C, Semmes OJ, Wright G L Jr, Drake RR, Vlahou A (2004) Pharmacoproteomic analysis of prechemotherapy and postchemotherapy plasma samples from patients receiving neoadjuvant or adjuvant chemotherapy for breast carcinoma. Cancer 100:1814–1822
9. Miguet L, Bogumil R, Decloquement P, Herbrecht R, Potier N, Mauvieux L, Van DA (2006) Discovery and identification of potential biomarkers in a prospective study of chronic lymphoid malignancies using SELDI-TOF-MS. J Proteome Res 5:2258–2269
10. Yip TT, Lomas L (2002) SELDI ProteinChip array in oncoproteomic research Technol Cancer Res Treat 1:273–280
11. Petricoin EF, Ardekani AM, Hitt BA, Levine PJ, Fusaro VA, Steinberg SM, Mills GB, Simone C, Fishman DA, Kohn EC, Liotta LA (2002) Use of proteomic patterns in serum to identify ovarian cancer. Lancet 359:572–577
12. Ostergaard M, Rasmussen HH, Nielsen HV, Vorum H, Orntoft TF, Wolf H, Celis JE (1997) Proteome profiling of bladder squamous cell carcinomas: Identification of markers that define their degree of differentiation. Cancer Res 57:4111–4117
13. Anderson NL, Anderson NG (2002) The human plasma proteome: History, character, and diagnostic prospects. Mol Cell Proteomics 1:845–867
14. Anderson NL, Polanski M, Pieper R, Gatlin T, Tirumalai RS, Conrads TP, Veenstra TD, Adkins JN, Pounds JG, Fagan R, Lobley A (2004) The human plasma proteome: A nonredundant list developed by combination of four separate source. Mol Cell Proteomics 3:311–326
15. Issaq HJ, Conrads TP, Prieto DA, Tirumalai R, Veenstra TD (2003) SELDI-TOF MS for diagnostic proteomics. Anal Chem 75:148A–155A
16. Chapman K (2002) The ProteinChip Biomarker System from Ciphergen Biosystems: A novel proteomics platform for rapid biomarker discovery and validation. Biochem Soc Trans 30:82–87
17. Langbein S, Lehmann J, Harder A, Steidler A, Michel MS, Alken P, Badawi JK (2006) Protein profiling of bladder cancer using the 2D-PAGE and SELDI-TOF-MS technique. Technol Cancer Res Treat 5:67–72
18. Melle C, Ernst G, Schimmel B, Bleul A, Koscielny S, Wiesner A, Bogumil R, Moller U, Osterloh D, Halbhuber KJ, von Eggeling F. (2004) A technical triade for proteomic identification and characterization of cancer biomarkers. Cancer Res 64:4099–4104
19. Pusch W, Flocco MT, Leung SM, Thiele H, Kostrzewa M (2003) Mass spectrometry-based clinical proteomics. Pharmacogenomics 4:463–476
20. Fung ET, Weinberger SR, Gavin E, Zhang F (2005) Bioinformatics approaches in clinical proteomics. Expert Rev Proteomics 2:847–862

Chapter 14
The Applicability of a Cluster of Differentiation Monoclonal Antibody Microarray to the Diagnosis of Human Disease

Peter Ellmark, Adrian Woolfson, Larissa Belov, and Richard I. Christopherson

Abstract Recent advances in antibody microarray technology have facilitated the development of multiplexed diagnostic platforms. Highly parallel antigen expression data obtained from these arrays allow disease states to be characterized using protein patterns rather than individual protein markers. The development of an antibody microarray platform of general applicability requires careful consideration of the array content. The human cluster of differentiation (CD) antigens constitute a promising candidate set, being united by their common expression at the leukocyte cell surface and the fact that the majority perform critical functions in the human immune response. The diagnostic potential of a microarray, containing 82 cluster of differentiation monoclonal antibodies (DotScan microarrays) has been demonstrated for a variety of infectious and neoplastic disease states, including HIV, many acute and chronic leukemias, and colorectal cancer. It is likely that these microarrays will have more general utility that extends to other pathological categories, including autoimmune, metabolic, and degenerative diseases.

Keywords antibody microarrays, CD, human cluster of differentiation antigens, cell surface antigens, disease signatures, protein profiles, fluorescence multiplexing, immunophenotype, pharmacogenomics

1 Introduction

The improvement of diagnostic procedures is an important goal for biotechnology and should lead to high-order parallel assay systems based exclusively on multiple molecular criteria from microarrays. Increasingly rapid, sensitive, and specific tests coupled with improved prognostic stratification will reduce health care costs and increase therapeutic success for a broad range of diseases. Although single-protein biomarkers, such as prostate-specific antigen (PSA), carcino-embryonic antigen (CEA), and cancer antigen 125 (CA-125), have wide clinical applicability, the use of unitary markers typically is limited by their low specificity and sensitivity. Such markers are more useful for monitoring the

From: Methods in Molecular Biology, Vol. 439: *Genomics Protocols: Second Edition*
Edited by Mike Starkey and Ramnath Elaswarapu © Humana Press Inc., Totowa, NJ

response to therapeutic interventions than in helping determine the initial diagnosis. One method for improving the sensitivity and specificity of biomarker-based diagnosis is to use assays where multiple protein biomarkers are analyzed simultaneously [1, 2]. Antibody microarrays constitute one of several technologies that enable the multiple parallel analyses of proteins derived from clinical samples (for a review, see[3–5]). Monoclonal antibody microarrays have the advantage of requiring tiny amounts of antibodies, high sensitivity, amenability to automation and high-throughput analysis, and the potential for hundreds to thousands of antigens to be analyzed in parallel [4, 5].

Antibody microarrays were pioneered by MacBeath, Haab, and their associates [6, 7] and subsequently were applied to the development of novel diagnostic methods [8, 9]. More recently, monoclonal antibody microarrays have been utilized to obtain larger data sets from tissue and blood samples [10–16]. In addition to diagnostic applications, antibody microarray technology is emerging as a useful tool in target discovery and proteomics. One such application has been the comparison of proteome expression signatures in malignant and normal samples [4].

It should be possible to construct antibody microarrays that target proteins selected from the entire human proteome (estimated at approximately 100,000 primary proteins), however, for reasons of economy and practicality, the content of the array may be restricted to a defined set of targets. Ideally, the target set should have differential expression in a range of different disease states, to make the test applicable to more than one pathological process. The human cluster of differentiation antigens constitute one such candidate set, being united by their common expression at the surface of leucocytes where the majority perform functions critical to the human immune response [3]. As there is no a priori method for defining candidate sets, selection and validation is empirical and determined from testing for unique patterns against a broad panel of pathologies. Whereas CD antigens have proven to be useful in a broad range of disease situations, alternative approaches may be equally valuable [11].

The human CD antigens are a well-characterized set of molecules expressed at the cell surface of leucocytes and other cell types. As the function of leucocytes is restricted principally to immunity, changes in the expression of CD antigens provide the basis for a general-purpose diagnostic platform. The parallel measurement of multiple CD antigens contrasts with the more limited surface immunophenotyping provided by flow cytometry-based techniques. Limited CD antigen profiling combined with morphological, cytochemical and cytogenetic data forms the basis of the classification system for human leukemias. CD antibody microarray-based diagnosis (DotScan microarrays) has been shown to concur with established diagnostic methods used in acute myeloid leukemia [16], HIV[10], and colorectal cancer [14]. The extensive immunophenotype obtained with DotScan microarrays may enable subclassification of these diseases, for example, into drug responders and nonresponders, and promises to facilitate the identification of new disease entities.

2 Materials

2.1 Preparing Leukemia and Lymphoma Samples

1. Histopaque-1077 (Sigma-Aldrich Pty. Ltd., St. Louis, MO).
2. Phosphate-buffered saline (PBS; Sigma-Aldrich).
3. K3EDTA vacuette tubes (Greiner Bio-one, Catalogue number 455036), or lithium heparin (BD vacutainer, catalogue number 367885).

2.2 Generation of Single-Cell Suspensions from Solid Tumors

1. Hanks buffer (Sigma-Aldrich).
2. Collagenase (Sigma-Aldrich) 2 % (w/v) in PBS, stored for weeks at −20 °C.
3. Roswell Park Memorial Institute 1640 (RPMI) medium (Sigma-Aldrich).
4. Stainless steel sieve, a standard tea sieve may be used.
5. 50 μm Filcon filter membrane (BD, Franklin Lakes, NJ).
6. 0.1 % (w/v) DNAse in PBS (Sigma-Aldrich), stored for weeks at −20 °C.

2.3 Capture And detection of Cells on DotScan Microarrays

1. DotScan microarrays, (Medsaic Pty Ltd, Eveleigh, NSW, Australia).
2. 1 mM EDTA in PBS (Sigma-Aldrich).
3. Roswell Park Memorial Institute 1640 (RPMI) medium (Sigma-Aldrich).
4. Heat-inactivated human AB serum (Sigma-Aldrich) (*see* **Note 1**).
5. Fetal bovine calf serum (FCS; Gibco, Grand Island, NY).
6. 4 % (v/v) formaldehyde in PBS (Sigma-Aldrich). Make up fresh as required. Formaldehyde is very toxic and should be handled with care.
7. Humidity chamber suitable for one to six slides. It is important that the slides are incubated horizontally.
8. Bovine serum albumin (BSA; Sigma-Aldrich, catalogue number A3902).
9. Fluorophore-labeled cell-type-specific antibody or antibodies for detection of captured cells.
10. Blocking reagent (e.g., serum) from the same species as the antibodies used in item 8.
11. Medsaic DotScan array scanner for optical scanning of DotScan arrays or Molecular Imager FX Pro (Bio-Rad, Hercules, CA) or Typhoon 8600 Variable Mode Imager (Amersham-Pharmacia, Castle Hill, NSW, Australia) for scanning DotScan arrays stained with fluorescently labeled antibodies.
12. Medsaic DotScan software.

3   Methods

The DotScan microarray platform is based on the capture of whole live cells on nitrocellulose microarrays containing 82 monoclonal antibodies with specificities against different CD antigens. Using whole cells for the detection of CD antigens is advantageous, since cell surface markers often are hydrophobic and aggregate when analyzed in solution.

3.1   Preparation of Cell Samples

Obtaining a homogeneous single-cell suspension is essential. Mononuclear leucocytes are purified from samples of peripheral blood, as red cells interfere with the capture of leukocytes. For samples obtained from solid tumor tissue, it is essential to remove cell aggregates that interfere with cell capture and fluorescent scanning, resulting in false positives. Furthermore, the proteolytic and mechanical tissue disaggregation results in DNA release, making the solution viscous and hampering the cell capture. Therefore, cell suspensions are treated with DNAse.

3.1.1   Leukemias and Lymphomas from Peripheral Blood

1. Collect the blood sample in a K3EDTA vacuette tube, or a lithium heparin vacutainer.
2. Dispense 3 mL of histopaque-Ficoll into a centrifuge tube.
3. Carefully overlay 5 mL of blood onto the histopaque-Ficoll.
4. Centrifuge at 800 g (without brakes) for 15 min at room temperature in a swing-out bucket rotor.
5. Remove the top layer (the plasma fraction) from the tube.
6. Carefully transfer the mononuclear leucocytes from the Ficoll/plasma interface to a 10-mL centrifuge tube.
7. Wash the cells with 10 mL of PBS and centrifuge at 400 g for 5 min at room temperature. If necessary, use lysis buffer and/or DNAse to remove red blood cells and/or DNA (see **Note 2**).
8. Resuspend the cell pellet in 10 mL of PBS.
9. Centrifuge at 400 g for 5 min at room temperature.
10. Resuspend the cell pellet in 10 mL of PBS.
11. Stain 10 μL of the cells with trypan blue and count the number of living cells using a light microscope in a manual counting chamber (Bürker chamber).
12. Centrifuge at 400 g for 5 min at room temperature to pellet the cells. Proceed immediately to Sect. 3.2.1.

3.1.2   Preparation of Single-Cell Suspensions from Solid Tumors

1. Immediately following the biopsy, place the tissue samples in a cold Hanks buffer to maintain cell viability, which is essential for cell capture (*see* **Note 3**).
2. Cut the samples into 2 mm strips, or 2 × 2 mm cubes.
3. Incubate the tissue with 2 % (w/v) collagenase in RPMI 1640 at 37 °C for 1 h.
4. Force the semidigested tissue through a fine stainless steel sieve to disaggregate it into a cell suspension.
5. Remove cell aggregates with a 50-μm Filcon filter membrane to obtain a suspension of single viable cells.
6. Add 0.1 % (w/v) DNAse to cells for 20 min at room temperature to digest DNA from lysed cells (*see* **Note 2**).
7. Centrifuge at 400 g for 5 min at room temperature.
8. Resuspend the cell pellet in 10 mL of PBS.
9. Centrifuge at 400 g for 5 min at room temperature.
10. Resuspend the cell pellet in 4–10 mL of PBS.
11. Stain 10 μL of the cells with trypan blue and count the number of living cells using a light microscope in a manual counting chamber.
12. Centrifuge at 400 g for 5 min at room temperature to pellet the cells. Proceed immediately to Sect. 3.2.2.

3.2   Capturing Cells on DotScan Microarrays

Cell capture on antibody dots of the microarray is an active process involving live cells, where CD antigens "cap" to the interface between the cells and dots. Evidence for such a time-dependent capping process has been obtained from experiments in which cells were captured by CD antibodies immobilized on Biacore chips (L. Bransgrove and R. I. Christopherson, unpublished). It therefore is essential that cells are viable for capture on the DotScan microarray.

3.2.1   Leukaemia Cells

1. Resuspend the leukocytes (4–6×10^6 cells; *see* **Note 3**) in 300 μL of PBS containing 1 mM EDTA.
2. Prewet the microarray with PBS and carefully wipe away excess PBS around the nitrocellulose surface.
3. Carefully apply the cell suspension to the microarray.
4. Incubate the slide in a humid chamber at room temperature for 30 min.
5. Gently wash off unbound cells with PBS.
6. Fix the DotScan array for ≥20 min in PBS containing 4% (v/v) formaldehyde.

3.2.2 Single-Cell Suspension from Solid Tumors

1. Resuspend the single cell suspension (4×10^6 cells; *see* **Note 3**) in 300 μL of RPMI containing 10% (v/v) FCS.
2. Prewet the array in PBS and carefully wipe away excess PBS around the nitrocellulose surface.
3. Add the cell suspension to the array.
4. Incubate the array in a humid chamber at 37 °C for 1 h (*see* **Note 4**).
5. Gently wash off unbound cells with PBS.
6. Fix the DotScan array for 15 min in PBS containing 4% (v/v) formaldehyde.

3.3 Detection of Captured Cells

Captured cells can be detected using either an optical scanner, where the average optical density of the spot is measured (see the example in Fig. 14.1), or by staining the cells with a fluorescently labeled antibody and using a laser scanner with the appropriate excitation and emission filters. The latter method allows for simultaneous detection of several cell populations captured on a single DotScan microarray.

3.3.1 Optical Scanning

1. Wash the fixed DotScan microarray twice in PBS for 20 s each.
2. Carefully remove excess fluid from the microarray surface and thoroughly wipe the back of the slide until clean and dry. The nitrocellulose surface must not be allowed to dry out during scanning.
3. Scan the slides using a Medsaic DotScan array scanner.

3.3.2 Fluorescence Multiplexing

1. Wash the fixed array twice in PBS for 20 s each (*see* **Note 5**).
2. Block the slides by the addition of 300 μL of 1% (w/v) BSA in PBS containing 5 μg/mL mouse IgG serum (*see* **Note 6**).
3. Incubate at room temperature for 30 min in a humid chamber.
4. Apply 1–5 μg/mL of fluorophore-labeled antibody or antibodies (*see* **Note 5**) in 150 μL of PBS containing 1% (w/v) BSA and 5 μg/mL mouse IgG serum. If there are several different antibodies directed at different antigens, the spectral properties of the combination of fluorophores must be considered (*see* **Note 7**).
5. Incubate in the dark in a humid chamber at room temperature for 30 min.
6. Wash the microarray by immersion in PBS at least five times (keep it immersed for 20 s for two of the immersions). Keep away from light.

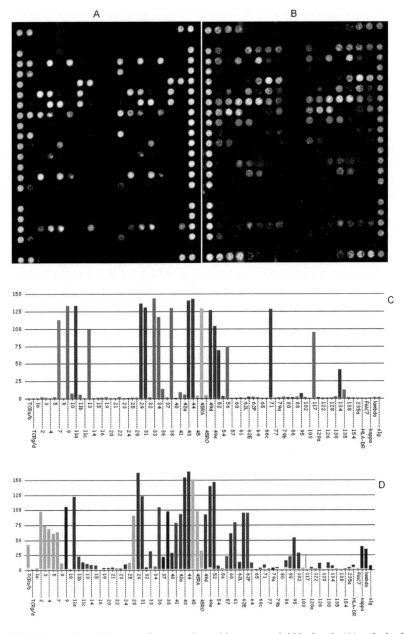

Fig. 14.1 Mononuclear leukocytes from a patient with acute myeloid leukemia (A) and a healthy control sample (B) captured on CD antibody microarrays. The DotScan microarray consists of two duplicate arrays, and the signal intensities are averaged by the DotScan software to produce the corresponding bar charts for acute myeloid leukemia (C) and a healthy control (D)

7. Scan the slides using a laser scanner. Select appropriate laser wavelengths and filter settings for the particular set of fluorophore-labeled antibodies used (*see* **Note 7**).

3.4   Identification of Disease Signatures

Dot patterns of cells captured on the DotScan microarray are quantified using Medsaic DotScan software. The software quantifies the intensities of each dot from the digital image file and compiles expression profiles as bar charts showing the average intensities above background on an 8-bit grayness scale from 1 to 256. For images generated with scanners other than the Medsaic scanner, for example, files from fluorescent multiplexing assays, the image file needs to be converted to a bitmap-file format prior to analysis with the DotScan software. To identify expression signatures associated with disease, a set of matched samples from healthy and diseased tissues (or healthy and sick individuals) should be analyzed, preferably with at least 20 samples per group.

3.5   Principal Component Analysis of Patient Data

The data obtained from DotScan analysis is highly (up to 82) multidimensional, and for visualization purposes, it may be projected onto two or three dimensions using principal component analysis (PCA). PCA is a method for reducing the dimensionality of a data set by performing a covariance analysis among factors and is suitable for data sets in multiple dimensions. For high-dimensional data, it is preferable to perform principal component analysis first utilizing a large number of components, followed by a discriminant analysis that improves separation. The immunophenotypic data then can be visualised in a two- or three-dimensional plot (see the example in Fig. 14.2), using the first two or three discriminant functions and used to identify sample populations and outliers in the sample groups.

3.6   Target Recognition for Therapeutic Antibodies

Monoclonal antibodies constitute the most rapidly growing class of human therapeutics and, currently, the second largest class of drugs after vaccines [17]. The majority of the monoclonal antibodies currently approved by the FDA are directed toward CD antigens or other cell surface antigens [17], such as Rituxan (Rituximab, Mabthera, anti-CD20), Mylotarg (anti-CD33), and Campath-1H (anti-CD52). It has been estimated that more than 150 other monoclonal antibodies against different targets are in various stages of clinical trial. For example, the following therapeutic

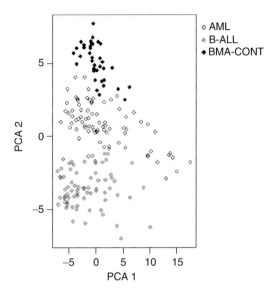

Fig. 14.2 Two-dimensional principal component analysis of CD antibody microarray data from bone marrow aspirates of acute myeloid leukemia (AML, $n=75$), B-cell acute lymphoblastic leukemia (B-ALL, $n=73$), and healthy controls (BMA-CONT, $n=36$). The symbols denote individual samples, the axes represent scores in the principal component space for the first and second principal components

antibodies target surface antigens overexpressed on colorectal cancer, Edrecolomab targets EpCAM (epidermal cell adhesion molecule), huAb A33 binds to the antigen A33, and Cetuximab binds to the epidermal growth factor receptor (EGFR). Inclusion of more therapeutic antibodies on the DotScan microarray would provide a more extensive immunophenotype and enable selection of appropriate therapeutic antibodies for patient-specific treatment.

4 Notes

1. The DotScan microarray may require heat-inactivated human AB serum (2%) at the steps in Sect. 3.2.1 to work optimally when analyzing acute myeloid leukemias or other cells with high levels of Fc receptors.
2. If the Ficoll gradient does not remove the red blood cells sufficiently, the remaining erythrocytes can be removed using lysis buffer (0.15 M ammonium chloride, 10 mM potassium hydrogen carbonate, 0.1 mM EDTA). If the cell suspension contains too much DNA (from lysed cells), it will be too viscous to allow proper cell binding. The cell suspension can be treated with DNAse (0.1% [w/v] in PBS for 20 min at room temperature). The cells must be alive to be captured on the DotScan microarray. Cells that have been frozen at −80 °C in dimethylsulfoxide/fetal calf serum may be used if the viability is retained.

3. For solid tissue samples, it is important to minimize the time between surgery and analysis, as the cells must be alive to be captured onto the DotScan microarray.

4. Cell suspensions generated from solid tissue are incubated in 37 °C for 1 h on the arrays to allow for regeneration of CD antigens that may have been affected by the collagenase/DNAse treatment.

5. If the DotScan microarray is to be stained with a fluorescent antibody, the fixation time should be minimized, as several CD antigens are fixation sensitive. If possible, choose an antibody or CD marker that can be used for staining fixed tissue sections (this information generally is provided by the antibody supplier). As a note of caution, data from spots that correspond to the same antibody as the staining antibody (i.e., the CD3 dot from a CD3-staining experiment) may be affected by the capping effect described in Sect. 3.2.

6. It is advisable to block the DotScan slides before staining to avoid the nonspecific binding of the fluorescently labeled antibodies to the fixed cells. Species-specific antibodies, or serum, should be used for this purpose, such as mouse IgG or mouse serum for blocking the slides prior to staining with a fluorescently labeled mouse antihuman anti-CD antigen antibody.

7. Fluorescence multiplexing using several antibodies with different fluorophores must be done with matching combinations of fluorophores. A combination of Alexa488, Alexa532, and Alexa647 works well with most scanners. However, both the wavelengths of the lasers and the bandwidth of the available filters have to be considered.

References

1. Schoniger-Hekele M, Muller C (2006) The combined elevation of tumour markers CA 19-9 and CA 125 in liver disease patients is highly specific for severe liver fibrosis. Dig Dis Sci 51:338–345

2. Louhimo J, Finne P, Alfthan H, Stenman UH, Haglund C (2002) Combination of HCGbeta, CA 19-9 and CEA with logistic regression improves accuracy in gastrointestinal malignancies. Anticancer Res 22:1759–1764

3. Woolfson A, Ellmark P, Chrisp JS, Scott MA, Christopherson RI (2006) The application of CD antigen proteomics to pharmacogenomics. Pharmacogenomics 7:759–771

4. Borrebaeck CA (2006) Antibody microarray-based oncoproteomics. Expert Opin Biol Ther 6:833–838

5. Lal SP, Christopherson RI, dos Remedios CG (2002) Antibody arrays: An embryonic but rapidly growing technology. Drug Discov Today 7:S143–S149

6. Haab BB, Dunham MJ, Brown PO (2001) Protein microarrays for highly parallel detection and quantitation of specific proteins and antibodies in complex solutions. Genome Biol 2 RESEARCH0004

7. MacBeath G, Schreiber SL (2000) Printing proteins as microarrays for high-throughput function determination. Science 289:1760–1763

8. Knezevic V, Leethanaku, C, Bichsel VE, Worth JM, Prabhu VV, Gutkind JS, Liotta LA, Munson PJ, Petricoin EF III, Krizman DB (2001) Proteomic profiling of the cancer microenvironment by antibody arrays. Proteomics 1:1271–1278

9. Belov L, de la Vega O, dos Remedios CG, Mulligan SP, Christopherson RI (2001) Immunophenotyping of leukemias using a cluster of differentiation antibody microarray. Cancer Res 61:4483–4489

10. Woolfson A, Stebbing J, Tom BD, Stoner KJ, Gilks WR, Kreil DP, Mulligan SP, Belov L, Chrisp JS, Errington W, Wildfire A, Erber WN, Bower M, Gazzard B, Christopherson RI, Scott MA (2005) Conservation of unique cell-surface CD antigen mosaics in HIV-1-infected individuals. Blood 106:1003–1007

11. Ellmark P, Ingvarsson J, Carlsson A, Lundin SB, Wingren C, Borrebaeck CA (2006) Identification of protein expression signatures associated with H. pylori infection and gastric adenocarcinoma using recombinant antibody microarrays. Mol Cell Proteomics 5:1638–46

12. Miller JC, Zhou H, Kwekel J, Cavallo R, Burke J, Butler EB, Teh BS, Haab BB (2003) Antibody microarray profiling of human prostate cancer sera: Antibody screening and identification of potential biomarkers. Proteomics 3:56–63

13. Orchekowski R, Hamelinck D, Li L, Gliwa E, vanBrocklin M, Marrero JA, Vande Woude GF, Feng Z, Brand R, Haab BB (2005) Antibody microarray profiling reveals individual and combined serum proteins associated with pancreatic cancer. Cancer Res 65:11193–11202

14. Ellmark P, Belov L, Huang P, Lee CS, Solomon MJ, Morgan DK, Christopherson RI (2006) Multiplex detection of surface molecules on colorectal cancers. Proteomics 6:1791–1802

15. Belov L, Huang P, Barber N, Mulligan SP, Christopherson RI (2003) Identification of repertoires of surface antigens on leukemias using an antibody microarray. Proteomics 3:2147–2154

16. Belov L, Mulligan SP, Barber N, Woolfson A, Scott M, Stoner K, Chrisp JS, Sewell WA, Bradstock KF, Bendall L, Pascovici DS, Thomas M, Erber W, Young GAR, Wiley JS, Juneja S, Wierda WG, Green AR, Keating MJ and Christopherson RI (2006) Classification of human leukemias and lymphomas using extensive immunophenotypes obtained by cell capture on an antibody microarray. Brit J Haem 135:184–197

17. Carter PJ (2006) Potent antibody therapeutics by design. Nat Rev Immunol 6:343–357

Chapter 15
Protein Profiling Based on Two-Dimensional Difference Gel Electrophoresis

Gert Van den Bergh and Lutgarde Arckens

Abstract The recent introduction of two-dimensional fluorescent difference gel electrophoresis has enabled the large screening of differential protein expression levels with higher confidence and greater sensitivity than using the classical two-dimensional electrophoresis approach. With this technology, multiple protein samples can be labeled with up to three fluorescent dyes. These labeled protein samples are mixed and applied on the same two-dimensional electrophoresis gel, subsequently scanned, and analyzed by specialized software. The possibility to run two or more protein samples on a single gel as well as the introduction of an internal standard on each gel drastically reduces the gel-to-gel variability and results in higher levels of certainty with regard to the differential character of the expressed proteins.

Keywords two-dimensional electrophoresis, two-dimensional difference gel electrophoresis (DIGE), CyDyes, proteomics, fluorescence

1 Introduction

Numerous technological developments, such as differential display PCR [1], suppressive subtractive hybridization [2], and cDNA microarrays [3], have enabled the rapid and in-depth analysis of differential mRNA expression levels of large sets of gene products in parallel. Notwithstanding the wealth of information these approaches provide regarding the molecular mechanisms underlying physiological processes or diseases, proteins are the ultimate effector molecules of the cell. The quantitative study of the proteome therefore should be given high priority, which is true for two more reasons. First, quantitative variations in mRNA levels are not always correlated with corresponding changes in the expression of proteins [4], and second, many proteins contain posttranslational modifications that are not detectable at the transcriptional level at all but are essential for their activation and functional regulation.

Large-scale quantitative analysis and comparison of protein expression levels, also called *functional proteomics*, generally is performed by two different approaches: gel-free or gel-based. In gel-free proteomics, proteins or enzymatically cleaved peptides derived from proteins are separated, quantified, and identified by liquid

From: Methods in Molecular Biology, Vol. 439: *Genomics Protocols: Second Edition*   211
Edited by Mike Starkey and Ramnath Elaswarapu © Humana Press Inc., Totowa, NJ

chromatography and mass spectrometry; whereas in gel-based proteomics, proteins are first separated and quantified by two-dimensional electrophoresis, followed by mass spectrometric identification of the differentially expressed proteins. While gel-free proteomics is better able to detect hydrophobic proteins, such as membrane proteins, gel-based proteomics methods can easily detect posttranslationally modified proteins. Moreover, it is easier to compare a large number of experimental conditions by two-dimensional electrophoresis than by mass spectrometry.

Since comparative gel-based proteomics depends on the quantitative separation and visualization of a large fraction of the expressed proteins, traditional two-dimensional electrophoresis currently is not the optimal choice, as it suffers from a rather high level of variability among different gels. Moreover, traditional staining methods, such as Coomassie brilliant blue or silver staining, provide either a low sensitivity or a limited linear range, impairing the quantitative aspect of protein visualization. The introduction of two-dimensional difference gel electrophoresis (2-D DIGE), a multiplexing approach for two-dimensional electrophoresis based on fluorescent protein visualization, proved to be of enormous importance for the renaissance of gel-based proteomics [5–8]. With this technique, proteins are labeled with up to three spectrally distinct fluorescent dyes that have no discernable influence on the electrophoretic mobility of the proteins. Two to three labeled samples therefore can be run on a single 2-DE gel. The protein spot patterns are visualized by sequentially scanning the gel at the excitation and emission wavelengths of each of the dyes used. The quantitative analysis of differential protein expression levels is performed by using appropriate software, a process facilitated by the relative comparison of protein expression levels obtained from the same gel, reducing the gel-to-gel variability inherent to 2-DE. This variability can be further reduced by using an internal standard, a mixture of protein samples from every experimental condition, which is applied on every gel [7]. As a result, the matching of spots and their quantification across different gels is greatly facilitated and enables 2-D DIGE to be employed as a high-throughput screening method for the detection of differential protein expression. Although recently, new dyes have been developed that enable saturation labeling of the proteins and, therefore, provide a much higher sensitivity than the minimal labeling dyes [9], in this chapter we describe only the standard 2-D DIGE approach using an internal standard and minimal labeling dyes.

2 Materials

2.1 Sample Preparation

1. 2-DE lysis buffer: $7M$ urea, $2M$ thiourea, 4% CHAPS, $40\,mM$ Tris-base. Store in 1mL aliquots at $-20\,°C$ or use fresh. Do not allow warming above room temperature, since this can result in the degradation of urea to isocyanate, leading to the carbamylation of proteins. Add $40\,\mu L$ of complete protease inhibitor (Roche, Basel, Switzerland) to 1 mL of aliquot prior to use (stock solution: 1 tablet in 2mL HPLC-grade water. Store for a maximum of 1 month at $-20\,°C$ (*see* **Note 1**).

2. Bath sonicator (Model 5510, Branson Ultrasonics, Danbury, CT).
3. PBS buffer: 0.1 M NaH$_2$PO$_4$, pH 7.4.
4. Protein Assay Kit (Bio-Rad, Hercules, CA).
5. ELISA plate reader (Labsystems Multiskan RC, Thermo Electron Corporation, Brussels, Belgium).

2.2 Experimental Setup and Protein Labeling with Fluorescent Dyes

1. pH indicator strips (Merck-Eurolabo, Leuven, Belgium).
2. Cy2, Cy3, and Cy5 minimal labeling dyes (GE Healthcare Life Sciences, Uppsala, Sweden).
3. Dimethylformamide (DMF), water free (Merck-Eurolabo).
4. 10 mM lysine solution (Acros Organics, Geel, Belgium).

2.3 Isoelectric Focusing

1. Immobilized pH gradient (IPG) strips, such as the Immobiline DryStrips (GE Healthcare Life Sciences).
2. Destreak rehydration solution with 0.5% pharmalytes (GE Healthcare Life Sciences).
3. IPG strip reswelling tray (GE Healthcare Life Sciences).
4. Parrafin oil as cover fluid (Merck-Eurolabo).
5. IPGphor cup loading manifold (GE Healthcare Life Sciences).
6. IPGphor IEF unit (GE Healthcare Life Sciences).
7. 2X lysis buffer: 7 M urea, 2 M thiourea, 4% CHAPS, 2% Dithiothreitol (DTT, Applichem, Darmstadt, Germany), 40 mM Tris-base. Store in 1mL aliquots at −20 °C or use fresh.

2.4 SDS-PAGE

1. Gel casting equipment (GE Healthcare Life Sciences).
2. Low-fluorescence glass plates (GE Healthcare Life Sciences).
3. Vacuum grease.
4. Acrylamide gel solution: For a 12.5% T gel solution, prepare 900 mL of the following solution, enough for casting 14 gels: 375 mL acrylamide stock solution (30% T, 2.6% C, BioRad Hercules, CA), 225 mL Tris buffer (1.5 M Tris-HCl, pH 8.8), 290 mL water and 9 mL 10% (w/v) SDS (Sigma-Aldrich, Bornem, Belgium).
5. TEMED (Bio-Rad) and ammoniumpersulphate (APS, Sigma-Aldrich).
6. 100 mL of displacing solution: 0.375 M Tris-HCl, pH 8.8, 50% (v/v) glycerol (Acros Organics), bromophenol blue (Merck-Eurolabo).
7. Water-saturated butanol (Merck-Eurolabo).

8. Two liter of 10X SDS running buffer: 250 m*M* Tris, 1.92 *M* glycine, 1% SDS, approximately pH 8.3. The pH of this solution should not be adjusted. Dilute to obtain 1X and 2X SDS running buffer.
9. 100 mL agarose sealing solution, containing 0.5% agarose in 2X SDS running buffer and a few grains of bromophenol blue.
10. Equilibration solution A: 50 m*M* Tris-HCl, pH 6.8, 6 *M* urea, 30% (v/v) glycerol, 2% (w/v) SDS and 1% (w/v) DTT.
11. Equilibration solution B: 50 m*M* Tris-HCl, pH 6.8, 6 *M* urea, 30% (v/v) glycerol, 2% (w/v) SDS and 2.5% (w/v) iodoacetamide.

2.5 Gel Imaging and Data Analysis

1. Fluorescent gel imager capable of the detection of Cy2, Cy3, and Cy5 with high resolution and high sensitivity (e.g., Ettan DIGE Imager or Typhoon 9400 Variable Mode Imager, GE Healthcare Life Sciences, or equivalent).
2. Image analysis software (DeCyder Differential Analyis Software, GE Healthcare Life Sciences, or equivalent).
3. Fixation solution: 50% (v/v) methanol, 5% (v/v) acetic acid in water.

2.6 Protein Identification

1. Coomassie blue G-250 protein staining solution (Bio-Rad) or other protein staining solutions.
2. Sterile razor blades or a spot picking robot (e.g., GE Healthcare Life Sciences).
3. Laminar flow hood.
4. Acetonitrile (ACN), trifluoroacetic acid (TFA), formic acid (FA) (Merck Eurolabo, Leuven, Belgium).
5. Digestion solution: 100 ng modified porcine trypsin (Promega, Leiden, The Netherlands, sequencing grade), 12.5 m*M* NH_4HCO_3, and 5% (v/v) ACN in water.
6. ZipTip C_{18} reversed-phase chromatography pipette tips (Millipore, Bedford, MA).
7. Mass spectrometer (Waters, Milford, MA).

3 Methods

3.1 Sample Preparation

1. Collect fresh or frozen cells or tissue; for example, from a small area of mammalian brain, the neocortex. To this end, slice the brain in 200 μm thick sections

on a cryostat. Collect the brain tissue by cutting gray matter from the 200 μm thick sections (approximately 160 mm² (±32 mm³)). Always cut the tissue on dry ice to prevent protein degradation.

2. Rapidly transfer the collected tissue to 100 μL of ice-cold lysis solution. This volume should be changed according to the volume of the cells or tissue to be lysed (*see* **Note 3**).

3. Homogenize brain tissue on ice. Avoid excessive foaming of the lysis buffer to prevent protein loss.

4. Sonicate in a bath sonicator for approximately 1 min at room temperature.

5. Put protein sample at room temperature for 1 h, to allow for complete solubilization of the proteins in the lysis buffer.

6. Sonicate again for 1 min at room temperature.

7. Clear the protein lysate by centrifugation for 20 min at 13,000 g at 4 °C.

8. Dialyze the supernatant against water for 2–3 h, to decrease the salt concentration of the protein sample.

9. Place the supernatant in dialysis tubing with a cutoff of 500 Da.

10. Determine the protein concentration of the dialysate.

11. Make a 2 μg/μL stock solution of ovalbumin and prepare a standard dilution series, adding 400 μL of a solution containing 500, 300, 100, 50, 25, and 12.5 μg ovalbumin in PBS and 300 μL 0.1 M NaOH.

12. Prepare the samples to be measured by mixing 1 μL of protein sample, 30 μL 0.1 M NaOH, and 69 μL PBS buffer.

13. Place 20 μL of the samples and ovalbumin standard in triplicate on a flat-bottomed 96-well plate. Add 200 μL of Bio-Rad protein assay solution (diluted 1/10 in HPLC-grade water), read the absorbance at 595 nm with an ELISA plate reader, and determine protein concentrations.

14. Store the protein samples at −70 °C in aliquots to reduce freeze-thaw cycles.

3.2 Experimental Setup and Protein Labeling with Fluorescent Dyes

Because of the high cost and complexity of 2-D DIGE, the design of a 2-D DIGE experiment is of utmost importance. When comparing two experimental conditions, it is possible to directly analyze both samples on a single gel to determine relative protein expression levels. However, in most cases where multiple samples have to be compared, it is advantageous to include an identical internal standard on every gel alongside the experimental samples, since this standard can be used to normalize the experimental samples across different gels (by dividing the volume of every spot on each gel by the volume of the same spot from the internal standard), thereby eliminating gel-to-gel variability. But, even in a comparison of two samples, it is useful to include an internal standard, because of the requirement to run several biological replicates on different gels to obtain valid statistical information.

It is recommended that Cy2 routinely be used for the labeling of the internal standard. Also, always ensure that one half of the biological replicates of an experimental condition are labeled with Cy3 and the other half with Cy5, to prevent dye-specific labeled proteins showing up falsely as differentially expressed. In addition, always try to pair the samples under investigation at random, to prevent any experimental bias due to technical reasons. An example of an experimental setup with three conditions is given in Table 15.1.

1. Prepare the internal standard. This standard is made of equal aliquots from every protein sample to be investigated in the experiment. Make a larger amount of the internal standard than strictly necessary (50 µg of protein for each gel to be run) to compensate for gels that have to be repeated for technical reasons or additional experimental conditions that are added late in the experimental design.
2. Check the pH of the protein samples and internal standard. The optimal pH for the labeling reaction is between pH 8.0 and 9.0. Adjust the pH only if it is higher than pH 9.0 or lower than pH 6.0.
3. Make a working solution of the CyDyes at a concentration of 400 pmol per µL, by diluting them in water-free DMF (*see* **Note 5**).
4. Mix 50 µg of protein sample with 400 pmol fluorescent dye (*see* **Note 6**).
5. Vortex sample and incubate for 30 min at 4 °C in the dark.
6. Add 1 mL of 10 mM lysine and incubate 15 min at 4 °C in the dark. This blocks the labeling reaction of the remaining free dyes and prevents cross-labeling after mixing the samples.
7. Mix the two samples labeled with Cy3 and Cy5 and the internal standard, labeled with Cy2, to be run on the same gel and store on ice.

3.3   Isoelectric Focusing

The isoelectric focusing method described here is based on the IPGphor IEF system, using broad pH gradient IEF strips to obtain a general overview of the proteins present in the samples.

Table 15.1 Example of an experimental setup for an experiment with three biological conditions (A–C) and four replicates (1–4); 50 µg of each sample is loaded onto each gel; the internal standard consists of equal amounts of protein from samples A1 to C4

Gel number	Cy3	Cy5	Cy2
1	A1	B1	Internal standard
2	C1	A2	Internal standard
3	B2	C2	Internal standard
4	A3	C3	Internal standard
5	B3	A4	Internal standard
6	C4	B4	Internal standard

1. To reconstitute the dried and frozen IPG strips to their original volume, they need to be rehydrated with destreak rehydration solution that at the same time provides the appropriate conditions to perform isoelectric focusing in (*see* **Note 7**). Remove the plastic cover sheets of the IPG strips and place them overnight (at least 10 h) with their gel surface down in the correct volume of rehydration buffer containing 0.5% IPG buffer (with the same pH gradient as the gels) in a IPG dry-strip reswelling tray. For a 24 cm gel, the required volume is 450 μL. Cover the strips with 2 mL of paraffin oil to prevent dehydration and carbon dioxide absorption.

2. Prepare the cup loading manifold for isoelectric focusing (IEF). Place the ceramic manifold tray on the surface of the IEF unit. Place the rehydrated IPG strips with their gel side up in the grooves of the manifold tray, with their anodic end pointing toward the marked direction. Place the sample cups on the strips and pipette about 20 μL of paraffin oil in the cups to check for leaks. If no leaks are found, the manifold tray and IPG strips should be covered as soon as possible with 100 mL paraffin oil to prevent drying of the strips (*see* **Note 8**).

3. Wet paper wicks evenly with 150 μL of distilled water and place them at the end of the IPG strips. One third of the paper wick should cover the end of the strip. Place the electrodes on the paper wicks, where the wicks and strips overlap.

4. Add an equal volume of 2X lysis buffer to the sample mixture. Pipette this mixture in the sample cups and close the lid. Make sure the sample is completely covered with paraffin oil.

5. Start the isoelectric focusing run. For a broad pH gradient on a 24 cm strip and 50 μg of protein per sample, the following protocol could be run: 3 h at a constant voltage of 300 V, 3 h at a constant voltage of 600 V, a gradual increase in voltage over 6 h to 1,000 V, a 3 h gradual increase toward 8,000 V, and a final 4 h focusing at a constant 8,000 V. The temperature should be maintained at 20 °C and the current at 50 μA/gel (*see* **Note 9**).

6. After the run, remove the IPG strips and proceed to the second dimension or store them at −80 °C between two plastic sheets for up to 3 months.

3.4   SDS-PAGE

3.4.1   Casting of Gels

3.4.1.1   Preparation of the Gel Caster

1. Wash each item (gel caster, glass plates, separator sheets, blank cassette inserts) with water and detergent and thoroughly rinse with tap water and HPLC-grade water. Glass plates can be further cleaned by washing them with ethanol. Dry glass plates with a lint-free cloth and keep away from dust. Any residual dust could result in artifactual spots on the gel images. Tilt the gel caster backward and fill it, starting with a separator sheet. Alternate between a gel cassette (large and small glass plate) and a separator sheet. If the caster

is not filled by the number of gel cassettes required, add additional blank cassette inserts and separator sheets until the stack in the caster is even with the edge of the caster.

2. Lubricate the sealing gasket of the gel caster lid with vacuum grease to ensure a liquidtight seal and close the lid.

3.4.1.2   Pouring of the Acrylamide Gel

1. Prepare an appropriate volume of acrylamide gel solution. For a full 14 gel set, this amounts to 0.9 L. Immediately prior to gel casting, add 250 μL TEMED and 250 mg APS in 1 mL of water to start the polymerization reaction.
2. Pour the gel solution in the gel caster through a funnel, taking care not to introduce any air bubbles. Fill to about 4 cm of the top of the small glass plates.
3. Remove the funnel and add displacing solution to the balance chamber in small volumes, until the bottom V-well of the gel caster is filled with this solution. The remaining acrylamide solution is forced into the gel cassettes to the final gel height. The required volume of displacing solution will be between 50 and 100 mL.
4. Immediately pipette 2 mL of butanol onto each gel to ensure completely level gel surfaces. Allow for overnight polymerization of the gels before running the second-dimension electrophoresis to complete the polymerization reaction. After that, the gels can be stored in sealed plastic bags at 4 °C for up to 1 week.

3.4.2   Equilibration of IPG Strips

1. Prepare 100 mL of a 0.5% (w/v) agarose sealing solution in 2X SDS running buffer, including a few grains of bromophenol blue. Dissolve the agarose by boiling.
2. To saturate the separated proteins with SDS, equilibrate the IPG strips with a solution containing SDS. Place the IPG strips in equilibration solution A for 15 min at room temperature.
3. Incubate the strips for another 15 min in equilibration solution B (5 mL per strip). The iodoacetamide in this solution will alkylate the free thiol groups of the reduced –SH groups and, more important, will help remove any remaining DTT, which could result in vertical streaking (*see* **Note 11**).

3.4.3   SDS-PAGE

Prepare the electrophoresis unit following the instructions of the manufacturer. The protocol here is for the Ettan Dalt 12 system.

1. Turn the pump valve to circulate and fill the separation unit with 7.5 L 1X SDS running buffer. Turn the pump on and set the temperature at 12 °C.
2. Prepare the gel cassettes for loading the IPG strips by rinsing the top of the gel with water and drain. Remove any residual polyacrylamide from the surface of the glass plates. Store the gel cassettes upside down until loading the IPG strips to prevent drying of the gel surface.
3. Remove one IPG strip from the equilibration solution, using forceps, and rinse it in fresh 2X SDS running buffer. Dry the strip by briefly placing it on its side on absorbent paper.
4. Put the SDS-PAGE gel in an upright position in the cassette rack, with the small glass plate in front.
5. Using two forceps, place the IPG strip with its plastic backing against the large glass plate just above the small glass plate and with the acidic side of the strip pointing to the left.
6. Gently slide the strip between the two glass plates using a small spatula or plastic spacer, pushing against only the plastic background and not making contact with the surface of the IPG strip or the SDS-PAGE gel, until it firmly rests against the SDS-PAGE gel. Make sure the gel surface of the IPG strip does not touch the small glass plate.
7. Pipette 2 mL of agarose solution on top of the IPG strip to seal it in place and allow it to solidify. Repeat steps 3–7 for every IPG strip.
8. Slide the gel cassettes in the separation unit and fill unused slots with blank cassette inserts.
9. Fill the upper buffer chamber with 2.5 L of 2X SDS running buffer.
10. Start the run at 40 mA for 30 min and continue with an overnight run at about 15 mA/gel at 20 °C. Running the second dimension overnight gives the best resolution and facilitates the scanning of the gels on the following day.
11. Stop the electrophoresis run when the bromophenol blue front reaches the lower edge of the gel. Leave the gels in the cooled running buffer in the apparatus until they are scanned.

3.5 Gel Imaging and Data Analysis

1. Remove the gel from the electrophoresis apparatus and clean with water and a lint-free cloth to remove dust particles from the glass plates. Transfer the gel to the gel scanner.
2. To determine the optimal exposure times for each fluorescent dye present in the gel, perform a test scan for a small portion of the first gel of the batch, scanning it at a 100 μm resolution. The maximal pixel value should be close to 65,000 counts for maximal sensitivity but should not exceed this value to avoid saturation. For laser based scanners, the voltage of the photomultiplier tubes has to be optimized for maximal sensitivity.

3. Scan each gel for each fluorescent dye with the exposure times as determined in [2] (see Fig. 15.1). The resolution is set at 100 µm and the correct dye chemistry setting should be checked (minimal labeling). We usually use the exposure settings as given in Table 15.2. After scanning, gels can be stored in fixation solution, if required.

4. Perform image analysis using the desired software. A very capable software package for the analysis of 2-D DIGE gel images is the DeCyder 2D, which is specifically designed with the use of an internal standard in mind. A typical analysis consists of the following two modules.

5. In the differential in-gel analysis (DIA) module, perform intragel analysis. On each gel, protein spots are codetected on the Cy2, Cy3, and Cy5 images and the Cy3 and Cy5 gel images are normalized with respect to the Cy2 image of the internal standard. Absolute protein abundances therefore are replaced by the normalized volume ratios of Cy3/Cy2 and Cy5/Cy2.

6. Perform intergel analysis using the biological variation analysis (BVA) module (Fig. 15.2). After automatic matching of all spots on each gel with the spots on a master gel, manually confirm or correct the matching for a large number of spots. Although taking some time, this step is required to achieve a correct statistical analysis.

7. Plot the relative abundances of all spots against the normalized internal standard and apply a student's t-test to generate a list of protein spots that are differentially regulated between the different experimental conditions.

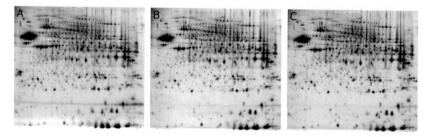

Fig. 15.1 Representative Cy2 (A), Cy3 (B), and Cy5 (C) images obtained from a 2-D DIGE gel, scanned with the Ettan DIGE imager. Two protein samples from a cat primary visual cortex were stained with Cy3 and Cy5, and an internal standard was labeled with Cy2

Table 15.2 Scan settings for the three CyDyes

CyDye	Excitation filter (nm)	Emission filter (nm)	Exposure time (s)
Cy2	480 ± 30	530 ± 40	1.0
Cy3	540 ± 25	595 ± 25	0.5
Cy5	635 ± 30	680 ± 30	2.0

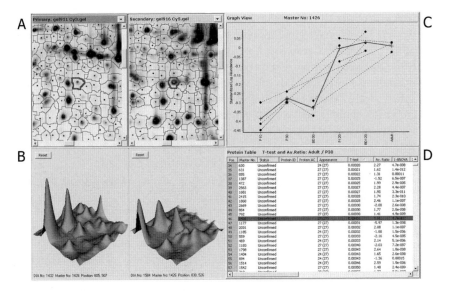

Fig. 15.2 Screen shot from the DeCyder Biological Variation Analysis workspace, showing the Protein Table view. This view is divided in four panes: (A) The image view shows Cy3, Cy5, or Cy2 images of the gels in the experiment, (B) A 3-D view shows a three-dimensional representation of the gel region around the selected spot from the gel images in (A). A graphical view (C) shows a graphical quantitative representation of the selected spot from all gels in the experiment. In this example, spot 1426 was selected in an experiment comparing protein expression levels of mammalian brain at different postnatal ages. Six experimental conditions were compared containing three data points each. The thick black line connects the means of each experimental condition, while the dashed lines each connect two points that were run on a single 2-D DIGE gel, labeled with either Cy3 or Cy5. (D) The table view shows the statistical data from the experiment, with the selected spot indicated by the gray bar

3.6   Protein Identification

The identification of the proteins present in the differentially regulated spots is a procedure that consists of several steps. These steps ultimately depend on the type and brand of mass spectrometer to which one has access. Therefore, the description that follows should be adapted for each specific mass spectrometer. In general, a preparative 2-DE gel is run that is matched to the protein spots visualized by 2-D DIGE. The spots of interest are excised from the gel, either manually or with a spot-picking robot. The proteins in the excised spots are digested with trypsin and the resulting peptides are concentrated and purified. This is followed by analysis of these peptides by mass spectrometry and identification of the protein by protein identification search engines.

1. Run one or more preparative two-dimensional electrophoresis gels as described in Sects. 15.3.4 and 15.3.5, except that 500 µg or more of the sample of interest is loaded onto each gel (*see* **Note 12**). Optionally, label 50 µg of this sample

with one of the CyDyes to facilitate the matching of the preparative gels with the 2-D DIGE images. In this case, the gel has to be imaged prior to detection of the proteins by Coomassie blue staining, as described in Sect. 3.6.

2. Rinse the gels twice for 5 min with 200 mL of water.

3. Stain the gels with 200 mL of Coomassie blue staining solution overnight and subsequently destain by rinsing several times with water, until the blue spots are visible and the background is translucent. The gels can be stored in water for several days at 4 °C prior to spot excision (*see* **Note 13**).

4. Excise the gel spots of interest using a sterile razor blade or a spot-picking robot using a picking list generated in DeCyder (*see* **Note 14**).

5. Perform an enzymatic digestion of the proteins in the gel spots using trypsin.

6. Pool identical spots from different gels.

7. Wash the gel cubes twice for 15 min with milliQ water, followed by two 15 min rinses in 50% (v/v) acetonitrile (ACN) and two 5 min washes with milliQ water.

8. Dry the gel pieces in a SpeedVac vacuum centrifuge and incubate overnight at 37 °C in 40 µL digestion solution.

9. Extract the tryptic peptides from the gel pieces by two washes with 100 µL and 50 µL of a 5:95 ACN:water solution and sonication in a bath sonicator for 30 min.

10. Concentrate the supernatant using SpeedVac vacuum centrifugation.

11. Desalt and concentrate the tryptic peptides using ZipTip C$_{18}$ reversed-phase chromatography. Rinse the column by pipetting five times with 10 µL of 100% ACN, followed by five times 50% ACN and five times with 1% TFA in water.

12. Dilute the peptide mixture in 1% TFA in water and apply to the column by pipetting up and down. Wash the column five times with 10 µL of 1% TFA followed by five times with 10 µL of 1% FA.

13. Elute the peptides from the column with 2 µL of 50% ACN and 1% FA.

14. Analyze the peptide mixture by mass spectrometry, either by peptide mass fingerprinting on MALDI-TOF or by generating peptide sequence tags on a tandem mass spectrometer, possibly in combination with liquid chromatography.

15. Submit the ion mass lists to the protein identification search engines, such as MASCOT, to search the relevant protein databases to obtain an unambiguous protein identification.

4   Notes

1. All water used during sample preparation and two-dimensional electrophoresis should be HPLC grade or of similar quality, having a resistance of 18.2 MΩ.

2. We use 24 cm strips with pH gradients from pH 3–11 nonlinear for starting experiments. In later experiments, we sometimes use strips with narrower pH gradients (pH 4–7 or 6–9) to focus on a specific subset of proteins.

3. A literature search should be performed to use the most efficient protein extraction protocol for each cell or tissue type. For many tissues or cell types, reproducible procedures have been described. Further optimization might be necessary to obtain optimal results. Try to keep the extraction protocol as simple as possible to prevent protein loss or the introduction of nonbiological sources of variation.

4. We utilized a modified Bradford assay, as described [10], but any protein quantification method that is compatible with the contents of the lysis solution (high concentrations of urea, thiourea, CHAPS) can be used.

5. To prevent degradation of the reactive N-hydroxy-succinimidyl group of the dyes, storage of the dyes should always be done in water-free conditions. Therefore, always use fresh water-free DMF for diluting the CyDyes and store the diluted working solution up to 1 week maximum at −20 °C. For prolonged storage of the dye, evaporate all solvents using a SpeedVac evaporator and store at 4 °C or −20 °C on silica gel.

6. 400 pmol CyDye per sample is the amount of dye recommended by GE Healthcare Life Sciences. However, depending on the protein samples or to decrease the cost of the experiment, the amount of dye can be lowered to 200 pmol, with no appreciable effect on the sensitivity of the protein detection.

7. Instead of the destreak solution, proteins can be reduced during IEF by the addition of DTT to the rehydration solution. However, using destreak solution results in less horizontal streaking of the more basic proteins in the gel.

8. IEF also is possible in individual focusing trays. This enables the addition of the sample during IPG strip rehydration, by mixing the sample with the rehydration solution. This also makes it possible to place a low voltage across the rehydrating strip, which ensures a better uptake of high-molecular-weight proteins [11].

9. The total volt hours should be in the range of 40–50 kVh. The number of volt hours should be increased when narrower pH gradients are used or higher amounts of protein are applied. However, to prevent overfocusing and associated horizontal streaking artifacts, the total number of volt hours should never surpass 100 kVh.

10. We generally use 5 mL of equilibration solution per strip, although higher volumes could result in better equilibration and therefore less vertical streaking and better transfer of proteins from first to second dimension. The DTT in this solution reduces the disulfide bonds.

11. Do not exceed the given times of reduction and alkylation, as this could result in the loss of protein. Perform this step as close to running the second dimension as possible.

12. If the differences in expression levels between the spots are not large, it is possible to use the internal standard as the sample on the preparative gel(s), since only one sample then has to be run to excise all the differential spots. If differential spots are absent in several of the gels, it usually is better to separate the samples where these spots show the highest expression levels, to ease the identification of these spots.

13. Alternative protein stains are the fluorescent dyes SyproRuby or DeepPurple. The use of silver staining is discouraged, since this stain generally interferes with the mass spectrometric analysis, resulting in a lower chance of protein identification.
14. If cutting manually, do not cut the gel plugs too large, to avoid diluting the protein with nonprotein contaminants. Perform all steps in a dust-free environment (laminar flow hood) to prevent keratin contamination.

References

1. Liang P, Pardee AB (1992) Differential display of eukaryotic messenger RNA by means of the polymerase chain reaction. Science 257:967–971
2. Diatchenko L, Lau YF, Campbell AP, Chenchik A, Moqadam F, Huang B, Lukyanov S, Lukyanov K, Gurskaya N, Sverdlov ED, Siebert PD (1996) Suppression subtractive hybridization: A method for generating differentially regulated or tissue-specific cDNA probes and libraries. Proc Natl Acad Sci USA 93:6025–6030
3. Schena M, Shalon D, Davis RW, Brown PO (1995) Quantitative monitoring of gene expression patterns with a complementary DNA microarray. Science 270:467–470
4. Anderson L, Seilhamer J (1997) A comparison of selected mRNA and protein abundances in human liver. Electrophoresis 18:533–537
5. Ünlü M, Morgan ME, Minden JS (1997) Difference gel electrophoresis: A single gel method for detecting changes in protein extracts. Electrophoresis 18:2071–2077
6. Tonge R, Shaw J, Middleton B, Rowlinson R, Rayner S, Young J, Pognan F, Hawkin, E, Currie I, Davison M (2001) Validation and development of fluorescence two-dimensional differential gel electrophoresis proteomics technology. Proteomics 1:377–396
7. Alban A, David SO, Bjorkesten L, Andersson C, Sloge E, Lewis S, Currie I.(2003) A novel experimental design for comparative two-dimensional gel analysis: Two-dimensional difference gel electrophoresis incorporating a pooled internal standard. Proteomics 3:36–44
8. Van den Bergh G, Arckens L (2004) Fluorescent two-dimensional difference gel electrophoresis unveils the potential of gel-based proteomics. Curr Opin Biotechnol 15:38–43
9. Shaw J, Rowlinson R, Nickson J, Stone T, Sweet A, Williams K, Tonge R (2003) Evaluation of saturation labelling two-dimensional difference gel electrophoresis fluorescent dyes. Proteomics 3:1181–1195
10. Qu Y, Moons L, Vandesande F (1997) Determination of serotonin, catecholamines and their metabolites by direct injection of supernatants from chicken brain tissue homogenate using liquid chromatography with electrochemical detection. J Chromatogr B Biomed Sci Appl 704:351–358
11. Görg A, Obermaier C, Boguth G, Weiss W (1999) Recent developments in two-dimensional gel electrophoresis with immobilized pH gradients: Wide pH gradients up to pH 12, longer separation distances and simplified procedures. Electrophoresis 20:712–717

Chapter 16
Quantitative Protein Profiling by Mass Spectrometry Using Isotope-Coded Affinity Tags

Arsalan S. Haqqani, John F. Kelly, and Danica B. Stanimirovic

Abstract A key issue in proteomics is to quantify changes in protein levels in complex biological samples under different conditions. Traditional two-dimensional gel (2-DE) electrophoresis-based proteomic approaches are tedious and suffer from several limitations, including difficulties in detecting low abundant and insoluble proteins. Isotope-coded affinity tagging (ICAT®), one of the most employed chemical isotope labeling methods, can address many of the shortcomings of 2-DE. ICAT relies on the sensitivity of mass spectrometry (MS) to quantify relative protein abundance in a mixture of two differentially labeled protein samples. We describe here a detailed protocol for ICAT-based quantification of proteins in two or more biological samples, including sample preparation, ICAT labeling, fractionation and purification, and analysis by MS. For the MS analysis, we describe a "targeted" approach, which includes quantification of the samples using MS followed by selective identification of only the differentially expressed ICAT pairs using tandem MS (MS/MS). This approach gives more biologically relevant information than a data-dependent MS/MS analysis. We also describe the steps in data analysis, statistical analysis, and protein database searching.

Keywords quantitative protein profiling, isotope-coded affinity tag, tandem mass spectrometry

1 Introduction

A major goal of proteomics is to accurately and reproducibly quantify changes in protein levels to understand the basic biological responses as well as to discover biomarkers for the treatment and diagnosis of disease. Traditionally, quantitative proteomics has been accomplished using two-dimensional gel electrophoresis (2-DE)-based protein separation and quantification followed by mass spectrometry (MS)-based or tandem mass spectrometry (MS/MS)-based protein identification. The 2-DE-based quantification method has several shortcomings [1, 2], including low quantitative reproducibility and difficulties in detecting membrane proteins, low abundant proteins, very large and small proteins, and extremely acidic or basic proteins. MS-based quantitative proteomic methods that utilize stable isotope labels can address many of the shortcomings of 2-DE method, including detection of membrane proteins and low abundant proteins [1–4]. Isotope labeling techniques (e.g., ICAT®, SILAC™, ¹⁸O-incorporation) allow relative quantification of proteins in a single mixture containing two samples using isotopically distinguishable, but structurally identical, labels.

Isotope-coded affinity tagging (ICAT) is a widely used MS-based method for quantitative proteomic analysis due to its commercial availability and sensitivity. The ICAT method involves differential isotope labeling of proteins from two biological states followed by mixing the two samples to allow direct comparison of protein levels within a single experiment [2]. The ICAT reagent consists of two isotopic tags that are structurally identical except that one has a linker containing nine ^{12}C atoms (Light) and the other containing nine ^{13}C atoms (Heavy) (Fig. 16.1). Each tag also contains a biotin group and a thiol-reactive group (Fig. 16.1). The latter group labels the proteins at cysteine residues, allowing selective isolation of cysteine-containing peptides and, hence, reducing the sample complexity. Mixing the two labeled samples (Fig. 16.2A) minimizes the run-to-run variability between the samples, making ICAT also more quantitatively reproducible than the traditional 2-DE method. The combined sample usually is trypsin digested, and peptides are either cleaned up or fractionated using cation exchange chromatography (Fig. 16.2A). Subsequently, cysteine-containing peptides are purified using avidin affinity chromatography and the biotin affinity tag is cleaved (Fig. 16.2A). Quantitative information of the isotopically labeled peptides is obtained using nanoscale liquid chromatography combined with MS (nanoLC-MS), while peptide sequence information is obtained using tandem MS (nanoLC-MS/MS) (Fig. 16.2B).

2 Materials

2.1 Protein Samples

1. A known protein, such as bovine serum albumin (BSA), as a standard to test the accuracy of quantification using the protocol (*see* **Note 1**).
2. 100 μg of proteins from two or more biological samples (*see* **Note 2**). The samples may include proteins from cell cultures (e.g., whole cell extracts, subcellular fraction, or secreted fraction), whole tissue or subfractionated tissue extracts, or laser capture microdissection (LCM) samples.

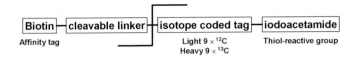

Fig. 16.1 Structure of cleavable isotope-coded affinity tag (ICAT) reagent

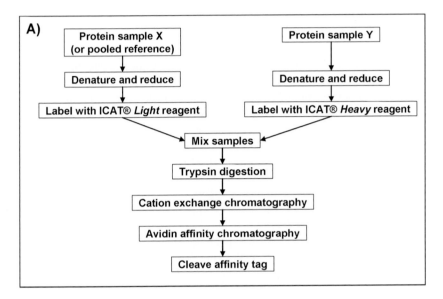

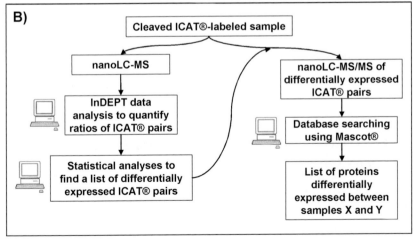

Fig. 16.2 Outline of a protocol for ICAT-based quantification of proteins in two biological samples: (A) Steps involved in ICAT labeling and purification. One of the samples may be a pooled reference if more than two samples are compared. (B) Targeted MS analysis of ICAT samples for quantification and identification of differentially expressed proteins. Steps involving computerized data analysis are indicated with a computer icon

2.2 Reagents for ICAT Sample Labeling, Purification, and Cleaving

1. Denaturing SDS buffer (DS buffer): 50 mM Tris-HCl, pH 8.5, 0.1% (w/v) SDS.
2. Cleavable ICAT Reagent Light and Heavy (Applied Biosystems, catalogue number 4339040).
3. Reducing reagent TCEP-HCl: 50 mM Tris(2-carboxyethyl) phosphine hydrochloride.
4. Trypsin Gold, Mass Spectrometry Grade (Promega, catalogue number V5280).
5. Cartridge holder (Applied Biosystems, catalogue number 4326688).
6. Cation exchange (CE) cartridge: POROS® 50 HS, 50-μm particle size (4.0 mm × 15 mm) (Applied Biosystems, catalogue number 4326695).
7. CE load buffer: 10 mM KH_2PO_4, pH 3.0, 25% (v/v) acetonitrile.
8. CE elute buffer: CE load buffer + 350 mM KCl.
9. CE clean buffer: CE load buffer + 1 M KCl.
10. CE storage buffer: CE Load buffer + 0.1% (w/v) NaN_3.
11. Avidin Affinity (AA) Cartridge (4.0 mm × 15 mm) (Applied Biosystems, catalogue number 4326694).
12. AA load buffer: 20 mM NaH_2PO_4, pH 7.2, 300 mM NaCl.
13. AA wash-1 buffer: 10 mM NaH_2PO_4, pH 7.2, 150 mM NaCl.
14. AA wash-2 buffer: 50 mM NH_4HCO_3, pH 8.3, 20% (v/v) methanol.
15. AA elute buffer: 25% (v/v) acetonitrile, 0.4% (v/v) trifluoroacetic acid.
16. AA storage buffer: AA Load buffer + 0.1% (w/v) NaN_3.
17. Cleaving reagent A (concentrated trifluoroacetic acid) (Applied Biosystems, catalogue number 4338543).
18. Cleaving reagent B (Applied Biosystems, catalogue number 4339052).
19. MS buffer: 5% (v/v) acetonitrile, 1% (v/v) acetic acid.
20. SpeedVac Concentrator SPD111V (Thermo Scientific, Waltham, MA).

2.3 Mass Spectrometry and Online Reverse-Phase Liquid Chromatography

1. A mass spectrometer (MS) equipped with an electrospray ionisation source (ESI) and capable of tandem MS (MS/MS) analysis; for example, a hybrid quadrupole time-of-flight Q-TOF™ Ultima (Waters, Millford, MA).
2. An online nanoflow liquid chromatography (nanoLC) system; for example, CapLC HPLC pump (Waters).
3. A reverse phase nanoLC column; for example, 75 μm × 150 mm PepMap C18 nanocolumn (Dionex/LC-Packings, San Francisco).

2.4 Software for Analyzing Data

1. Software for generating peak lists from the nanoLC-MS and nanoLC-MS/MS data, such as ProteinLynx™ (Waters).
2. Software for finding and quantifying ICAT pairs from nanoLC-MS data; for example, ProICAT Application Software (Applied Biosystems, catalogue number WC026995), ProteinLynx Global SERVER™ 2.2.5 (Waters), or InDEPT software (NRC-IBS).
3. Software for data normalization and statistical analysis, such as Statomics software using R language for statistical computing and graphics (www.dawningrealm.org/stats/statomics), or BioMiner™ software (NRC-IIT/NRC-IBS).
4. Software for identifying peptide sequences from the nanoLC-MS/MS data using database searching; for example, Mascot® version 2.1.0 (Matrix Science Ltd., London, UK), PEAKS (Bioinformatics Solutions Inc., Ontario, Canada), PeptideProphet™ (http://peptideprophet.sourceforge.net).

3 Methods

Although the ICAT reagent originally was designed to quantify protein differences between two samples, differences among *multiple* protein samples are desired under many conditions (e.g., multiple replicate or time-series experiments). For such experiments, we demonstrated the utility of a pooled reference control experimental design [3], which allows quantitative comparisons of ICAT-labelled peptides among multiple samples and provides a standard for reducing intraexperimental variations. We describe the use of a pooled reference control here and recommend using it for quantifying more than two samples, as it allows a more reliable comparison of data among experiments.

The commonly used MS-based method for analyzing ICAT samples is a "shotgun" approach that involves data-dependent nanoLC-MS/MS analysis. Such an approach includes an MS survey scan that recognizes ions above a certain set threshold then switches to an MS/MS mode to identify these ions. Although it has been successfully used to identify a large number of ICAT labeled peptides (5), this approach is limited to identifying abundant peptides, and most of the peptides it does identify are not differentially expressed (Fig. 16.3A). We therefore demonstrated the use of a "targeted" MS approach [3]. This approach involves two steps to the MS analysis (Fig. 16.3B). In the first step, samples are analyzed by nanoLC-MS to quantify ratios of all ICAT pairs and subsequently recognize differentially expressed peptides. In the second step, the differentially expressed peptides are selectively identified using targeted nanoLC-MS/MS analysis with an "include" list. Although it requires two steps, this approach gives a more biologically relevant result, since it focuses on identifying differentially expressed proteins (Fig. 16.3B). As a result, the targeted approach also allows identification of lower abundant peptides (Fig. 16.3B) and achieves higher sequence coverage of proteins and a higher peptide score (for the same peptide) than the shotgun approach. We therefore strongly recommend using the targeted MS approach and describe it here in detail.

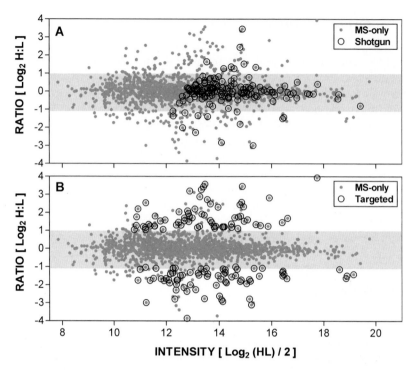

Fig. 16.3 Comparison of the "shotgun" and "targeted" MS/MS approaches for analysis of ICAT samples. All the ICAT pairs present in the sample (as identified by MS-only) are indicated by *filled gray circles* on ratio-intensity plots in panels A and B. The ICAT pairs for which MS/MS spectra were collected using the shotgun (A) or targeted (B) MS/MS approach are indicated by *open circles*. The ICAT sample consisted of a mixture of two differentially tagged biological samples. For "shotgun" analysis, data-dependent MS/MS was done. For "targeted" analysis, the sample was first analyzed by MS-only to quantify the ratios of all the ICAT pairs (H:L ratios). The sample then was reanalyzed by MS/MS with an "include" list of differentially expressed peptides to select only these for identification. A H:L ratio >2-fold or <0.5-fold is used here as differentially expressed (nongray area). Note that, using the shotgun method, MS/MS analysis is done only on abundant peptides, most of which are not differentially expressed, whereas the targeted approach allows identification of only differentially expressed as well as low abundant peptides. Plotted are $\log_2(H/L)$ vs. $\log_2(HL)/2$, where H and L are peak intensities of the heavy and light ICAT-labeled pairs, respectively

3.1 Protein Sample Preparation

1. Prepare each protein sample in DS buffer, at a concentration of ≥ 1.25 mg/mL (or $\geq 100\,\mu$g/$80\,\mu$L) (*see* **Notes 2 and 3**).
2. If necessary, concentrate or clean protein samples by acetone precipitation: Add 10 volumes of cold acetone and precipitate at $-20\,^\circ$C for 1 h (*see* **Note 4**). Pellet the protein by centrifugation at 5,000 g for 5 min and redissolve the pellet in DS buffer.

3. To extract protein from LCM samples, place the ethylene vinyl acetate membrane containing the captured tissue [3] in 80 μL of DS buffer and incubate with shaking at 65 °C for 1 h. Discard the membrane afterward.
4. For comparing more than two samples, pool all the control or untreated protein samples to generate a single reference control. Retain a separate aliquot of each treated or diseased protein sample and individually compare it with the pooled reference control in the subsequent steps.

3.2   Denaturing and Reducing the Samples

1. Add 80 μL of control or untreated sample (or pooled reference), containing 100 μg of protein in DS buffer to a new tube labeled A. If testing the quantitative accuracy of the protocol (see **Note 1**), this will be a known protein, such as BSA.
2. Add 80 μL of diseased or treated sample, containing 100 μg of protein in DS buffer to a new tube labeled B. If testing the quantitative accuracy of the protocol, this will be the same known protein (e.g., BSA) as in tube A.
3. Add 2 μL of reducing agent TCEP-HCl to each of tubes A and B.
4. Vortex to mix, and centrifuge at 10,000 g for 10 s to bring the solutions down to the bottom of the tubes.
5. Heat both tubes at 95 °C for 10 min to denature proteins.
6. Vortex to mix, and centrifuge at 10,000 g for 10 s.
7. Cool the tubes at room temperature for 2 min.

3.3   ICAT Labeling and Tyrpsin Digestion of Samples

1. Bring vials of cleavable ICAT Light and Heavy reagents to room temperature.
2. Centrifuge the vials at 10,000 g for 3 min to bring down all the reagents.
3. Add 20 μL of acetonitrile to each vial.
4. Vortex to mix, and centrifuge at 10,000 g for 10 s.
5. To the Light reagent vial, add all the contents of tube A (from Sect. 3.2).
6. To the Heavy reagent vial, add all the contents of tube B (from Section 3.2).
7. Vortex each vial to mix, and centrifuge at 10,000 g for 10 s.
8. Incubate each vial at 37 °C for 2 h.
9. Vortex each vial to mix, and centrifuge at 10,000 g for 10 s.
10. If testing the quantitative accuracy using a known protein, label three clean tubes: 1:1, 2:1, and 1:2. In the 1:1 tube, add 25 μL from both the Light and Heavy reagent vials. In the 2:1 tube, add 25 μL from the Light and 50 μL from the Heavy reagent vial. In the 1:2 tube, add 50 μL from the Light and 25 μL from the Heavy reagent vial. Otherwise, transfer all the contents from the Light

vial to the Heavy vial (this is your 1:1 sample). Keep the empty Light vial because it is needed later.

11. Resuspend a vial of 30 μg of trypsin in 200 μL of Milli-Q® water. Dissolve in 600 μL if testing for quantitative accuracy.

12. Tap the trypsin vial to mix, and centrifuge at 10,000 g for 10 s.

13. Transfer 200 μL of the trypsin to the empty Light vial (unless testing quantitative accuracy) then to the 1:1 sample (also to the 2:1 and 1:2 samples, if testing quantitative accuracy).

14. Tap the sample(s) to mix, and centrifuge at 10,000 g for 10 s.

15. Incubate at 37 °C for 12–16 h.

16. Vortex each vial to mix, and centrifuge at 10,000 g for 10 s.

17. Store samples at −20 °C for up to 2 weeks.

3.4 Cleaning up and Fractionating Digested Samples Using the CE Cartridge

1. Transfer the trypsin digested sample to a new tube, labeled PreCE, and add 2 mL of CE load buffer.

2. Check the pH. If it is not < 3.3, adjust by adding more CE load buffer.

3. Clean the CE cartridge by injecting 1 mL of CE clean buffer. Discard the flow-through.

4. Condition the CE cartridge by injecting 2 mL of CE load buffer. Discard the flow-through.

5. Slowly inject (~1 drop/s) all the contents of the PreCE tube. The flow-through may be collected, although it is not required in the subsequent steps.

6. Wash the cartridge by injecting 1 mL of CE load buffer. Discard the flow-through.

7. If fractionation is not needed, elute the peptides by slowly injecting (~1 drop/s) 0.5 mL of CE elute buffer and collect the eluted sample in a new tube.

8. If fractionation is needed, prepare a set of elute buffers (0.5 mL each) by mixing CE load buffers and salt-containing CE elute buffers at different ratios for a stepwise elution (*see* **Note 5**). For example, to collect six fractions, prepare buffers containing 4% (v/v), 8% (v/v), 12% (v/v), 16% (v/v), 20% (v/v), and 100% (v/v) CE elute buffer in CE load buffer. Elute peptides in a stepwise manner (from low to high CE elute buffer concentration) by slowly injecting (~1 drop/s) 0.5 mL of each elute buffer and collect the eluted fraction in a separate new tube.

9. Add 0.5 mL of AA load buffer to each eluted sample or fraction. Check the pH. If it is not 7, adjust by adding more AA load buffer. Store samples at −20 °C for up to a week.

10. If there are additional trypsin-digested samples, repeat steps 1–8.

11. Wash the cartridge by injecting 1 mL of CE clean buffer and 2 mL of CE storage buffer. Store the cartridge at 2–8 °C.

3.5 Affinity Purification of ICAT-Labelled Peptides and Cleavage of Affinity Tag

1. Inject 2 mL of AA elute buffer into the AA cartridge. Discard the flow-through.
2. Inject 2 mL of AA load buffer into the AA cartridge. Discard the flow-through.
3. Slowly inject (~1 drop/5 s) all the contents of the CE eluted sample (or fraction) into the AA cartridge. Collect the flow-through in a tube labeled Flow.
4. Inject 0.5 mL of AA load buffer. Collect in the Flow tube.
5. Inject 1 mL of AA wash-1 buffer to reduce the salt concentration. Discard the flow-through.
6. Inject 1 mL of AA wash-2 buffer to remove nonspecifically bound peptides. The first 0.5 mL of the flow-through may be collected and the remaining discarded.
7. Inject 1 mL of Milli-Q water. Discard the flow-through.
8. Slowly inject (~1 drop/5 s) 50 µL of AA elute buffer. Discard the flow-through.
9. Elute the peptides by slowly injecting (~1 drop/5 s) 750 µL of AA elute buffer and collect the eluted sample (or fraction) in a tube labeled AA elute.
10. If there are additional CE eluted samples (or fractions), repeat steps 1–9.
11. Wash the AA cartridge by injecting 2 mL of AA elute buffer and 2 mL of AA storage buffer. Store the cartridge at 2–8 °C.
12. Evaporate each AA eluted sample to dryness (usually takes 4–5 h) in a SpeedVac.
13. In a new tube, prepare the final cleaving reagent by combining cleaving reagent A and cleaving reagent B in a 95:5 ratio. Vortex to mix, and centrifuge at 10,000 g for 10 s.
14. Transfer 90 µL of freshly prepared cleaving reagent to each evaporated AA eluted sample.
15. Vortex to mix, and centrifuge at 10,000 g for 10 s.
16. Incubate at 37 °C for 2 h.
17. Centrifuge at 10,000 g for 10 s.
18. Evaporate to dryness (usually takes 0.5–1 h) in a SpeedVac.
19. Dissolve each sample in 300 µL of MS buffer.

3.6 NanoLC-MS and Data Analyses

1. Inject 5–10 µL of each sample in MS buffer onto the nanoLC online to Q-TOF Ultima MS. Separate peptides by gradient elution (5–75% (v/v) acetonitrile, 0.2% (v/v) formic acid in 90 min, 350 nL/min) and analyze in the MS-only mode.
2. Generate peak lists as a list of text files (usually ~4500–5000 files per 90 min run) for each nanoLC-MS run using ProteinLynx. Each file contains mass/charge (m/z) and intensity information for all the peaks in a scan.

3. Find the charge states, elution range (or single ion chromatogram), and an integrated intensity value for each ion using InDEPT software.
4. Find ICAT pairs as all the coeluting ion pairs separated by 9 Da/z (one cysteine) or 18 Da/z (two cysteine) and quantify their Heavy:Light (H:L) ratios, where z is the charge-state on the ICAT-labeled peptides. An example of an ICAT pair from a nanoLC-MS spectrum is shown in Fig. 16.4A.
5. Plot the H:L ratios on Ratio-Intensity plots to visualize the distribution of H:L ratios for all the ICAT pairs. Examples of these plots are depicted in Figs. 16.3 and 16.5. It may be necessary to normalize the H:L ratios of all the ICAT pairs (using global median normalization) to correct for possible unequal mixing or labeling of the two samples. To do this, calculate the median (M) of the H:L ratios and the expected ratio (E). The expected ratio is the ratio at which the Heavy and Light sample volumes were mixed in Sect. 3.3 (e.g., $E = 1$ for 1:1, $E = 2$ for 2:1, $E = 0.5$ for 1:2). Normalization is recommended if M is >5% different from E. Normalize each H:L ratio as follows: Normalized ratio = (Original ratio) × E/M. An example of the data before and after normalization is shown in Fig. 16.5.
6. Find a list of differentially expressed ICAT pairs. If only one replicate is being analyzed, a simple cutoff, such as 1.5-fold or 2-fold, may be used (*see* **Note 6**). However, we strongly recommend doing multiple replicate experiments and performing statistical tests such as t-tests or SAM analysis as used in microarray data analysis (*see* **Note 7**). A combination of cutoff fold value (e.g., 1.5-fold) and p-value from a statistical test (e.g., p<0.05) is a more reliable way of finding differentially expressed ICAT pairs.
7. For each sample (or fraction), generate an "include" list, containing the list of differentially expressed ICAT pairs in a format accepted by the MS instrument. Usually, the list includes m/z values, charge state, and retention times of the peptide to be analyzed.

3.7   NanoLC-MS/MS Analysis of Differentially Expressed Peptides and Database Searching

1. Inject another 5–10 µL of each sample in MS buffer onto the nanoLC online to Q-TOF Ultima MS.
2. Separate peptides by gradient elution (as in Sect. 3.6.2) and analyze in the MS/MS mode using the "include" list corresponding to that sample. Set to "only analyze the ions in the include list."
3. Generate a peak list as a single text file (usually a .pkl file from Waters instruments) for each nanoLC-MS/MS run using ProteinLynx. Each file contains m/z and charge values of all the precursor ions and their corresponding m/z and intensity values of fragment ions.
4. Submit each file to the Mascot search engine or any other probability-based search engine. The following parameters may be used: (1) specify trypsin enzymatic cleavage with two possible missed cleavages, (2) allow variable modifications

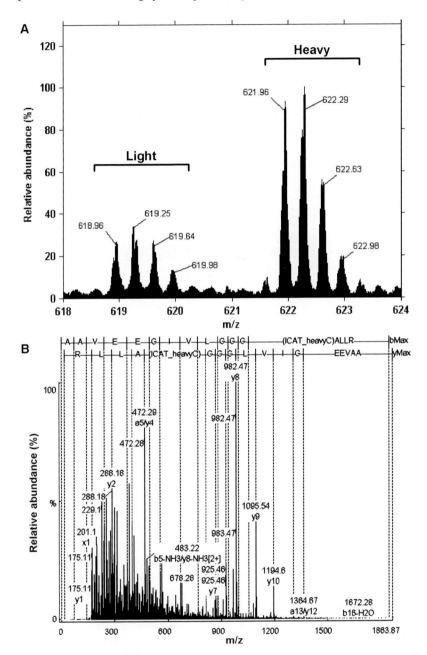

Fig. 16.4 Quantification and identification of an ICAT-labeled peptide: (A) MS spectrum showing a differentially expressed ICAT pair. Two triply charged peptides, 618.96 and 621.96, are potential ICAT pairs since they differ in molecular mass by 3 Da (or 9/charge). The pairs have a H:L ratio of 2.52. (B) MS/MS spectrum of the heavy ICAT pair (621.96) from panel A. Targeted MS/MS analysis confirmed that the peptide is labeled with heavy ICAT reagent and identified it as AAVEEGIVLGGGC*ALLR, belonging to heat shock protein 60

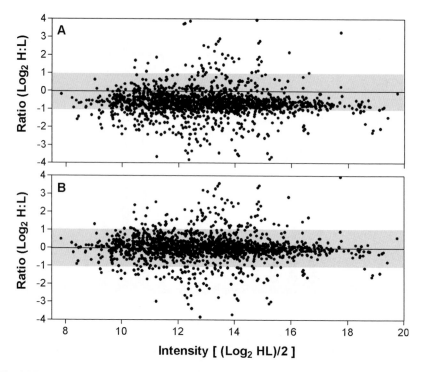

Fig. 16.5 Normalization of H:L ratios of ICAT pairs. Shown are ratio-intensity plots before (A) and after (B) global median normalization. Note that, before normalization, the ratios of ICAT pairs are "off-centered" and consequently significantly more down-regulation (>21%) than up-regulation (3%) of peptides is observed. After normalization, the data is more centered and equivalent numbers of up- (7%) and down-regulated (7%) peptides are observed. A H:L ratio >2-fold or <0.5-fold is used here as indicating differential expression (nongray area). Plotted are $\log_2(H/L)$ vs. $\log_2(HL)/2$, where H and L are peak intensities of Heavy and Light ICAT-labeled pairs, respectively

for cleavable ICAT Light (+227.13 Da) and Heavy (+236.16 Da) at the cysteine residues (recommended), (3) set parent ion tolerance ≤0.5 Da, (4) set fragment ion tolerance ≤0.2 Da. It is recommended that searches be done with and without restriction to cysteine-containing peptides. An ion identified as an ICAT peptide in a restricted search but identified as another peptide with a higher score in an unrestricted search should be removed.

5. For large-scale experiments, we strongly recommend using additional statistical analyses that indicate or establish a measure of identification certainty (e.g., PeptideProphet software) or allow a determination of the false positive rate (FPR). To calculate the FPR, do the searches using a composite database consisting of a forward and reversed (or randomized) protein database (*see* **Note 8**). The FPR can be estimated using the formula: $FPR = (2 \times N_{rev})/(N_{rev} + N_{fwd})$, where N_{rev} is the number of peptides identified (after filtering) from the reverse database, and N_{fwd} is the number of peptides identified (after filtering) from the forward database. Usually, a FPR of <2.5% is acceptable (*see* **Note 9**).

6. Follow the guidelines described on the websites of several proteomic journals (including *Molecular and Cellular Proteomics*, at www.mcponline.org, and the *Journal of Proteome Research*, at pubs.acs.org/journals/jprobs) for the analysis and documentation of peptide and protein identifications (established by LC-MS/MS) when reporting or publishing results.

3.8 Validation of an Identified List of Proteins and Their Significance

To have confidence in a large list of differentially expressed proteins identified by ICAT, validation of a few of the results using alternative biochemical and molecular methods is necessary. For confirmation of protein level changes, ELISA, Western blot, or *in-situ* immunochemistry may be used. To obtain physiologically meaningful information from a *large* list of the differentially expressed proteins identified, many high-throughput analyses (such as genomics and proteomics) rely on various bioinformatics tools to identify patterns in the results by categorizing or clustering the proteins using various methods. The proteins may be categorized by biological processes or molecular functions using the Panther classification system (www.pantherdb.org) or gene ontology database (www.geneontology.org). For experiments involving time-course series or clinical samples, various clustering methods may be used [3, 6].

4   Notes

1. Before running a complex sample for the first time, we recommend measuring the quantitative accuracy of the protocol with a well-characterized known protein (e.g., BSA) or a mixture of several known proteins. This protein(s) must contain multiple cysteine residues and be free of contaminants (*see* **Note 3**). Quantitative accuracy is the degree to which an observed H:L ratio matches the expected H:L ratio. Usually, an acceptable value for quantitative accuracy is <25% of expected values. We have observed that >95% of ICAT pairs show H:L ratios within 20% of expected values. If the H:L ratios consistently deviate too much from the expected values, normalization of the ratios may be required.

2. If the amount of sample is not limited (e.g., samples from cell cultures or tissue extracts), 100–150 µg of protein from each sample should be used. The higher the amount of protein used, the more low-abundant proteins are detected, although very high-abundant proteins may be less quantitative (due to signal saturation). If the amount of sample is limited (e.g. LCM samples), as little as 1 µg of protein may be used [3]. Samples with <1 µg of protein may not give reliable results. It is recommended that equivalent amounts of the samples being compared (e.g., control vs. treated or diseased) are used to minimize the need to normalize the data.

3. It is essential that each protein sample is free of potential contaminants, such as reducing agents and high levels of salt, acid, detergents, and denaturants, as these may interfere with the subsequent steps. Reducing agents, such as dithiothreitol or beta-mercaptoethanol, can react with the iodoacetamide group of ICAT reagents and compete with protein labeling. High amounts of detergents or denaturants, such as SDS and urea, can inactivate trypsin activity. High salt, acid, or detergent concentrations can prevent peptides binding to the CE cartridge. If sample quantities are not limited, these contaminants may be removed by dialysis against the DS buffer or by acetone precipitation.

4. Precipitate overnight if the quantity of protein is low. Incubate at 4 °C instead of −20 °C if the protein is present in urea.

5. If the sample to be analyzed is highly complex in nature (e.g., whole cell or tissue extract), peptide detection and quantification may not always be accurate due to an overabundance of overlapping and coeluting peaks. Fractionation can be used to reduce the complexity of the original sample and usually increase the number of unique peptides detected. The number of fractions depends primarily on the complexity of the investigation. Here, we describe the protocol for CE chromatography because it is widely used for peptide fractionation and also allows for the removal of SDS and unreacted ICAT reagents from the samples. The stationary phase surface of a CE column displays a negatively charged functional group (sulfopropyl for the column described here), which binds with and retains peptide ions of positive charge. The peptides can be eluted with increasing salt concentrations (KCl) in either a continuous or stepwise manner.

6. A simple fold cutoff of 1.5 implies that any peptide with a H:L ratio above 1.5 or below 1/1.5 (or 0.67) is assumed to be differentially expressed. However, the cutoff should be chosen appropriately, because its value depends on the distribution of the H:L ratios and reproducibility of the system. For example, when a control biological sample is labeled with Heavy and Light ICAT reagents separately and mixed at 1:1 ratio, <2% of the ICAT pairs show H:L ratios >1.5 or <0.67 [3].

7. ICAT pairs in multiple replicates must first be matched (e.g., using InDEPT software). The H:L ratios of each ICAT pair then are placed in a $m \times n$ matrix, where m is the number of ICAT pairs and n is the number of replicates. Statistical analysis may be carried out on the matrix using the SAM test [8] (and Biominer or Statomics software) to generate a p-value and false discovery rate for each ICAT pair. Typically, ICAT pairs satisfying all the following criteria may be chosen as differentially expressed: 1.5-fold cutoff (*see* **Note 6**), p-value <0.05, and false discovery rate <10%.

8. To create a reverse database, each protein sequence in the database is precisely reversed in order of the amino acid sequence so that the C-terminus becomes the N-terminus. A randomized database can be created by randomly shuffling the sequence of each protein. Each sequence in a reversed or randomized database is given a unique identifier (to differentiate it from the forward sequences) and added to the end of the forward database. The final composite database containing both forward and reversed, or randomized, sequences is used for the searches.

Table 16.1 Major differences between 2-DE and ICAT-based proteomic methods

	2-DE	ICAT
Quantitative reproducibility[a]	28% CV	12% CV
Detection limit	1–4 ng (SYPRO Ruby)	<0.5 ng
Insoluble/membrane proteins	Not easily detectable	Detectable
Protein pI range	pI 4–11	pI 4–12
	(pI 7–11 is problematic)	
Amount of protein sample required	>0.2 mg pH 4–7	<0.1 mg
	>0.2 mg pH 6–11	(ideal for LCM)
Keratin contamination	High	Very low
Cysteine-free proteins[b]	Detectable	Not detectable
Posttranslational modifications detection	Possible	Not easy

[a] The coefficient of variance (CV) values are from [4].The lower the CV, the more reproducible the method is.
[b] Cysteine-free proteins amount to approximately 3% of the proteins in mammals and 14% in *E. coli*.

9. If the FPR is large, further filtering of the data should be done to reduce false positives (e.g., by reducing parent or fragment ion tolerances or by increasing ion scores cutoff) and maximizing the number of peptides.
10. Our recent detailed comparison of 2-DE and ICAT in the same cellular model [4] demonstrated that ICAT and 2-DE are complementary techniques (summarized in Table 16.1). ICAT showed superior quantitative reproducibility over 2-DE and was more suitable for detecting small, large, basic, hydrophobic, and secreted proteins [4]. On the other hand, 2-DE was more reliable for the identification of posttranslational modifications. The sensitivity of the ICAT method enables a more comprehensive interrogation of the proteome including detection of low-abundant proteins such as signaling molecules, cytokines, and transcription factors [2]. We have successfully used the ICAT technology to quantify and identify ischemia-induced temporal changes in protein expression in minute samples of brain vessels dissected from brain sections acquired using LCM [3]. Due to the limited amount of protein (~1 μg) present in LCM-extracted samples, gel-free approaches are not suitable for this application. We recently used the ICAT method to identify pattern changes in serum protein biomarkers in patients suffering from severe head injury [6]. Several low-abundant brain-specific proteins (including neuron-specific enolase, amyloid beta A4 protein precursor, α-spectrin, and cleaved microtubule-associated protein tau), which may be of diagnostic or prognostic value, were identified in the serum of these patients [6]. Since ICAT selectively labels cysteine residues, it cannot analyze cysteine-free proteins, which correspond to 3% of the proteins in mammals, and cannot easily analyze protein posttranslational modifications or protein isoforms (Table 16.1). In conclusion, the comparative ICAT protein profiling is amenable to identifying and quantifying changes in rare and low-abundant proteins in a variety of biological samples, including cell cultures, small tissue samples, and patient body fluids.

References

1. Aebersold R, Mann M (2003) Mass spectrometry-based proteomics. Nature 422:198–207
2. Gygi SP, Rist B, Gerber SA, Turecek F, Gelb,MH, Aebersold R (1999) Quantitative analysis of complex protein mixtures using isotope-coded affinity tags. Nat Biotechnol 17:994–999
3. Haqqani AS, Nesic M, Preston E, Baumann E, Kelly J, Stanimirovic D (2005) Characterization of vascular protein expression patterns in cerebral ischemia/reperfusion using laser capture microdissection and ICAT-nanoLC-MS/MS. FASEB J 19:1809–1821
4. Haqqani AS, Kelly J, Baumann E, Haseloff RF, Blasig IE, Stanimirovic DB (2007) Protein markers of ischemic insult in brain endothelial cells identified using 2D gel electrophoresis and ICAT-based quantitative proteomics. J Proteome Res 6:226–239
5. Han DK, Eng J, Zhou H, Aebersold R (2001) Quantitative profiling of differentiation-induced microsomal proteins using isotope-coded affinity tags and mass spectrometry. Nat Biotechnol 19:946–951
6. Haqqani AS, Hutchison JS, Ward R, Stanimirovic DB (2007) Protein biomarkers in serum of pediatric patients with severe traumatic brain injury identified by ICAT-LC-MS/MS. J Neurotrauma 24:54–74
7. Ingebrigtsen T, Romner B (2003) Biochemical serum markers for brain damage: A short review with emphasis on clinical utility in mild head injury. Restor Neurol Neurosci 21:171–176
8. Tusher VG, Tibshirani R, Chu G (2001) Significance analysis of microarrays applied to the ionising radiation response. Proc Natl Acad Sci USA 98:5116–5121

Chapter 17
Quantitative Protein Profiling by Mass Spectrometry Using Label-Free Proteomics

Arsalan S. Haqqani, John F. Kelly, and Danica B. Stanimirovic

Abstract "Gel-free," or mass spectrometry (MS)-based, proteomics techniques are emerging as the methods of choice for quantitatively comparing proteins levels among biological proteomes, since they are more sensitive and reproducible than two dimensional gel (2-DE)-based methods. Currently, the MS-based methods utilize mainly stable isotope labels (e.g., ICAT®, iTRAQ™) that enable easy identification of differentially expressed proteins in two or more samples. "Label-free" MS-based methods would alleviate several limitations of the labeling methods, provided that relative quantitative profiling of proteins among multiple MS runs is achievable. However, comparisons of multiple MS runs of highly complex biological samples are very challenging and time consuming. To alleviate this problem, several laboratories and MS vendors have developed software for computer-assisted comparisons of multiple label-free MS runs to allow profiling of differentially expressed proteins. In this chapter, we describe the use of custom-developed MatchRx software in quantitative comparison of multiple label-free MS runs. We also describe details of sample preparation, fractionation, statistical analysis, and protein database searching for label-free comparative quantitative proteomics, as well as the application of a "targeted" MS approach, which includes quantification of the samples using MS followed by selective identification of only the differentially expressed peptides using tandem MS (MS/MS).

Keywords label-free quantitative protein profiling, tandem mass spectrometry

1 Introduction

The quantitative proteomic profiling of complex biological samples for the purposes of target and biomarker discovery is a key challenge in proteomics. The coupling of nanoscale liquid chromatography (nanoLC) to mass spectrometry (MS) enables the online separation of complex mixtures of digested proteins (peptides) prior to MS analysis, thus bypassing the need for 2-DE gels

From: Methods in Molecular Biology, Vol. 439: *Genomics Protocols: Second Edition*
Edited by Mike Starkey and Ramnath Elaswarapu © Humana Press Inc., Totowa, NJ

for protein separation. To add a quantitative dimension to nanoLC-MS experiments, isotope-labeling methods have been developed that introduce stable isotope tags to proteins via metabolic (e.g., SILAC™), enzymatic labeling (e.g., using ^{18}O water for trypsin digestion), or via chemical reactions using isotope-coded affinity tags (ICAT® [1–4]), or similar reagents. These methods achieve relative quantification of proteins in a single mixture containing two samples using isotopically distinguishable but structurally identical labels. Such "gel-free" proteomic methods have emerged as sensitive and reproducible approaches for comparing protein levels among complex biological proteomes [5].

Although MS-based methods are more sensitive and reproducible than gel-based methods, the use of isotopic labeling introduces several limitations. Labeling with stable isotopes often is very expensive, and some labeling procedures involve complex sample manipulations and introduce artifacts. The label may interfere with the MS/MS analysis, making it difficult to sequence the peptide. Labeling methods also make the nanoLC-MS spectra more complex due to the presence of additional isotopic peaks, which often overlap and complicate peak detection and quantification. Methods such as ICAT circumvent this complexity by selectively targeting peptides containing a specific amino acid (e.g., cysteine). Proteome coverage is reduced as a consequence and information on posttranslational modifications and protein isoform specificity is lost. Metabolic labeling, on the other hand, is applicable only to metabolically active samples (usually cell cultures) and not suitable for tissue extracts, biopsies and body fluids, and other *ex-vivo* biological samples. Therefore, even though the quantification step has been made easier, methods using isotopic labels have several pitfalls.

"Label-free" MS-based quantitative methods would be an attractive alternative to isotopic labeling if they could be used to quantitatively compare protein expression in two or more samples. However, the comparison of multiple LC-MS runs of highly complex biological samples, which may contain thousands of proteins, is very challenging and virtually impossible to do manually. Several laboratories have developed software to assist in analysis of label-free samples, including SpecArray [6], 2DICAL [7], and MatchRx (our laboratory); DeCyder MS software is commercially available from GE Healthcare. Although these software tools use distinct algorithms for data analyses, the underlying principles involved in the label-free sample analyses are very similar and include first finding or matching then quantifying peptides in different nanoLC-MS runs. In this chapter, we describe the main steps involved in the sample analyses achieved using the MatchRx software.

The steps involved in the label-free nanoLC-MS protocol are outlined in Fig. 17.1. Proteins from two or more samples to be compared are isolated and trypsin digested into peptides. A chromatography step usually is included to further reduce the complexity of the sample. We routinely use cation exchange chromatography, since it also allows removal of SDS from the samples. Each

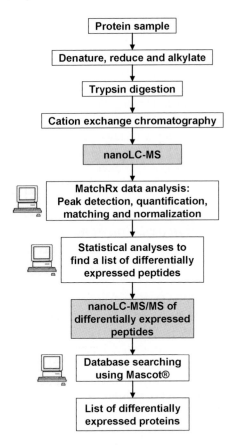

Fig. 17.1 Protocol outline for label-free MS-based proteomics. Steps involving computerized data analysis are indicated with a computer icon. Steps involving mass spectrometry are highlighted in gray

sample is then separately analyzed by nanoLC-MS. The label-free software extracts the quantitative intensity values of peptides from each nanoLC-MS run. The intensities of the same peptide observed in two or more separate runs are compared to determine their relative abundance. To address the concerns of lack of linearity and reproducibility among nanoLC-MS runs, quantification is performed in conjunction with data normalization procedures and retention time correction. Differentially expressed peptides are identified using statistical analysis, and the sequence information of each peptide is obtained using tandem MS (nanoLC-MS/MS) and database searching.

2 Materials

2.1 Protein Samples

Samples may include proteins from cell cultures (e.g., whole cell extracts, subcellular fractions, or secreted fractions), whole tissue or subfractionated tissue extracts, minute samples procured by biopsy or laser-capture microdissection (LCM), and body fluids (serum, CSF, etc.). 100 μg of proteins from two or more biological samples is required for comparison (*see* **Note 1**).

2.2 Sample Preparation and Fractionation

1. Denaturing SDS buffer (DS buffer): 50 mM Tris-HCl, pH 8.5, 0.1% (w/v) SDS.
2. 0.4 M dithiothrietol (DTT). Freshly prepare ~250 mM in Milli-Q water.
3. 1 M iodoacetamide (IAA). Freshly prepare ~500 mM in Milli-Q water and protect from light.
4. Trypsin Gold, mass spectrometry grade (Promega, catalogue number V5280).
5. Cartridge holder (Applied Biosystems, catalogue number 4326688).
6. Cation exchange (CE) cartridge: POROS® 50 HS, 50-μm particle size (4.0 mm × 15 mm) (Applied Biosystems, catalogue number 4326695).
7. CE load buffer: 10 mM KH_2PO_4, pH 3.0, 25% (v/v) acetonitrile.
8. CE elute buffer: CE load buffer + 350 mM KCl.
9. CE clean buffer: CE load buffer + 1 M KCl.
10. CE storage buffer: CE load buffer + 0.1% (w/v) NaN_3.
11. MS buffer: 5% (v/v) acetonitrile, 1% (v/v) acetic acid.
12. SpeedVac Concentrator SPD111V (Thermo Scientific, Waltham, MA).

2.3 Mass Spectrometry and Online Reverse Phase Liquid Chromatography

1. A mass spectrometer equipped with an electrospray ionization source (ESI) and capable of tandem MS (MS/MS) analysis, such as a hybrid quadrupole time-of-flight Q-TOF™ Ultima (Waters, Millford, MA).
2. An online nanoflow liquid chromatography (nanoLC) system, such as a CapLC HPLC pump (Waters).
3. A reverse phase nanoLC column, such as a 75 μm × 150 mm PepMap C18 nano-column (Dionex/LC-Packings, San Francisco).

2.4 Software for Analyzing Data

1. Software for generating peak lists from the nanoLC-MS and nanoLC-MS/MS data, such as ProteinLynx (Waters).
2. Software for quantifying peak abundance from nanoLC-MS data; for example, ProteinLynx Global SERVER™ 2.2.5 (Waters) or MatchRx software (NRC-IBS, Canada).
3. Software for data normalization and statistical analysis, such as Statomics software using R language for statistical computing and graphics (www.dawning-realm.org/stats/statomics) or BioMiner™ software (NRC-IIT/NRC-IBS, Canada).
4. Software for identifying peptide sequences from the nanoLC-MS/MS data using database searching; for example, Mascot® version 2.1.0 (Matrix Science Ltd., London), PEAKS (Bioinformatics Solutions Inc., Ontario, Canada), PeptideProphet™ (http://peptideprophet.sourceforge.net).

3 Methods

3.1 Protein Sample Preparation

1. Prepare each protein sample in DS buffer. A concentration of ≥2 mg/mL (or ≥100 μg/50 μL) is recommended (*see* **Notes 1 and 2**).
2. If necessary, concentrate or clean protein samples by acetone precipitation: add 10 volumes of cold acetone and precipitate at −20 °C for 1 h (*see* **Note 3**). Pellet the protein by centrifugation at 5,000 g for 5 min and redissolve the pellet in DS buffer.
3. To extract protein from LCM samples, place the ethylene vinyl acetate membrane containing the captured tissue [4] in 80 μL of DS buffer and incubate with shaking at 65 °C for 1 h to extract proteins. Discard the membrane afterward.

3.2 Reducing, Alkylating, and Trypsin Digesting the Samples

For trypsin digestion to reach completion, proteins require denaturation (in SDS) and disulfide bond reduction (using DTT). To prevent aggregation after trypsin digestion, alkylation of the reduced thiols (using IAA) is recommended although not required (*see* **Note 4**).

1. Add 50 μL of each sample (containing 100 μg of protein) to a fresh tube.
2. Add freshly prepared DTT to a final concentration of 4 mM.
3. Vortex to mix, and centrifuge at 10,000 g for 10 s to collect the solution in the bottom of each tube.

4. Incubate at 95 °C for 10 min to reduce the disulfide bonds of cysteine residues in proteins.
5. Vortex to mix, and centrifuge at 10,000 g for 10 s to collect the solution in the bottom of each tube.
6. Cool at room temperature for 2 min.

Steps 7–9 for alkylating the cysteine residues in proteins are recommended but not required (*see* **Note 4**).

7. Add freshly prepared IAA to each sample to a final concentration of 10 mM.
8. Incubate at room temperature for 20 min in the dark.
9. Vortex to mix, and centrifuge at 10,000 g for 10 s to collect the solution in the bottom of each tube.
10. Add 50 μL of 100 ng/μL trypsin (in Milli-Q water) to each sample.
11. Tap each sample tube to mix, and centrifuge at 10,000 g for 10 s to collect the solution in the bottom of each tube.
12. Incubate at 37 °C for 12–16 h.
13. Vortex to mix, and centrifuge at 10,000 g for 10 s to collect the solution in the bottom of each tube.
14. Store samples at −20 °C for up to 4 weeks.

3.3 Fractionating and Cleaning up Digested Samples Using a CE Cartridge

1. Transfer the trypsin digested sample to a new tube labeled PreCE, and add 2 mL of CE load buffer.
2. Check the pH. If it is not < 3.3, adjust by adding more CE load buffer.
3. Clean the CE cartridge by injecting 1 mL of CE clean buffer. Discard the flow-through.
4. Condition the CE cartridge by injecting 2 mL of CE load buffer. Discard the flow-through.
5. Slowly inject (~1 drop/s) all the contents of the PreCE tube. The flow-through may be collected, although it is not required in the subsequent steps.
6. Wash the cartridge by injecting 1 mL of CE load buffer. Discard the flow-through.
7. If fractionation is not needed, elute the peptides by slowly injecting (~1 drop/s) 0.5 mL of CE elute buffer and collect the eluted sample in a new tube.
8. If fractionation is needed, prepare a set of elute buffers (0.5 mL each) by mixing CE load buffers and salt-containing CE elute buffers at different ratios for a stepwise elution (*see* **Note 5**). For example, to collect five fractions, prepare buffers containing 20% (v/v), 40% (v/v), 60% (v/v), 80% (v/v), and 100% (v/v) CE elute buffer in CE load buffer. Elute peptides in a stepwise manner (from low to high CE elute buffer concentration) by slowly injecting (~1 drop/s) 0.5 mL of each elute buffer and collect the eluted fraction in a separate new tube.

9. If there are additional trypsin-digested samples, repeat steps 1-8.
10. Wash the cartridge by injecting 1 mL of CE clean buffer and 2 mL of CE storage buffer. Store the cartridge at 2–8 °C.
11. Evaporate each sample to dryness in a SpeedVac.
12. Dissolve each sample in 300 µL of MS buffer.
13. Store samples at −20 °C for up to 4 weeks.

3.4 NanoLC-MS Analysis and Data Conversion

1. Inject 5–10 µL of each sample in MS buffer onto the nanoLC online to Q-TOF Ultima MS.
2. Separate peptides by gradient elution (5–75% (v/v) acetonitrile, 0.2% (v/v) formic acid for 90 min, 350 nL/min), and analyze in the MS-only mode.
3. Acquire data between mass/charge (m/z) 400 and 1,600 with 1.0 s scan duration and 0.1 s interscan interval (for a 90 min run, this translates into 4,700–4,900 scans per run; i.e., one scan is approximately equivalent to 1.1 s).
4. In between samples, inject a 40-min blank run (MS buffer only) to prevent cross-contamination.
5. For each sample, convert the raw nanoLC-MS data file into text format using ProteinLynx or other appropriate software from the vendor. Usually, one text file is generated for each scan (i.e., 4,700–4,900 text files per run). Each file contains m/z and intensity information for all the peaks in a scan.

3.5 MatchRx Software and Statistical Analysis

Several steps are involved in the analysis of each nanoLC-MS run and matching multiple runs using MatchRx software (Fig. 17.2). Although these steps are automated, validation of results (from global images) usually is required by the user, to have higher confidence in the results. Once multiple runs are matched using MatchRx, the results can be analyzed using available statistical software to identify differentially expressed proteins, or find patterns in the data.

1. *Global image generation step.* Generate a "2DE gel-like" image for each run (Fig. 17.3A and B). The global images of runs allow visual examination and validation of subsequent steps in the data analysis. Each "spot" on the image corresponds to a peptide ion. The image contains three-dimensional information for each peptide: (1) the x-axis corresponds to m/z values, (2) the y-axis corresponds to the retention (or elution) time, and (3) the volume of the spot corresponds to peptide abundance.
2. *Peak detection step.* Identify the isotopic pattern of peptide peaks and calculate the charge states. Validate this visually (Fig. 17.3C).

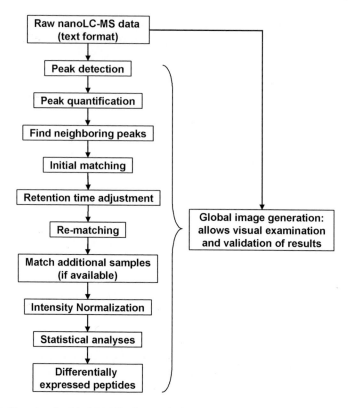

Fig. 17.2 Steps involved in MatchRx data analysis

3. *Peak quantification step*. Quantify the level of each peptide by finding its elution profile and integrating the intensities. Calculate and subtract background
   levels. Validate this visually (Fig. 17.3D).
4. *Neighboring peak finding step*. For each peptide peak, identify up to ten additional peaks within its surroundings (on a global image). These are useful in
   aligning peptides in two runs in the matching steps.
5. *Run matching step*. Once the peak detection and quantification steps are completed for each nanoLC-MS run, select the runs to be compared and match them
   using the MatchRx software as described in steps 6 and 7.
6. *Initial matching of two nanoLC-MS runs*. In this step, peptides from two runs
   are considered matched if they have the same charge, their *m/z* values are within
   0.2 Da, their neighboring peaks align, and their retention time values are
   within 200 s. This initial step of matching allows identification of the relative
   shift in retention times between two runs (Fig. 17.4A and B). Usually the shifts
   are not linear. Note that a large retention time tolerance is used, which may
   introduce false matches. These are reduced in the second step of matching.
7. *Rematching of two nanoLC-MS runs*. Once the shifts in retention times are identified, they are adjusted for one of the samples and matching is done again

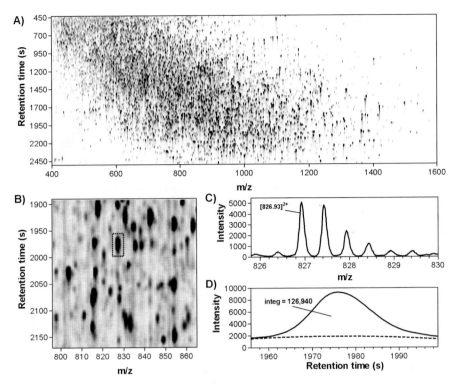

Fig. 17.3 Examination of a label-free nanoLC-MS run using MatchRx software. (A) Global image of a nanoLC-MS run of a biological sample. Each spot represents a peptide ion. More than 2,000 spots are present in the run. (B) Close-up of a global image. The image contains three-dimensional information about each peptide: m/z (x-axis), retention time (y-axis), and abundance or intensity (z-axis and spot volume). Note the variability in the abundance of different peptides. A peptide of $m/z = 826.93$ is highlighted for quantification (see subsequent panels). (C) A spectrum showing individual isotopic pattern of the peptide highlighted in panel B. Note that the peptide is doubly charged. (D) Elution profile of the peptide highlighted in panel B. Shown is the intensity of the peptide (solid line) and the calculated background intensity (dotted line). Background-subtracted integrated intensity is also indicated (integ). This value is used for relative quantification of the peptide among multiple samples

(Fig. 17.4C and D). In this step, matching is done as in the previous step except that a more stringent tolerance for retention time is used, that is, $\pm 10\,\text{s}$. This dramatically reduces the number of false positive matches. An example of a matched peptide between two biological samples that shows differential expression is shown in Fig. 17. 5.

8. *Matching more than two nanoLC-MS runs.* If more than two samples are to be compared, first match each consecutive sample using the double-step matching of two runs as just described. MatchRx then combines all the matches into one matrix (Fig. 17.6). This allows for easy importation of the data into a statistical software package such as BioMiner, to find patterns in the data using various types of clustering methods (Fig. 17.6 [2, 4]).

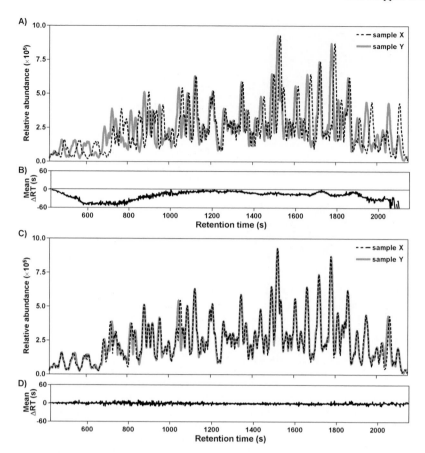

Fig. 17.4 Matching of two label-free nanoLC-MS runs and retention time adjustments: (A) Total ion chromatograms, showing the elution of peptides from nanoLC-MS of two biological samples X and Y. Note that the chromatogram of sample X appears to be shifted relative to sample Y. (B) Identification of the relative shift in retention time (RT) of sample X to sample Y after the first level of matching. Note that the shift is not linear. (C) Total ion chromatogram after adjustment of RT of peptides in sample X. (D) Relative shift in RT of sample X to sample Y after the second level of matching. In panels B and D, mean DRT (i.e., RTY – RTX) of the matched peptides for each RTY are plotted, where RTY and RTX are RT values of peptides from sample Y and X, respectively

9. *Normalization step*. This is one of the most important steps in the data analysis for label-free proteomics. It is necessary to normalize the intensities of the peptides in each sample (using median normalization) to correct for run-to-run variability and possible unequal injections of the samples. Only peptides that have been matched among all the samples are used for normalization. Calculate the median intensity of the matched peptides for each run (M_i) and for all the runs (M_a). Normalize the intensity of each peptide (matched and unmatched) as follows: Normalized intensity = (Original intensity) × M_a/M_i. An example of the data before and after normalization is shown in Fig. 17.7.

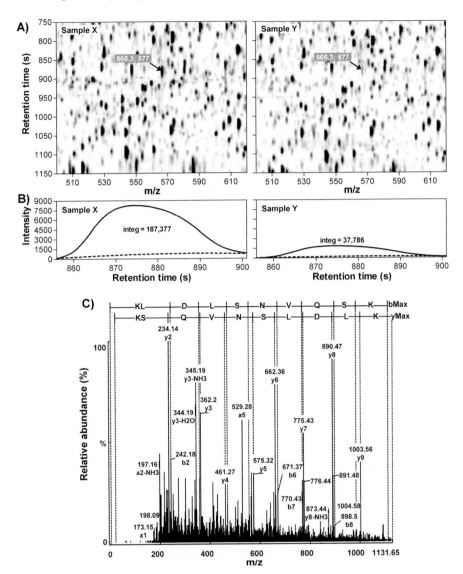

Fig. 17.5 Identification of a differentially expressed peptide: (A) Global images comparing two samples, X and Y. A peptide showing differential expression is indicated by an arrow corresponding to m/z = 566.33 and scan = 877. (B) Elution profiles of the peptide indicated in panel A. The peptide is 4.8-fold higher in sample X than Y. (C) MS/MS spectra of the peptide quantified in panel A. The peptide is identified as KLDLSNVQSK, corresponding to microtubule-associated protein tau

m/z	z	Control 1	Control 2	Control 3	Treated 1	Treated 2	Treated 3	Control 1	Control 2	Control 3	Treated 1	Treated 2	Treated 3
		RT	RT	RT	RT	RT	RT	INTEG	INTEG	INTEG	INTEG	INTEG	INTEG
652.09	3	920	922	916	924	917	910	453,506	544,253	489,251	488,023	442,142	460,282
747.14	3	1019	1015	1015	1020	1004	1008	524,450	542,372	550,136	295,011	295,146	271,498
737.79	3	997	1005	1005	1005	990	991	404,821	392,547	394,400	301,790	296,883	289,343
723.96	2	1259	1268	1256	1263	1260	1258	309,847	306,212	316,857	375,468	394,333	376,524
654.38	3	991	982	979	972	979	987	302,863	275,383	280,805	225,412	275,543	242,754
615.03	3	822	823	827	816	828	808	220,722	221,710	235,504	222,081	267,183	249,090
738.46	3	1107	1115	1115	1122	1119	1120	222,065	228,279	230,831	220,083	228,822	237,450
701.81	2	1008	1010	1004	1006	995	1011	201,743	201,637	201,583	246,420	247,110	246,664

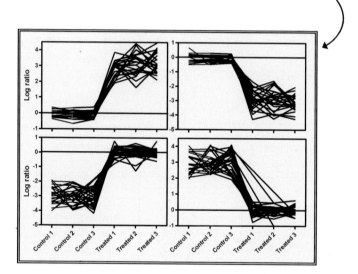

Fig. 17.6 A portion of a matrix generated by MatchRx from matching multiple samples. The normalized integrated intensity values may be converted to ratios and used for clustering or other statistical analyses to find patterns in the data. Interesting peptides are then identified by MS/MS

10. A list of differentially expressed peptides can then be found using BioMiner or related software. If only two samples are being compared (e.g., control vs. treated), a simple cutoff ratio of intensity (treated/control) of 1.5-fold or 2-fold may be used (*see* **Note 6**). However, we strongly recommend doing multiple replicate experiments and performing statistical tests such as *t*-tests or SAM analysis [8] as used in microarray data analysis (*see* **Note 7**). A combination of cutoff fold value (e.g., 1.5-fold) and *p*-value from a statistical test (e.g., $p < 0.05$) is a more reliable way of finding differentially expressed peptides.

11. Visually examine and validate the differentially expressed peptides on global images of the runs (see Fig. 17.5).

12. For each run, generate an "include" list, containing the list of differentially expressed peptides in a format accepted by the MS instrument. Usually, the list includes *m/z* values, charge state, and retention times of the peptides to be analyzed.

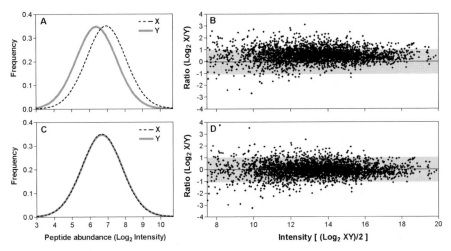

Fig. 17.7 Normalization of data before (A, B) and after (C, D) global median normalization. Shown are the distributions of peptide intensities (A, C) and ratio-intensity plots (B, D) before and after global median normalization of two biological samples, X and Y. Note that, before normalization, the overall intensities of the peptides in sample X are higher than sample Y (A). As a result, peptide ratios are "off-centered" and consequently a lot more up-regulation (>19%) than down-regulation (2%) of peptides is observed in X relative to Y (B). After normalization of intensities, the distribution of intensities in the two samples is almost identical (C). As a result, the ratios are more "centered" since an equivalent number of up- (5%) and down-regulated (6%) peptides are observed (D). The distributions in panels A and C are t location-scale distributions of the intensities created using MATLAB® v. 7.0 software. Ratio-intensity plots in panels B and D are $Log_2(X/Y)$ vs. $Log_2(XY)/2$, where X and Y are peak intensities of peptides from two samples. A ratio >2 or <0.5 is used here as being indicative of differential expression (nongray area)

3.6   NanoLC-MS/MS Analysis of Differentially Expressed Peptides and Database Searching

1. Inject another 5–10 µL of each sample in MS buffer onto the nanoLC online to the Q-TOF Ultima MS.
2. Separate peptides by gradient elution (as in Sect. 3.4, item 2) and analyze in the MS/MS mode using the "include" list corresponding to that sample. Set to "only analyse the ions in the include list" or a similar parameter.
3. Generate a peak list as a single text file (usually a .pkl file from Waters instruments) for each nanoLC-MS/MS run using ProteinLynx. Each file contains m/z and charge values of all the precursor ions and their corresponding m/z and intensity values of fragment ions.
4. Submit each file to the Mascot search engine or any other probability-based engine. The following parameters may be used: (1) Specify trypsin enzymatic cleavage with two possible missed cleavages, (2) allow variable modification for oxidation (+15.99 Da) at the methionine residues, (3) allow variable modification for carbamidomethyl (iodoacetamide derivative; +57.02 Da) at the cysteine

residues if proteins were alkylated, (4) set parent ion tolerance $\leq 0.5\,\text{Da}$, (5) set fragment ion tolerance $\leq 0.2\,\text{Da}$. An example of peptide identified by Mascot is shown in Fig. 17.5C.

5. For large-scale experiments, we strongly recommend using additional statistical analyses that indicate or establish a measure of identification certainty (e.g., using PeptideProphet software) or allow a determination of the false-positive rate (FPR). To calculate the FPR, perform the searches using a composite database consisting of a forward and reversed (or randomized) protein database (*see* **Note 8**). The FPR can be estimated using the formula $\text{FPR} = (2 \times N_{rev})/(N_{rev} + N_{fwd})$, where N_{rev} is the number of peptides identified (after filtering) from the reverse database and N_{fwd} is the number of peptides identified (after filtering) from the forward database. Usually, a FPR of $<2.5\%$ is acceptable (*see* **Note 9**).

6. Follow the guidelines described on the web sites of several proteomic journals, including *Molecular and Cellular Proteomics* www.mcponline.org) and the *Journal of Proteome Research* (http://pubs.acs.org/journals/jprobs), for the analysis and documentation of peptide and protein identifications (established by LC-MS/MS) when reporting or publishing results.

3.7 Validation of the Identified List of Proteins and Their Significance

To have confidence in a large list of differentially expressed proteins identified by label-free proteomics, validation of a few of the results using alternative biochemical and molecular methods is necessary. For confirmation of protein level changes, ELISA, Western blot, or in-situ immunochemistry may be used. To obtain physiologically meaningful information from a *large* list of the differentially expressed proteins identified, many high-throughput analyses (such as genomics and proteomics) rely on various bioinformatics tools to identify patterns in the results by categorizing or clustering the proteins using various methods. The proteins may be categorized by biological processes or molecular functions using a Panther classification system (www.pantherdb.org) or gene ontology database (www.geneontology.org). For experiments involving time-course series or clinical samples, various clustering methods may be used [2, 4].

4 Notes

1. If the amount of sample is not limited (e.g., samples from cell cultures or tissue extracts), $100–150\,\mu\text{g}$ of protein from each sample should be used. The higher the amount of protein used, the more low-abundant proteins are detected, although very high-abundant proteins may be less quantitative (due to signal saturation). If the amount of sample is limited (e.g., LCM samples), as little as $1\,\mu\text{g}$ of protein may be used [2]. Samples with $<1\,\mu\text{g}$ protein may not give reliable results. It is recommended that equivalent amounts of the samples being

compared (e.g., control vs. treated or diseased) are used to minimize the need to normalize the data.

2. It is essential that each protein sample is free of potential contaminants, such high levels of salt, acid, detergents, and denaturants, as these may interfere with the subsequent steps. High amounts of detergents or denaturants, such as SDS and urea, can inactive trypsin activity. High salt, acid, or detergent concentrations can prevent peptides binding to the CE cartridge. If sample quantities are not limited, these contaminants may be removed by dialysis against the DS buffer or by acetone precipitation.

3. Precipitate overnight if the quantity of protein is low. Incubate at 4 °C instead of −20 °C if the protein is present in urea.

4. If proteins are not alkylated, cysteine-containing peptides may not be detected, since they may aggregate by reforming disulfide bonds after trypsin digestion. Although alkylation of protein is recommended, the majority of the proteins are detectable even if the alkylation step is omitted. This is because cysteine-containing peptides amount to only a small portion of all the peptides (<15% for mammals and <9.5% for *E. coli*).

5. If the sample to be analyzed is highly complex in nature (e.g., whole cell or tissue extract), peptide detection and quantification may not always be accurate, due to an overabundance of overlapping and coeluting peaks. Fractionation can be used to reduce the complexity of the original sample and usually increases the number of unique peptides detected. The number of fractions depends primarily on the complexity of the investigation. Here, we describe the protocol for CE chromatography because it is widely used for peptide fractionation and also allows for the removal of SDS and unreacted IAA from the samples. The stationary phase surface of a CE column displays a negatively charged functional group (sulfopropyl, for the column described here), which binds with and retains peptide ions of positive charge. The peptides can be eluted with increasing salt concentrations (KCl) in either a continuous or stepwise manner.

6. A simple fold cutoff of 1.5 implies that any ratio above 1.5 or below 1/1.5 (or 0.67) is assumed differentially expressed. However, the cutoff should be chosen appropriately, because its value depends on the distribution of the ratios and reproducibility of the system. For example, when a complex biological sample is run consecutively and compared, <5% of the peptides show intensity ratios >1.5, or <0.67.

7. Statistical analysis may be carried out on the matrix produced by MatchRx using the SAM test [8] to generate a p-value and a FPR for each peptide identified using BioMiner or other software. Typically, peptides satisfying all the following criteria may be chosen as differentially expressed: 1.5-fold cutoff (*see* **Note 6**), p-value < 0.05, and FPR < 10%.

8. To create a reverse database, each protein sequence in the database is precisely reversed in order of the amino acid sequence so that the C-terminus becomes the N-terminus. A randomized database can be created by randomly shuffling the sequence of each protein. Each sequence in a reversed or randomized database is given a unique identifier (to differentiate it from the forward sequences) and added to the end of the forward database. The final composite database containing both forward and reversed, or randomized, sequences is used for the searches.

9. If the FPR is large, further filtering of the data should be done to reduce false positives (e.g., by reducing parent or fragment ion tolerances or by increasing ion score cutoffs) and maximizing the number of peptides.

10. We successfully applied the label-free comparative proteomics procedures described here to analyze differential protein expression in brain synaptosomes isolated from animals subjected to focal cerebral ischemia. In these studies (unpublished work), we compared the label-free nanoLC-MS with the ICAT method. In this application, both methods exhibited high quantitative reproducibility, quantitative accuracy, and a wide dynamic range. However, the label-free method identified more than five times more peptides and proteins than the ICAT method. In addition, the number of peptides per protein (proteome coverage) was significantly higher by the label-free method; the label-free method also showed higher peptide scores and identified several cysteine-free proteins. We recently used the label-free method described here for the absolute quantification of "therapeutic" proteins injected into animals in various organs and tissues (unpublished data), an attractive application for studies of biodistribution of injected biologics. In conclusion, the label-free nanoLC-MS-based method is a sensitive and reproducible technique that alleviates many of the limitations of isotopic labeling methods and, when coupled to appropriate bioinformatics tools, permits fast quantitative profiling of a large number of proteins from complex biological samples for the purposes of differential expression measurements and biomarker discovery.

References

1. Gygi SP, Rist B, Gerber SA, Turecek F, Gelb MH, Aebersold R (1999) Quantitative analysis of complex protein mixtures using isotope-coded affinity tags. Nat Biotechnol 17:994–999
2. Haqqani AS, Nesic M, Preston E, Baumann E, Kelly J, Stanimirovic D (2005) Characterisation of vascular protein expression patterns in cerebral ischemia/reperfusion using laser capture microdissection and ICAT-nanoLC-MS/MS. FASEB J 19:1809–1821
3. Haqqani AS, Kelly J, Baumann E, Haseloff RF, Blasig IE, Stanimirovic DB (2007) Protein markers of ischemic insult in brain endothelial cells identified using 2D gel electrophoresis and ICAT-based quantitative proteomics. J Proteome Res 6:226–239
4. Haqqani AS, Hutchison JS, Ward R, Stanimirovic DB (2007) Protein biomarkers in serum of pediatric patients with severe traumatic brain injury identified by ICAT-LC-MS/MS. J Neurotrauma 24:54–74
5. Aebersold R, Mann M (2003) Mass spectrometry-based proteomics. Nature 422:198–207
6. Li XJ, Yi EC, Kemp CJ, Zhang H, Aebersold R (2005) A software suite for the generation and comparison of peptide arrays from sets of data collected by liquid chromatography-mass spectrometry. Mol Cell Proteomics 4:1328–1340
7. Ono M, Shitashige M, Honda K, Isobe T, Kuwabara H, Matsuzuki H et al (2006) Label-free quantitative proteomics using large peptide data sets generated by nanoflow liquid chromatography and mass spectrometry. Mol Cell Proteomics 5:1338–1347
8. Tusher VG, Tibshirani R, Chu G (2001) Significance analysis of microarrays applied to the ionizing radiation response. Proc Natl Acad Sci USA 98:5116–5121

Chapter 18
Using 2D-LC-MS/MS to Identify *Francisella tularensis* Peptides in Extracts from an Infected Mouse Macrophage Cell Line

John F. Kelly and Wen Ding

Abstract Two dimensional nano-high-performance liquid chromatography (nanoHPLC) coupled directly to a high-resolution tandem mass spectrometer (2D-nLC-MS/MS) is an excellent method for analyzing very complex peptide mixtures, especially when the quantity of sample available for analysis is severely limited. We describe here a relatively simple 2D-nLC-MS/MS approach that we often use to characterize complex peptide mixtures, such as those produced by the proteolytic digestion of protein extracts. A peptide mixture is resolved in the first dimension by stepped elution from a strong cation exchange (IEC) column and in the second dimension by reverse phase (RP) nanoHPLC chromatography prior to electrospray ionization. The peptide ions are analyzed by automatic tandem mass spectrometry in a hybrid quadrupole time-of-flight mass spectrometer (Q-TOF). In this chapter, we illustrate this approach by way of an example featuring analyses of peptides extracted from a mouse macrophage cell line infected with the live vaccine strain of *Francisella tularensis*.

Keywords two-dimensional liquid chromatography-tandem mass spectrometry, *Francisella tularensis*, automated database searching, MHC, mouse macrophage cells

1 Introduction

Complex peptide samples, such as those produced by proteolytic digestion of protein extracts, contain thousands of different peptides species and pose an enormous challenge to characterize thoroughly. The peptides in these mixtures can vary widely in amino acid composition, molecular weight, and abundance and are often posttranslationally modified. One-dimensional reverse-phase nanoHPLC (<1 µL/min flowrate) coupled with tandem mass spectrometry (1D-nLC-MS/MS) is an effective and sensitive method for analyzing simpler mixtures, such as those produced by in-gel proteolytic digestion of gel-separated proteins, where the number of peptide species present is relatively low (tens or

From: Methods in Molecular Biology, Vol. 439: *Genomics Protocols: Second Edition*   257
Edited by Mike Starkey and Ramnath Elaswarapu © Humana Press Inc., Totowa, NJ

hundreds of different peptide species). However, more sophisticated separation methods are required when the peptides present in a sample number in the thousands and tens of thousands. It often is possible to prefractionate the sample off line using an orthogonal separation technique (i.e., ion exchange) prior to 1D-nLC-MS/MS. However, this approach does not lend itself well to the analysis of minute quantities of samples (i.e., samples obtained by laser-capture microdissection or peptides extracted from MHC complexes, etc.), as peptide losses can be significant. In these instances, it usually is more appropriate to employ a two-dimensional chromatography-tandem mass spectrometry (2D-LC-MS/MS) technique.

We define 2D-LC-MS/MS as the direct coupling of two orthogonal methods of separation to each other and to a mass spectrometer capable of fragmenting and sequencing peptides. Numerous examples of 2D-LC-MS/MS are to be found in the literature and a few of these are described at the end of this chapter (*see* **Note 1**). Here, we present a straightforward 2D-nLC-MS/MS technique that can be fashioned from a standard 1D-nLC-MS/MS arrangement by making a single modification. An outline of the 2D-nLC-MS/MS experiment is presented in Fig. 18.1. While not the most sophisticated 2D-LC-MS/MS arrangement, it nevertheless has proven very effective in sequencing and identifying thousands of peptides. Furthermore, the entire process of analysis, from sample injection to database searching, can be automated, a distinct advantage when analyzing multiple samples.

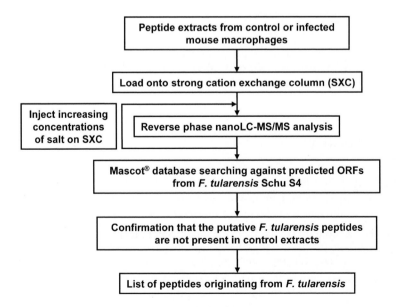

Fig. 18.1 Protocol for the 2D-LC-MS/MS described in this chapter

In this chapter, we demonstrate the utility of the 2D-nLC-MS/MS technique for the analysis of peptides extracted from a mouse macrophage cell line infected with the live vaccine strain (LVS) of *Francisella tularensis. F. tularensis* is a human intracellular pathogen and the causative agent of tularemia. Perceived problems associated with the current live vaccine strain of *F. tularensis* means that an opportunity exists to develop more clinically acceptable vaccines against this pathogen. Our objective was to use 2D-nLC-MS/MS to identify *F. tularensis* peptides that could bind to the host's major histocompatibility complexes (MHC) during experimental infection. Other groups have used similar approaches to identify MHC peptide ligands from microbial intracellular pathogens [1, 2].

2 Materials

2.1 Preparation of Peptide Extracts

The peptide extracts were prepared by Drs. M. Telepnev and A. Sjöstedt (Umeå University, Sweden). Approximately 10^8 J774 mouse macrophage cells were grown in Dulbecco's modified Eagle medium media (DMEM) containing 10% (v/v) fetal calf serum (FCS) and infected with *F. tularensis* LVS at a multiplicity of infection of 500 per macrophage cell. Two hours later, all extracellular bacteria were washed away with DMEM without FCS, and fresh medium containing 5 µg/mL of gentomycine was added. The cells were incubated for a further 15 h, harvested, resuspended in 10 mL of 0.1% (v/v) trifluoroacetic acid, and sonicated for 1 min. The suspension was centrifuged at 100,000 g for 30 min. The clear supernatant was split into three equal portions and lyophilized overnight. Control J774 cells were treated in the same manner, but were not infected with LVS.

2.2 Equipment and Reagents

1. A strong cation exchange column, such as 300 µm i.d.×5 cm polysulfoethyl aspartamide strong cation exchange column (PolyLC Inc., Columbia, MD).
2. A reverse phase nanoHPLC column, such as 75 µm i.d.×15 cm PepMap™ C18 nanoHPLC column (Dionex/LC Packings, San Francisco).
3. A reverse phase peptide trap, such as 300 µm i.d.×0.5 cm PepMap C18 traps (Dionex/LC Packings, San Francisco).
4. 3,000 MWL Centricon® centrifugal filters (Millipore, Nepean, ON, Canada).
5. Solvent A: 0.2% (v/v) formic acid in deionized water (HPLC-grade reagents, solution degassed before use).

6. Solvent B: 0.2% (v/v) formic acid in acetonitrile (HPLC-grade reagents, solution degassed before use).
7. Solvent C: 0.2% (v/v) formic acid in 5% (v/v, aqueous) acetonitrile (HPLC-grade reagents, solution degassed before use).
8. Salt solution 1: 10 mM KH_2PO_4, 1 mM NH_4OAc.
9. Salt solution 2: 10 mM KH_2PO_4, 2 mM NH_4OAc.
10. Salt solution 3: 10 mM KH_2PO_4, 5 mM NH_4OAc.
11. Salt solution 4: 10 mM KH_2PO_4, 10 mM NH_4OAc.
12. Salt solution 5: 10 mM KH_2PO_4, 25 mM NH_4OAc.
13. Salt solution 6: 10 mM KH_2PO_4, 50 mM NH_4OAc.
14. Salt solution 7: 10 mM KH_2PO_4, 100 mM NH_4OAc.
15. Salt solution 8: 10 mM KH_2PO_4, 200 mM NH_4OAc.
16. Salt solution 9: 10 mM KH_2PO_4, 1 M NH_4OAc.

2.3   Instrumentation

1. A nanoHPLC system. In this instance, we used a CapLC capillary liquid chromatography system (Waters, Millford, MA), modified for nanoflow delivery.
2. A mass spectrometer with MS/MS capabilities. In this instance, we used a Q-TOF Ultima hybrid quadrupole/time-of-flight mass spectrometer (Waters).

2.4   Software for the Analysis of Peptide MS/MS Spectra

1. Software for generating peak lists from the nanoLC-MS/MS spectral data, such as ProteinLynx™ (Waters).
2. Software for identifying proteins using peptide MS/MS spectra, such as Mascot®, version 1.9 (Matrix Science Ltd, UK).
3. Protein sequence databases: a *F. tularensis* database; the NCBI nr protein sequence database.

3   Methods

3.1   Preparation of the Peptide Extracts for 2D-LC-MS/MS Analysis

Redissolve the peptide extracts in 900 μL of 0.1% (v/v) formic acid and filter through a 3,000 MWL Centricon centrifugal filter to remove large molecular weight proteins and peptides. Utilize 10% of each filtered sample for each set of 2D-nLC-MS/MS experiments.

3.2   2D-LC-MS/MS Instrument Setup

The column and flow stream configurations for injection and analysis are illustrated in Fig. 18.2, and are based on that described by Hughes [3]. The instrument setup is very similar to that for 1D-nLC-MS/MS except that a strong cation exchange capillary column is inserted in the sample flow stream prior to the ten-port valve. The direction of the liquid flow streams (sample and HPLC gradient) is governed by the orientation of the ten-port switching valve. The entire instrument arrangement (sample injection, nanoHPLC, switching port, and mass spectrometer) is automatically controlled through MassLynx™, the software provided by the manufacturer (Waters). This allows easy automation of the entire analysis.

3.3   2D-LC-MS/MS

1. Place the LC-MS system into inject mode (Fig. 18.2a), such that sample flow is directed through the cation exchange column and the 0.5 cm C_{18} peptide trap before being sent to waste (*see* **Note 2**). Inject the filtered peptide extracts (~10% of total) into a 30 µL/min flow stream of solvent C.

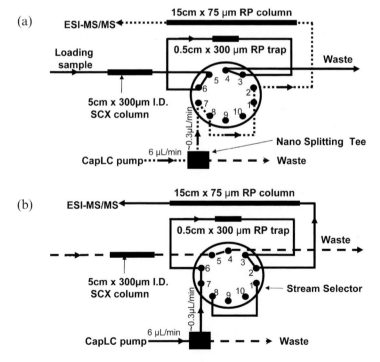

Fig. 18.2 2D-nanoLC-MS/MS experiment setup for sample loading and salt injection (a) and peptide analysis (b). The configuration is based on that described by Hughes [3]

2. Rinse the two trapping columns with the sample flow stream (solvent C, 30 μL/min) for 10 min to ensure that all soluble contaminants are removed.

3. Place the LC-MS system into peptide analysis mode (Fig. 18.2b), such that the peptide trap is brought online with the reverse phase nanocolumn and the Q-TOF mass spectrometer. Analyze the peptides that pass through the ion exchange column but are retained on the peptide trap by nanoLC-MS/MS as described in the following steps.

4. Use the following HPLC gradient conditions to separate peptides prior to elution and MS/MS analysis: (1) 95% solvent A + 5% solvent B for 5 min, (2) gradually increase to 95% solvent B over 100 min, (3) maintain at 95% solvent B for 5 min before gradually reducing back to 5% solvent B over 5 min (flow rate through the trap and nanocolumn ~ 300 nL/min).

5. With the electrospray voltage at 4 kV (and without using nebulizing gas), set the Q-TOF Ultima to automatically acquire MS/MS spectra on doubly, triply, and quadruply charged ions (*see* **Note 3**). An example of a nLC-MS/MS chromatogram acquired during this analysis is presented in Fig. 18.3.

6. Switch the LC-MS system back to inject mode (Fig. 18.2a). Inject 10 μL of salt solution 1 and wait 10 min before switching back to analysis mode (Fig. 18.2b). Perform the nanoLC-MS/MS analysis as described in steps 4 and 5.

7. Repeat step 6 using salt solutions 2–9.

3.4 Database Searching Using Mascot and the Identification of *F. tularensis* Peptides

1. These experiments were performed before a fully sequenced genome of *F. tularensis* was publicly available. Therefore, the peptide MS/MS spectra were searched against annotated (by BLAST sequence similarity search against bacterial subset of the Genbank nonredundant protein database) contigs representing the partially completed *F. tularensis* Schu S4 genome (http://artedi.ebc.uu.se/Projects /Francisella).

2. The raw peptide MS/MS spectra in each 2D-nLC-MS/MS file were converted to peak lists (.pkl files) using MassLynx software. Each file contained the *m/z* value and charge state of the precursor ions as well as the *m/z* and intensity values of their fragment ions.

3. The peak list files were submitted to Mascot (*see* **Note 4**) and searched against the *F. tularensis* protein sequence database using the following parameters: (1) no cleavage agent was specified, (2) the parent ion mass tolerance was set at ≤0.5 Da, (3) the fragment ion tolerance was set to ≤0.3 Da, (4) variable oxidation of methionine (+ 15.99 Da), and (5) deamidation of asparagine (+1 Da) were enabled. In theory, only those peptides derived from *F. tularensis* proteins would be successfully identified. Each potential hit was examined by hand to ensure good correlation of the fragment ion data with the theoretical match. An example of a *F. tularensis* peptide MS/MS spectrum is presented in Fig. 18.4.

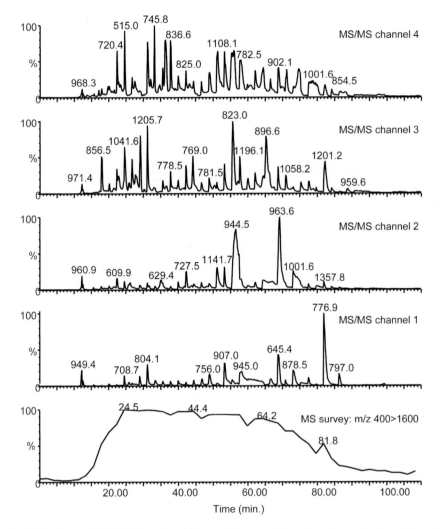

Fig. 18.3 2D-nanoLC-MS/MS analysis of the peptide extract from the spleen cell culture infected with *F. tularensis*. The MS survey total ion chromatogram (TIC) is shown in the lower panel, while four peptide MS/MS TICs acquired during the same run are presented in the upper panels. Up to eight peptide ions can be selected from each survey scan for MS/MS analysis. Some of the peaks in the MS/MS TICs have been annotated with the *m/z* value of the peptide precursor ions

3.5 Confirmation That the Putative F. tularensis Peptides Are Exclusively Expressed in the Infected Cell Line

1. In theory, peptides arising from *F. tularensis* proteins should be observed only in the 2D-nLC-MS/MS files of the infected cell line extracts. Therefore, the ion chromatograms for each putative *F. tularensis* peptide identified in Sect. 3.4, step 3 were

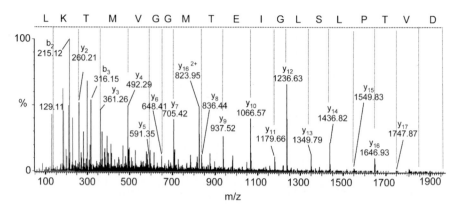

Fig. 18.4 MS/MS spectrum of the doubly protonated ion at *m/z* 981.57 (MH$_2^{2+}$) of an MHC-like peptide originating from the DNA K protein of *F. tularensis* (gi: 89144343). This peptide was observed in the extract of the LVS-infected spleen cells only. The sequence of the peptide is DVTPLSLGIETMGGVMTKL. The peptide fragment ions observed in this MS/MS spectrum are predominantly y ions (fragment ions originating from the C-terminus)

extracted from the survey scans of the infected and control 2D-nLC-MS/MS files. Only those peptides with no correlating peak in the control were retained.

2. The amino acid sequence of the putative *F. tularensis* peptides were blasted against all murine entries in the NCBI nr protein sequence database. Only those peptides that could not be matched reasonably were retained.

4   Notes

1. Examples of other 2D-nanoHPLC applications:

   a. MudPIT. An innovative method of online two-dimensional chromatography, Multidimen-sional Protein Identification Technology (MudPIT) was developed by John Yates III and his colleagues [4–7]. The technique consists of packing a fused-silica microcapillary (100 μm i.d. × 360 μm o.d.) first with reverse phase particles (i.e., 10 cm of 5-μm Zorbax Eclipse XDB-C$_{18}$ from Agilent Technologies, Palo Alto, CA) then with cation exchange beads (i.e., 4 cm of 5-μm Partisphere strong cation exchange resin from Whatman, Clifton, NJ). Prior to packing, the tip of the column is pulled into a fine point (~2 μm) using a microcapillary puller, thus trapping the particles inside the microcolumn without the need of a frit and ensuring a stable electrospray at submicroliter flow rates. The sample (i.e., a proteolytic digest of a complex protein extract) is loaded onto the biphasic column, which then is linked directly to a quaternary HPLC pump system capable of nanoflow delivery (150–250 nL/min) and a mass spectrometer capable of rapid MS/MS analysis (i.e., an ion trap). Using a fully automated multistep chromatography procedure, the peptides are sequentially eluted off the ion exchange beads, separated by reverse phase chromatography, and analyzed by nESI-MS/MS. The process consists of the application of a standard reverse phase gradient followed by a wash with a salt solution (i.e., ammonium acetate in 5% acetonitrile, 0.012% HFBA, 0.5% acetic acid). The concentration of the salt solution increases with each iteration of this cycle (i.e., 2.5, 5, 10, 25, 50 100, 250, and 500 mM ammonium acetate). The last salt wash is followed by a final reverse phase gradient run to complete the analysis. The peptide MS/MS spectra then are searched against an appropriate protein or nucleotide sequence database using an automated database searching engine such as SEQUEST™. Using this method, it is possible to identify thousands of proteins expressed in a sample [4]. Furthermore, the inventors demonstrated that MudPIT could be used to perform quantitative proteomics analysis using stable isotope labels[5, 6]. The samples, in this case yeast, were grown either on ^{14}N-minimal media or ^{15}N-enriched minimal media, combined, and analyzed using MudPIT.

b. Liquid isoelectric focusing and reverse phase chromatography of proteins. David Lubman and his coworkers described a procedure for isoelectric focusing and reverse phase chromatography of proteins [8–10]. The method consists of resolving and fractionating complex protein mixtures by liquid isoelectric focusing on a device such as the mini-Rotofor system (Bio-Rad). Each fraction is then analyzed by nonporous silica (NPS) reverse phase chromatography coupled with an electrospray mass spectrometer (NPS-RP-ESI-MS). NPS reverse phase media are preferred, as this reduces the problems associated with protein separation on porous media, such as poor peak shape and low recovery of proteins from the column. This technique has the advantage that both separation techniques are liquid based and, provided certain detergents and denaturing reagents are avoided, entirely compatible with one another and with ESI-MS. Although this is not true 2D-LC-MS/MS, it is nevertheless a fine example of the use of two orthogonal separation techniques to attain a quality of separation not possible by either technique on its own.

2. In our experience, this trapping arrangement captures the large majority of peptides, with only the smallest, most acidic, and most hydrophilic getting through to waste. The majority of peptides are retained on the cation exchange column, although some, likely the most acidic ones, pass through and are retained on the C_{18} peptide trap.

3. The mass spectrometer surveys the incoming ions in MS-only mode (m/z 400–1,600, one scan per second), analyzes the spectra "on-the-fly," and selects up to eight of the most intense multiply charged ions for MS/MS analysis. Four MS/MS spectra are acquired per precursor ion (m/z 50 > 2,000, 1 s per spectrum) and the collision offset is automatically chosen by the mass spectrometer based on their size and charge state.

4. Some of the peak list files were so large that Mascot, version 1.9, could not handle them (although this has not been a problem with later versions of Mascot). On these occasions, the peak list files were divided in two and sometimes three smaller .pkl files and resubmitted to Mascot.

5. Using 1D-nLC-MS/MS, we acquired hundreds of peptide MS/MS spectra in the cell extracts and managed to identify 4 *F. tularensis* peptides with a high degree of confidence. However, by using this simple 2D-nLC-MS/MS technique, we acquired approximately five times more peptide MS/MS spectra and confidently identified 27 *F. tularensis* peptides, many of which appeared to be excellent MHC peptide candidates. Modern 2D-nanoHPLC systems offer a greater degree of sophistication and control over every step in the analysis, which can result in increased peptide resolution, greater sensitivity, and less carryover from one fraction to another (*see* **Note 1**). In addition, a faster scanning mass spectrometer (i.e., an ion trap) would likely have acquired MS/MS spectra on more peptide ions, albeit with a significant loss in ion resolution (*see* **Note 7**). Therefore, it is entirely likely that many less-abundant *F. tularensis* peptides were overlooked in this analysis. Nevertheless, this experiment demonstrates the real potential of 2D-nLC-MS/MS for analyzing highly complex peptide samples.

6. 2D-nLC instruments. Many modern nanoHPLC systems are capable of true, online 2D-LC-MS/MS. Typically, these instruments are equipped with multiple switching valves and three or four pumps capable of both capillary and nanoflow solvent delivery. Although more expensive than standard one-dimensional nanoHPLC instruments, these 2D systems allow fuller control of the first and second dimensions of chromatography. For example, instead of eluting the peptides from the ion exchange column using relatively small volumes of salt solution introduced via the sample injection system (as described in this chapter), a 2D-HPLC system can perform true gradient ion exchange chromatography in the first dimension, fractionate or trap the eluting peptides using multiple peptide traps, and analyzse each fraction in turn by nanoHPLC-MS/MS. This degree of sophistication ensures greater run-to-run reproducibility and less carryover of peptides from one fraction to another due to insufficient washing of the first dimension column.

7. Mass spectrometers. Mass spectrometers capable of fast, "on-the-fly" MS/MS spectral acquisition are very well suited for the type of experiments described in this chapter. Modern ion trapping instruments, for example, can acquire MS/MS spectra at up to 100 times the rate of the Q-TOF instrument used here. This enhanced acquisition speed makes it possible to delve deeper into complex samples and analyze more of the less-abundant peptides. However, this speed comes with a marked reduction in ion resolution, which can make it difficult to match the fragment ions in a peptide MS/MS spectrum to the appropriate protein sequence in a database. Furthermore, the high-resolution instruments, although slower, have a distinct advantage when it comes to *de novo* sequencing of peptides (i.e., when a peptide MS/MS spectrum cannot be matched to a known protein in a database and the amino acid sequence must be determined from the fragment ions in the spectrum). Some hybrid mass spectrometers offer the

advantage of rapid MS/MS analysis as well as high ion resolution. For example, using a hybrid ion trap/Fourier transform ion cyclotron resonance mass spectrometer (IT-FTMS), it is possible to obtain ultrahigh-resolution and mass accuracy on peptide ions in the FTMS while rapidly acquiring their MS/MS spectra in the ion trap, a very powerful combination for the type of experiments described here.

Acknowledgments We thank Dr. Wayne Conlan (NRC-IBS) and Dr. Anders Sjöstedt (Umeå U.) for their guidance and support during this work. We also thank Dr. John Nash, Simon Foote, and Ken Chan (all from NRC-IBS) for their assistance with the informatics associated with this project.

References

1. Brockman A (1999) Rapid Communications in Mass Spectrometry 13:1024–1030
2. Flyer D. (2002) Infection and Immunity 70:2926–2932.
3. Hughes C (2001) Presented at the 49th American Society for Mass Spectrometry conference, Chicago, May 27–31
4. Link AJ, Eng J, Schieltz DM, Carmack E, Mize GJ, Morris DR, Garvik BM, Yates JR (1999) Direct analysis of protein complexes using mass spectrometry. Nature Biotechnology 17:676–682
5. Washburn MP, Wolters D, Yates JR (2001) Large-scale analysis of the yeast proteome by multidimensional protein identification technology. Nature Biotechnology 19:242–247
6. Washburn MP, Ulaszek R, Deciu C, Schieltz DH, Yates JR (2002) Analysis of quantitative proteomics data generated via multidimensional protein identification technology. Analytical Chemistry 74:1650–1657
7. Washburn MP, Ulaszek RR, Yates JR (2003) Reproducibility of quantitative proteomic analysis of complex biological mixtures by multidimensional protein identification technology. Analytical Chemistry 75:5054–5061
8. Lubman DM, Kachman MT, Wang H, Gong S, Yan F, Hamler RL, O'Neil KA, Buchanan NS, Barder TJ (2002) Two-dimensional liquid separations-mass mapping of proteins from human cancer cell lysates. Journal of Chromatography B 782:183–196
9. Kachman MT, Wang H, Schwartz DR, Cho KR, Lubman DM (2002) A 2-D liquid separations/mass mapping method for interlysate comparison of ovarian cancers. Analytical Chemistry 74:1779–1791
10. Hamler RL, Zhu K, Buchanan NS, Kreunin P, Kachman MT, Miller FR, Lubman DM (2004) A two-dimensional liquid-phase separation method coupled with mass spectrometry for proteomics studies of breast cancer and biomarker identification. Proteomics 4:562–577

Chapter 19
Baculovirus Expression Vector System: An Emerging Host for High-Throughput Eukaryotic Protein Expression

Binesh Shrestha, Carol Smee, and Opher Gileadi

Abstract The increasing demand for production and characterization of diverse groups of recombinant proteins necessitates the analysis of several constructs and fusion tags in a variety of expression systems. The challenge is to screen multiple clones quickly for the desired properties. When using a eukaryotic system, such as baculovirus-mediated expression in insect cells, the total time required and the volume of culture needed to obtain reasonable results are limiting factors. This chapter focuses on addressing these issues by describing rapid small-scale expression as a mode of screening. The method allows the rapid identification of the best clone before scaling-up and the production of heterologous protein.

Keywords high-throughput (HT), baculovirus, insect cell, functional titration, plaque assay

1 Introduction

Baculoviruses are a diverse group of viruses isolated from different insect hosts. Most of the baculoviruses isolated so far are host specific, and the majority of them are isolated from the larvae of butterflies and moths (Phylum Arthropoda, Class Insecta, Order Lepidoptera). Baculoviruses are rod shaped, 40–50 nm in diameter, and 200–400 nm in length [1]. The length of the capsid can extend, accommodating larger double-stranded DNA [2, 3]. The parallel skills developed to enable the growth and multiplication of cell lines derived from fall armyworm, *Spodoptera frugiperda* (Sf9, Sf21), and cabbage looper, *Trichopulsia ni* (HighFive) made it possible to propagate baculoviruses in cell culture [4]. This opened the door to extensive studies on the molecular biology of baculoviruses [5]. The model virus in these studies was *Autographa californica* multiple nuclear polyhedrosis virus (AcMNPV). Most expression vectors used in the baculovirus expression vector system (BEVS) at present are based on AcMNPV infection of *S. frugiperda*. *Bombyx mori* nuclear polyhedrosis virus (BmNPV)

From: Methods in Molecular Biology, Vol. 439: *Genomics Protocols: Second Edition*
Edited by Mike Starkey and Ramnath Elaswarapu © Humana Press Inc., Totowa, NJ

has been used extensively for the production of heterologous protein in silkworms [6].

The BEVS exploits the characteristics of the baculovirus and its mode of infection in cultured insect cells. Therefore, it is important to understand the infection cycle to optimize the level of foreign gene expression. During the infection cycle, the parental baculovirus in cell culture displays three distinct phases: early, late, and very late. During the early phase, viral attachment occurs followed by penetration of viral genetic material, which reprograms the host cells for virus gene expression and shuts off host gene expression [7]. The nucleus of the infected host cell then synthesizes viral DNA (nucleocapsid) in the late phase. The viral DNA is enveloped by budding through the plasma membrane, commonly known as the *budded virus* (BV). Logarithmic production of BV occurs between 12 to 20 h post infection [8, 9]. A decline in host protein synthesis is observed at 18 h post infection [10–12]. The infective viral titer remains very high during this phase. The very late phase (24–36 h post infection) is specific to viral gene expression. During this time, large numbers of occlusion bodies containing viruses are produced. In this phase, each virus acquires an envelope within the nucleus along with a crystalline protein matrix, a polyhedrin. The virus produced in this phase is commonly known as the *occluded virus* (OV). BV is more infectious than OV.

The polyhedrin produced in the very late phase is under the control of a strong promoter called *polh*, which is responsible for the formation of OV [13]. The *polh* promoter is responsible for very high levels of expression during the later stages of infection [14]. The gene encoding the polyhedrin is not essential for the replication of the virus. This gene may be replaced by a foreign gene under the control of the same promoter *polh*. Production of the recombinant protein is achieved following infection of insect cells or insect larvae with the engineered virus. Another promoter used in the BEVS is *p10*, which has been less extensively characterized but appears to be regulated in a fashion similar to the *polh* promoter [15, 16].

The increasing demand for BEVS over the past 20 years resulted in significant developments in using baculovirus as a prominent tool in molecular biology. This led to the commercialization of the system, making it easier to use, especially for those with little experience in eukaryotic expression systems. The advancement of the system is reviewed by Kost et al. [17]. The wealth of information obtained from genome projects increased demand for the production and characterization of diverse groups of proteins. The most efficient approach is to analyze several constructs and fusion tags in a variety of expression systems [18]. Therefore, the present day challenge is to quickly screen multiple clones for desired properties, such as expression level, posttranslation modification, solubility, localization of protein, and activity. Although *Escherichia coli* (*E. coli*) is the first choice to achieve this using a high-throughput (HT) platform, it has limitations when expressing eukaryotic proteins. The alternative HT system to complement the *E. coli* expression system is BEVS.

One of the most commonly used systems in BEVS is the Bac-to-Bac system, which uses site-specific transposition of an expression cassette into the baculovirus genome maintained in *E. coli* [19]. Improvements to the throughput of recombinant eukaryotic protein in insect cells has been described by several authors [18, 20–23]. This chapter provides a detailed methodology to achieve efficient high-throughput cloning and expression of eukaryotic proteins in insect cells. An overview of the work flow is presented in Fig. 19.1. The rapid small-scale expression study described here yields sufficient amounts of protein to unravel the desired properties of heterologous proteins, thus addressing the challenges mentioned previously. This method is based on the Bac-to-Bac system, however, it can easily be adapted to other systems.

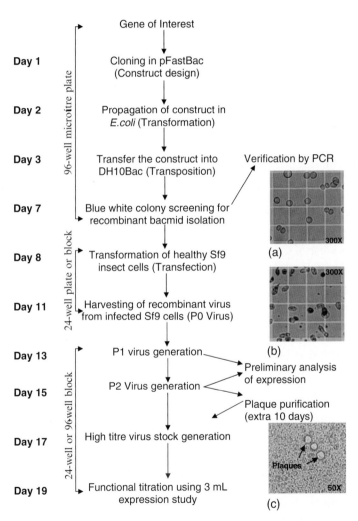

Fig. 19.1 An overview of the steps involved in baculovirus-mediated cloning and expression of gene in insect cell. The images observed under the microscope help to identify (a) the healthy cells, (b) infected cells, and (c) the plaque formed by viruses

2 Materials

2.1 Microorganisms and Cell Line

1. *E. coli* MACH1 (Invitrogen, C8620-03).
2. *E. coli* DH10Bac (Invitrogen, 10361-012).
3. Sf9 insect cells SFM adapted (Invitrogen, 11496-015).
4. High-Five insect cells, SFM adapted (Invitrogen, 10486-025).

2.2 Enzymes

1. Platinum® Pfx DNA polymerase (Invitrogen, 11708-039).
2. Benzonase (HC, 250 units/μL) (Novagen, 71205-3).

2.3 Chemicals and Kits

1. Cellfectin reagent (Invitrogen, 10362-010)
2. EDTA-Free Protease Inhibitor Cocktail Set III (Merck Biosciences, 539134)
3. 0.4% Trypan Blue Stain (Invitrogen, 15250-061)
4. Oligonucleotides (Invitrogen, 10297-117)

   M13 forward: 5′ GTTTTCCCAGTCACGAC 3′.
   M13 reverse 5′ CAGGAAACAGCTATGAC 3′.

5. 96-Well DNA Purification Kits for PCR Products (Qiagen 28183 or Millipore LSKMPCR50).
6. Plasmid Miniprep Kit (Qiagen 16191 or Millipore LSKP09624).
7. Sybr-Safe DNA gel stain (Invitrogen S33102): use at 1:30,000 dilution in agarose gels.
8. Low melting-point agarose (Invitrogen, 15517-014).
9. Solutions P1 and P2 (Qiagen 19051 and 19052, respectively).
10. Isopropanol (Fisher Scientific, P/7500/17)
11. Ethanol (Fisher Scientific, E/0650DF/17)
12. Potassium acetate (FisherBiotech, BP364-500)

2.4 Plasticware

1. 96-well thin-walled PCR plates (ABgene,AB-0900).
2. Adhesive, heat-resistant film (ABgene, AB-0558).
3. 96 deep-well blocks, 2.2-mL (Thomson, 951652B).
4. 96-well filter plates (Chromacol, CLS-370011, or Thomson, 921673B).

5. Porous adhesive film for 96-well blocks (Thomson, 899410).
6. 250-, 500-, and 1,000-mL flasks with vented cap (Corning 431144, 431145, and 431147, respectively).
7. 500 mL sterile centrifuge tubes (Corning, 431123).

2.5   Instruments

1. 96-well thermocycler with heated lid (Biometra).
2. Micro-express Glas-Col shaker (8 × 96-well blocks).
3. Large, chilled shaker-incubators.
4. Inverted microscope (Axiovert 25, CarlZeiss).
5. Haemocytometer, improved Neubauer (VWR International, 720-0104).

2.6   General Stock Solutions and Media

1. Penicillin/streptomycin (Invitrogen, 15140-122).
2. Kanamycin (Apollo Scientific Ltd., BIK0126).
3. Tetracycline (Fluka/BioChemika, 87128).
4. Gentamycin (Sigma-Aldrich, 46305)
5. Antibiotic stock solutions (1000X):

   a) 50 mg/mL kanamycin in water.
   b) 10 mg/mL tetracycline in ethanol.
   c) 7 mg/mL gentamycin in water.

6. Bluo-gal (Invitrogen, 15519-028): 100 mg/mL in DMSO.
7. Foetal bovine serum (Invitrogen, 10082-147).
8. Heparin (Sigma-Aldrich, H3400-100).
9. L-glutamine (Invitrogen, 25030-024).
10. CHAPS (AG Scientific Inc., C-1019): 10% (w/v).
11. 5 M NaCl (Fluka, 71376).
12. 5 M Imidazole, pH 7.5 (Fluka, 56750): dissolve imidazole in 80% of the final volume, titrate with concentrated HCl to the desired pH, then adjust the volume with water, filter, and sterilize. Store at room temperature.
13. Glycerol (Fisher Scientific, G/0650/17): prepare 60% (w/v) stock, autoclave, and store at room temperature.
14. Isopropanol (Fisher Scientific, P/7500/17)
15. 1 M HEPES (Calbiochem, 391340): dissolve HEPES in 80% of the final volume, titrate with 5 M NaOH to pH 7.5, adjust the volume with water, autoclave, and store at room temperature.
16. 0.5 M TCEP (Tris[2-carboxyethyl] phosphine hydrochloride) (Fluorochem, BV1300): dissolve the solid in water and store in 1 mL aliquots at −20 °C.

17. IPTG (Apollo Scientific, BIMB1008): dissolve the solid in water to prepare 40 mg/mL stock, filter-sterilize and store in 1 mL aliquots at −20 °C).
18. TE buffer: 10 mM Tris-HCl, pH 7.5 and 1 mM EDTA.
19. Sf-900 II (1X) liquid with L-glutamine (Invitrogen, 10902-153).
20. Grace's insect medium (1X) (Invitrogen, 11595-030).
21. Express SFM (Invitrogen, 10486-025).
22. Virkon (Appleton Woods, GC383).
23. Decon90 (Decon Laboratories Ltd.)

2.7 Solutions and Media for Cloning

1. LB medium: 1% tryptone, 0.5% yeast extract, 1% NaCl.
2 SOC: 2% Bacto-tryptone, 0.5% yeast extract, 0.05% NaCl, 2.5 mM KCl, 10 mM MgCl$_2$, 20 mM glucose, pH 7.0. MgCl$_2$ and glucose added after autoclave.
3. 2X LB: LB medium prepared in 0.5 volume.
4. LB-agar plates: prepare LB medium, before autoclaving add 1% (w/v) bacto-agar, and autoclave. After autoclaving, let the medium cool to 55–60 °C, add antibiotics and other additives (if required), mix well, and pour into 60-mm petri dishes (approximately 10 mL/dish). Prepare LB agar plates containing 50 µg/mL kanamycin, 7 µg/mL gentamycin, 10 µg/mL tetracycline, 100 µg/mL bluo-gal, and 40 µg/mL IPTG to select for DH10Bac transformants. These plates can be stored at 4 °C for up to 4 weeks, covered to prevent exposure to light.

2.8 Buffers for Protein Purification

The following solutions are made by combining general stock solutions and water; no pH adjustment is needed, but it is good practice to check and maintain the pH between 7.5 and 8.0.

1. Suspension buffer: 50 mM HEPES, pH 7.5, 0.5 M NaCl, 5% glycerol.
2. Lysis buffer: 50 mM HEPES, pH 7.5, 0.5 M NaCl, 5% glycerol, 1% CHAPS, 10 mM imidazole, 1 µL of antiprotease cocktail per mL buffer, 1 mM TCEP and 20 U Benzonase per mL buffer, added before use.
3. Wash buffer: 50 mM HEPES, pH 7.5, 0.5 M NaCl, 5% glycerol, 20 mM imidazole, and 1 mM TCEP.
4. Elution buffer: 50 mM HEPES, pH 7.5, 0.5 M NaCl, 5% glycerol, 500 mM imidazole, and 1 mM TCEP.
5. Gel filtration buffer: 10 mM HEPES, pH 7.5, 150 mM NaCl, 5% glycerol, 0.1 mM TCEP, added before use.

3   Methods

3.1   Growth and Maintenance of Insect Cell Lines

3.1.1   Initiation of Cell Cultures

Three insect cell lines widely used in baculovirus expression systems are Sf9, Sf21, and High Five. Sf9 and Sf21 cell lines are suitable for all insect cell–based work from transfection to expression. However, we suggest starting with the Sf9 cell line. The Sf21 and High Five cell lines are suitable to optimize the expression level once the expression in Sf9 is tested. For HT work, a suspension culture is preferred over the adherent culture; therefore, the protocol is designed for cell culture in suspension.

A full description of insect cell culture methods and BEVS is described in two laboratory manuals [6, 24]. Basically, strict aseptic technique is required to handle and manipulate all cell lines, that is, inside a sterile laminar flow hood (*see* **Note 1**). Anything taken inside the hood should be surface treated with 70% ethanol to decontaminate it, unless stated otherwise. All vessels and pipettes used with live cells or viruses must be decontaminated properly using autoclaving, Virkon (10% w/v) or diluted Decon (10% v/v). Wipe work surfaces with Virkon/bleach followed by 70% ethanol. Treat any spills immediately.

1. Before starting the manipulation, bring the Sf-900 II medium, fetal bovine serum (FBS), and antibiotic (penicillin/streptomycin) solution to 27 °C. Transfer them into a laminar flow hood. Supplement the Sf-900 II medium with a final concentration of 2% FBS (v/v), 50 U of penicillin and 50 μg of streptomycin per mL of medium. Mix by gently inverting the medium bottle three times. Prepare a 250-mL Erlenmeyer flask with a vented cap with 29 mL of supplemented Sf-900 II.
2. Take a cryotube containing frozen stock of Sf9 cells from a liquid nitrogen tank and thaw rapidly in a 37 °C water bath. When completely thawed, rinse the outer surface of the tube in 70% ethanol, and take it into the hood.
3. Transfer the contents of the cryotube into an Eppendorf tube. Spin the tube in a tabletop centrifuge at 500 g for 5 min. Discard the supernatant (the storage medium contains DMSO that may slow down cell growth). Resuspend the cell pellet in 1 mL of supplemented Sf-900 II medium. Use gentle pipetting in all cell handling steps.
4. Add 1 mL of resuspended cells from step 3 to the flask containing the medium from step 1. Incubate the culture at 27 °C with 150 rpm in a humidified shaker incubator (*see* **Note 2**).
5. After 24 h of incubation, take out 100 μL of culture sample and place in an Eppendorf tube. Add 10 μL of 0.4% Trypan blue stain, mix well and transfer a small volume into a hemocytometer. Count the cells using an inverted microscope. Count the number of live cells (appear white) and dead cells (appear blue). Calculate the total number of viable cells per mL of culture and, hence, the percentage viability. If the viability is less than 95%, the culture is not suitable for further work.

6. Keep a record of cell growth patterns for 24 h. Once the count reaches 1.8×10^6 to 2.0×10^6 cells per mL, passage the culture with supplemented Sf 900 II medium back to 0.8×10^6 to 1.0×10^6 cells per mL, and incubate as previously.

During growth and maintenance of cultures, examine them carefully for any signs of contamination. Bacteria, yeast, and filamentous fungus often contaminate the culture. Cloudiness of the overnight culture, a pungent smell in the incubator, and swollen and disordered insect cells are the primary signs of contamination. Contamination can be verified by microscopic observation of the samples under high magnification. However, viral contamination is hard to detect in the first instance. As the incubation continues, appearance of distorted cells and an increase in cell lysis are preliminary signs of unwanted viral contamination. All the contaminated cultures should be isolated from any others and disposed of safely.

3.1.2 Generation of Stock Cell Lines

Cell lines can be stored for long periods of time if grown and stored properly. We recommend backing up stocks every year in case of problems that may arise in the main stock. Preparation of ten vials of stock each with 1 mL of culture is described here. It can be expanded according to need (*see* **Note 3**).

1. Examine the Sf9 culture as described previously (see Sect. 3.1.1). The viability of the culture should be more than 98% and in exponential growth, that is, ~1×10^6 cells per mL, for the purpose of storage.
2. Prepare 10 mL of freezing medium by mixing 6 mL (60% v/v) Grace's insect medium, 3 mL (30% v/v) FBS, and 1 mL (10% v/v) DMSO. Bring the medium to room temperature.
3. Transfer 100 mL of cell culture (1×10^6 cells per mL) into two 50-mL Falcon tubes. Centrifuge at 500 g for 5 min. During this time, take 10 cryovials and mark each with the name of the cell line, the date, expected cell density (i.e., 1×10^7 cells) and original batch number from the purchaser. Keep the vials on ice.
4. Discard the medium safely after centrifugation. Resuspend the pellets from both tubes in 10 mL of freezing medium; they should give a count of ~1×10^7 cell per mL. Dispense 1 mL of the suspension into each cryovial.
5. Place the vials in a CryoCane and freeze gradually by placing at −20 °C for 1–2 h, transfer to −80 °C for 2 days and finally transfer to liquid nitrogen for long-term storage. Slow freezing is a critical step for the long-term viability of insect cells.
6. After 2 weeks, thaw one vial to check viability and absence of contamination.

3.2 Cloning the Gene of Interest

The cloning of a gene into the baculovirus expression vector is a two-step process. The first step involves the cloning of the gene into a transfer vector, such as pFastBac. This contains multiple cloning sites flanked by baculovirus DNA fragments to facilitate recombination into the baculovirus genome. The second step is the transfer of the cloned fragment into a baculovirus genome. This was done originally by cotransfection and recombination in insect cells. In the pFastBac system, the recombinant baculovirus genome is generated in bacteria using a coexpressed DNA transposition system.

When processing multiple genes in a structural genomics context, cloning based on restriction sites is problematic, since restriction sites used for cloning often are present within the gene sequence. Effective, generic alternatives include ligation-independent cloning (LIC) [25] and recombinase-based system (e.g., the Gateway system, Invitrogen).

In this chapter, use of transfer vector pFBOH-LIC-SGC is described, which is derived from the pFastBac vector by introducing LIC cloning sites. The cloning procedure used is based on method described by Stols et al. [26]. Primers for PCR include 14-base extensions that are identical to sequences at the cloning sites in the vector. Both vector and inserts are treated with T4 DNA polymerase, exposing complementary 14-nt single-stranded tails. The vector and inserts are then annealed and introduced into bacteria. We routinely clone the same fragments in parallel into compatible bacterial expression vectors to allow comparative expression studies.

The gene fragments in the transfer vector are propagated in a recA, endA E. coli strain, such as MACH-1, DH5α, or DH10B. The second cloning step is site-specific transposition. The construct cloned in pFBOH-LIC-SGC is transferred into E. coli DH10Bac which carries viral DNA [27]. The recombinant bacmid is reconstituted by recombination, and in the process, the lacZ cassette introduced into the system is disrupted, allowing the selection of clones by color [6]. This allows the HT generation of recombinant bacmid in E. coli. The procedure is described in detail:

1. Bring the DH10Bac selective medium plate (see Sect. 2.7) to room temperature. Prepare the media plates one day before starting the transposition experiment and allow them to dry at room temperature overnight. They can be stored for later use at 4 °C protected from light for up to 4 weeks.
2. Thaw the competent DH10Bac cells on ice. Dispense 20 μL into each well of a prechilled 96-well microtiter plate. Add ~0.3 μg or 2–3 μL of recombinant plasmid (construct). Mix gently using a multichannel pipette. Incubate on ice for 30 min.
3. Heat shock at 42 °C for 45 s. Return to ice and chill for 2 min. Take a 96-deep-well block and dispense 900 μL of prewarmed SOC medium into each well. Transfer the transformed bacterial suspension into the corresponding well of the block and cover with a porous film. Incubate at 37 °C at 700 rpm for 5 h. This incubation is essential for the transposition to occur.
4. Dilute the cultures (10 μL in 90 μL medium) in LB in a 96-well microtiter plate. Spread 50 μL of the diluted cultures on a LB-agar plate prepared in step 1. Store the rest of the culture at 4 °C.
5. Incubate plates at 37 °C overnight. Check the total number of colonies the next day. If there are more than 100 or less than 10 colonies, take the rest of the culture stored at 4 °C (from step 4), spread the appropriate dilution on another plate, and incubate at 37 °C until the blue/white colonies can be differentiated. This usually requires 48 h.

6. Streak two white colonies from each plate onto two fresh LB-agar plates containing antibiotics, bluo-gal and IPTG to verify the phenotype. Incubate for 48 h at 37 °C. White colonies contain the recombinant bacmid and, therefore, are selected for the isolation of recombinant bacmid DNA.

7. Take a 96-deep-well block. Dispense 1 mL of LB containing 50 µg/mL kanamycin, 7 µg/mL gentamycin, and 10 µg/mL tetracycline into each well. From a single colony confirmed as having a white phenotype on plates containing Bluo-gal and IPTG, inoculate the corresponding well of the 96-deep-well block. Incubate at 37 °C overnight, shaking at 700 rpm in a Glas-Col incubator. Add 330 µL of 60% (w/v) sterile glycerol to each well of the 96-deep-well block. Mix well and store at −80 °C.

3.3   High-Throughput Bacmid Production

1. Take a 96-well microtiter plate and dispense 50 µL of 2X LB medium into each well. The recombinant white colonies isolated after blue-white screening are used to isolate recombinant bacmid. Pick a white colony using the end of a pipette tip and suspend in the corresponding well of a 96-well microtiter plate. The suspension is used as an inoculum.

2. Take 2X 96-deep-well blocks to generate 2 mL of each culture for bacmid DNA miniprep. Dispense 1 mL 2X LB medium with 50 µg/mL kanamycin, 7 µg/mL gentamycin, and 10 µg/mL tetracycline into each well. Inoculate each well with 20 µL of inoculum, prepared in step 1, using a multichannel pipette. Cover the blocks with porous adhesive film. Incubate at 37 °C overnight at 700 rpm in a Glas-Col incubator.

3. Centrifuge the 96-deep-well blocks at 2,600 g for 20 min. Immediately decant the supernatant into a container for proper disposal. Invert and tap the blocks on absorbent paper to remove residual supernatant. Freeze the pellets at −20 °C for 30 min (the samples can be frozen for up to 24 h prior to processing).

4. Allow the pellets to thaw at room temperature for 15 min. Vortex the blocks vigorously until the cell pellets are dispersed. Add 250 µL of solution P1 (containing RNase A from Qiagen) to each well of the first block using a multichannel pipette. Seal the block with adhesive sealing film and mix by incubating in a Glas-Col incubator at 800 rpm for about 2 min. Use a multichannel pipette to resuspend the pellet, if necessary. Transfer the suspension to the corresponding wells of the second block. Seal and repeat the mixing process.

5. Add 250 µL of solution P2 (Qiagen) to each well. Seal the block with adhesive sealing film. Invert gently five times and incubate at room temperature for about 10 min.

6. Add 300 µL of $3 M$ potassium acetate pH 5.5, seal and mix gently, but thoroughly, by inverting five times. A thick white precipitate of protein and *E. coli* genomic DNA will form. Place the sample on ice for 20 min, followed by centrifugation at 2,600 g for 30 min at 4 °C.

7. Transfer the clear supernatant gently into a new 96-deep-well block. Centrifuge once again at 2,600 g for 30 min. (Note: It is difficult to separate the clear supernatant after the first centrifugation step. Great care should be taken while transferring the supernatant. A second centrifugation step helps obtain a cleaner supernatant.)

8. Take another 96-deep-well block. Dispense 0.8 mL of isopropanol into each well. Take 0.8 mL of clear supernatant from step 7 and add it to the corresponding well. Do not try to take all the supernatant. Seal the block and invert it five times to mix and incubate on ice for 30 min. The block can be left overnight at 4 °C, if necessary.

9. Centrifuge at 2,600 g for 30 min. Spray the outer surface of the block with 70% ethanol and take it inside the laminar flow hood (*see* **Note 4**). Discard the supernatant.

10. Add 500 µL of 70% (v/v) ethanol to each well. Gently tap the block to wash the pellet properly. Centrifuge at 2,600 g for 30 min. Take the block back inside the flow hood. Discard the supernatant and drain the last drops of ethanol by tapping the block on layers of blotting paper or tissue paper.

11. Allow the pellet to dry inside the flow hood. Check often to avoid overdrying. It takes about 15–20 min to dry. Add 50 µL of sterile TE buffer to resuspend the bacmid DNA. Separate 2 µL of each bacmid sample to measure the concentration and transfer the remaining to a microtiter plate for further use and storage. Make sure the absorbance ratio A260 nm/A280 nm is more than 1.7. Run a few samples on 1% (w/v) agarose gel containing CYBR Safe DNA gel stain.

12. Use 10 µL of the bacmid DNA just prepared in each well for each Sf9 transfection experiment.

3.4   Confirmation of Transposition by PCR on Bacmid DNA

It has been found that colony PCR, to confirm the transposition, does not work consistently. Therefore, PCR on bacmid DNA is recommended for verifying the presence of the gene of interest. Set up the PCR reaction in 96-well PCR plate(s). Prepare PCR master mix as shown in Table 19.1.

Table 19.1 PCR master mix

Stock solution	Concentration (100X)
10X Pfx Buffer	500 µL
10X PCR enhancer	500 µL
50 mM MgSO$_4$	100 µL
25 mM dNTP mix	50 µL
M13-Bac forward primer (100 µM)	9 µL
M13-Bac forward primer (100 µM)	9 µL
Water	1.4 mL
Platinum Pfx polymerase	20 µL

1. Aliquot 20 μL of PCR mix into each well of a 96-well PCR plate. Take the microtiter plate containing bacmid DNA. Use a multichannel pipette to add 1 μL of Bacmid DNA template into the corresponding well of the PCR plate.

2. Perform amplification of the target genes with initial denaturation for 3 min at 94 °C, followed by 30 cycles of 94 °C for 45 s, 50 °C for 45 s, and 72 °C for 5 min. The reaction is terminated after a final extension for 7 min at 72 °C.

3. Analyze 10 μL of amplification product on 1% (w/v) agarose gel (containing CYBR safe DNA gel stain) by electrophoresis. Insertion of the gene into the pFastBac plasmid results in a PCR product of 2,300 bp fragment of vector plus the size of the insert.

3.5   Transfection and Infection of Insect Cells

1. Grow Sf9 cells to mid log phase in Sf900-II medium supplemented with 2% (v/v) fetal bovine serum (see Sect. 3.1.1). The experimental setup is alternated between 24-deep-well blocks or 24-well TC plates and 96-deep-well blocks or 96-well microtiter plates, according to the required volumes at each step (see Note 5). To facilitate pipetting using multichannel pipettes, the convention shown in Fig. 19.2 is used in the transfer from 96-well to 24-well plates or blocks. Attach tips to every second position of a multichannel pipette for the transfer. The top four rows of the 96-well plate or block correspond to the first two in the 24-well block or plate and so forth. Alternatively, samples can be transferred one at a time.

2. Take mid-log phase Sf9 cells and dilute to 2×10^5 cells/mL. Dispense 1 mL (2×10^5 cells) into each well of four 24-well TC plates. Include one for cellfectin and the other for untreated cell controls. Incubate the plate at 27 °C for 1 h to allow cell attachment.

3. Bring the Grace's insect medium (serum free) and cellfectin to room temperature. Take a sterile flat-bottomed 96-well microtiter plate (that can hold ~200 μL of sample). Dispense 50 μL of Grace's insect medium into each well. Transfer 10 μL of recombinant bacmid DNA into each well and mix by shaking the plate gently or pipetting.

4. Mix cellfectin thoroughly by tapping the tube gently. In a 50-mL tube, mix 5 mL of Grace's insect medium and 0.3 mL of cellfectin. Mix thoroughly and add 50 μL to each well of the 96-well microtiter plate containing the recombinant bacmid DNA from step 3. Do not add cellfectin mixture to the cell control well, but add an equivalent amount of medium to it. Mix briefly by tapping the plates.

5. Cover the microtiter plate with adhesive film and incubate the mixture inside the laminar flow hood for 45 min. Dilute the solution by adding 100 μL of Grace's insect medium (serum free) to each well of the 96-well microtiter plate.

6. Take out the 24-well TC plate containing cells from step 2, aspirate the medium using a multichannel pipette. Wash once with 1 mL of Grace's insect medium. Remove the medium and overlay the cells with diluted lipid-DNA complexes

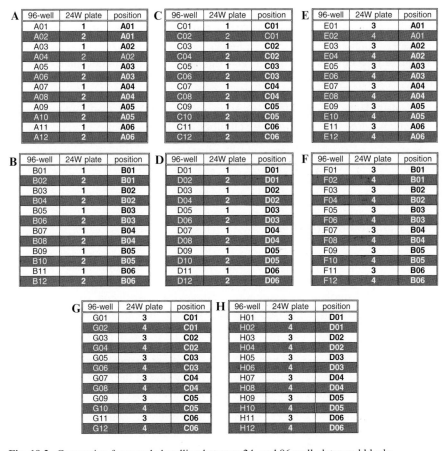

A

96-well	24W plate	position
A01	1	A01
A02	2	A01
A03	1	A02
A04	2	A02
A05	1	A03
A06	2	A03
A07	1	A04
A08	2	A04
A09	1	A05
A10	2	A05
A11	1	A06
A12	2	A06

C

96-well	24W plate	position
C01	1	C01
C02	2	C01
C03	1	C02
C04	2	C02
C05	1	C03
C06	2	C03
C07	1	C04
C08	2	C04
C09	1	C05
C10	2	C05
C11	1	C06
C12	2	C06

E

96-well	24W plate	position
E01	3	A01
E02	4	A01
E03	3	A02
E04	4	A02
E05	3	A03
E06	4	A03
E07	3	A04
E08	4	A04
E09	3	A05
E10	4	A05
E11	3	A06
E12	4	A06

B

96-well	24W plate	position
B01	1	B01
B02	2	B01
B03	1	B02
B04	2	B02
B05	1	B03
B06	2	B03
B07	1	B04
B08	2	B04
B09	1	B05
B10	2	B05
B11	1	B06
B12	2	B06

D

96-well	24W plate	position
D01	1	D01
D02	2	D01
D03	1	D02
D04	2	D02
D05	1	D03
D06	2	D03
D07	1	D04
D08	2	D04
D09	1	D05
D10	2	D05
D11	1	D06
D12	2	D06

F

96-well	24W plate	position
F01	3	B01
F02	4	B01
F03	3	B02
F04	4	B02
F05	3	B03
F06	4	B03
F07	3	B04
F08	4	B04
F09	3	B05
F10	4	B05
F11	3	B06
F12	4	B06

G

96-well	24W plate	position
G01	3	C01
G02	4	C01
G03	3	C02
G04	4	C02
G05	3	C03
G06	4	C03
G07	3	C04
G08	4	C04
G09	3	C05
G10	4	C05
G11	3	C06
G12	4	C06

H

96-well	24W plate	position
H01	3	D01
H02	4	D01
H03	3	D02
H04	4	D02
H05	3	D03
H06	4	D03
H07	3	D04
H08	4	D04
H09	3	D05
H10	4	D05
H11	3	D06
H12	4	D06

Fig. 19.2 Convention for sample handling between 24- and 96-well plates and blocks

(0.2 mL) from step 5. Add a further 0.2 mL of Grace's insect medium (serum free) to each well. Incubate cells for 5 h at 27 °C.

7. Remove the transfection mixture and add 0.7 mL of SF900 II insect medium containing 2% (v/v) FBS and antibiotics (50 U penicillin and 50 μg streptomycin per mL medium) to each well. Incubate cells at 27 °C for 72–96 h.

8. Look for signs of infection in the transfected cells 72 h post transfection, by comparing with the control cells under an inverted microscope. Confluent growth of cells will be seen in control wells, whereas areas of clearing will be prominent in wells with infected cells. Infected cells usually are larger and deformed or elongated compared to uninfected cells. Collect viruses when the cells are well infected; this may take up to 96 h or more (*see* **Note 6**).

9. Transfer the liquid contents from the 24-well tissue culture plate into a sterile 96-deep-well block and centrifuge at 1,500 g for 20 min at room temperature. Collect the clear supernatant in another sterile 96-deep-well block. This is the P0 viral stock. Store at 4 °C, protected from light.

10. Further amplification of the virus can be carried out in 24-deep-well blocks in suspension culture. Dispense 3 mL of Sf9 cells (2×10^6 cells/mL) into each well of four sterile 24-deep-well blocks.
11. Infect the cells with 120 µL of P0 virus stocks using a multichannel pipette. Cover the block with porous seal and incubate at 27 °C at 450 rpm for 48 h in a Glas-Col incubator.
12. Centrifuge the 24-well block at 1,500 g at 4 °C for 20 min. Collect the supernatant and store in a 2X 96-deep-well block as P1 virus stock.

Further, viral stocks (P2 virus and P3 virus) can be amplified as necessary by repeating this procedure (*see* **Note 7**). The virus stocks are stored at 4 °C, protected from light (*see* **Note 8**).

3.6 Plaque Purification

The first baculovirus plaque assay was described by Hink and Vail [28]. The method was modified by others in due course [29, 6]. However, the principle is based on the infection of cells with a very dilute virus suspension, so that each cell becomes infected by a single viral particle. The cells then are overlaid with a solid medium to limit the spread of virus particles. After several cycles of infection, a clearing around the cell, called a *plaque*, is formed. All viruses in an isolated plaque are derived from a single viral particle and so represent a clonal population. Plaque assay is the direct method for determining the viral titer of the stock. Plaque purification allows producing a pure, clonal viral stock (*see* **Note 9**).

1. Take a log phase Sf9 insect cell culture and dilute to obtain a suspension of 5×10^5 cells/mL. Aliquot 2 mL (1×10^6 cells) into each well of six-well tissue culture plates. Use duplicate plates for each target. Incubate the plates at 27 °C for 1 h in an incubator to allow the cells to attach.
2. During the 1-h incubation, prepare a serial dilution (10^{-1} to 10^{-8}) of the clarified baculovirus stock (either P2 virus or P3 virus stock) in Sf-900 II medium (use 0.5 mL of stock and dilute with 4.5 mL of the Sf-900 II medium). Perform the dilution in a sterile laminar flow hood.
3. After 1 h, observe the cell monolayer under the microscope. Sf9 cells should be attached and at ~50% confluency. Move the six-well tissue culture plates containing Sf9 cells to the sterile hood. Mark the wells in sequence with 10^{-4}, 10^{-5}, 10^{-6}, 10^{-7}, 10^{-8}, and blank (cells only). Remove the medium and replace each well with 1 mL of the appropriate virus dilution, using 1 mL of medium in the control well. Incubate for 1 h at room temperature to initiate the infection.
4. During the incubation period, prepare the plaque medium as follows: melt the 4% (w/v) agarose gel by heating in a microwave oven and place in a 40 °C water bath until ready to use, and prewarm an empty, sterile, 100 mL container, sterile pipettes and 1.3X Sf-900 II medium to 40 °C.
5. Prepare the plaquing medium. Transfer the 4% agarose, prewarmed sterile bottle and medium into a sterile laminar flow hood. Quickly combine 20 mL of 1.3X

Sf-900 II medium and 7 mL of the melted 4% agarose in an empty sterile bottle. Return the bottle of plaquing medium to the 40 °C water bath until ready to use.

6. Following the 1-h incubation of Sf9 cells and virus, remove the medium containing virus from the wells and replace with 2 mL of plaquing medium. Work as quickly as possible using aseptic technique. Allow the agarose to harden for 10–20 min at room temperature. Incubate the cells in a 27 °C humidified incubator for 7–10 days until plaques are visible and ready to count.

7. Observe the plaques under the microscope. Mark the clear plaques and use them for the amplification of recombinant virus stock. Pick each plaque using a sterile Pasteur pipette and transfer into 500 μL of Sf-900 II medium. Mix well by vortexing. Use 100 μL to infect mid-log phase Sf9 cells. Follow steps 10, 11, and 12 in Sect. 3.5. The virus thus obtained can be used as plaque purified stock.

3.7 Production and Storage of Recombinant Viruses

Large-scale production of virus often is necessary for the production of sufficient recombinant protein. It is recommended to keep a viral stock of 500 mL for each potential protein target. Whenever possible, amplify the plaque-purified virus to keep stocks maintained (*see* **Note 9**). The alternative approach would be to use either P2 or P3 virus stock (see Sect. 3.5). The further production of virus is a two-step procedure.

1. Infect 100 mL of Sf9 cells (2×10^6 cells/mL) with 200 μL of plaque-purified virus or 2 mL of P3 virus. Incubate at 27 °C at 150 rpm for 48 h. Centrifuge the culture at 1,500 g for 25 min. Separate the clear supernatant containing the virus and store at 4 °C in the dark. Analyse the pellet for possible expression.

2. Infect 500 mL of Sf9 cells with 15–20 mL of virus obtained as just described. Incubate and harvest the virus as specified.

Virus can be stored at 4 °C for up to 6 months in a stable condition. After 6 months, the titer may decrease. If long-term storage is necessary, aliquot the stock in smaller volumes, add 5% (v/v) FBS and store at −80 °C. Frequent freezing and thawing highly reduces the viral titer. Before carrying out any expression studies using the virus stock, it is strongly recommended that a functional titration is performed (see Sect. 3.9).

3.8 High-Throughput Small-Scale Expression Study

This protocol has been developed using 24-deep-well blocks and is designed to rapidly analyze the expression of 96 clones.

1. Take four 24-deep-well blocks. Dispense 3 mL of mid-log phase Sf9 cells (~2×10^6 cells per mL). Infect the cells with 120 μL of virus stock to be analyzed for expression. Incubate at 27 °C at 450 rpm in a Glas-Col incubator for 48 h.

2. Centrifuge the blocks at 1,500 g for 25 min. Discard the supernatant safely. Use the pellet for the extraction of protein. Store at −20 °C if using later.

3. Add 0.4 mL of Suspension buffer (see Sect. 2.8) to each well of the 24-deep-well blocks containing pellets. Shake at 800 rpm in a Glas-Col shaker for 5 min to resuspend. Do not incubate longer. Transfer to ice if incubation needs to be longer. During this time, prepare lysis buffer (see Sect. 2.8).

4. Make sure the pellet in the 24-deep-well block is resuspended completely; if not, use a pipette to resuspend the pellet.

5. Transfer the suspension from the 24-deep-well block into a 96-deep-well block using a multichannel pipette (see Fig. 19.2). Add 1 mL of lysis buffer to each well. Shake the suspension at 800 rpm in a Glas-Col shaker for 30 min. Check the lysed cells under a microscope. If lysis is not complete, freeze and thaw the cells a few times. Take 50 μL of the suspension, mix with 10 μL SDS-PAGE loading dye, heat at 98 °C for 10 min, spin at 10,500 g for 15 min in a benchtop centrifuge, and analyze the clear supernatant on SDS-PAGE (total fraction).

6. Centrifuge the block at 1,500 g for 30 min at 4 °C. Transfer the clear supernatant into another 96-deep-well block. Take 50 μL of the suspension, mix with 10 μL SDS-PAGE loading dye, heat at 98 °C for 10 min, spin at 10,500 g for 15 min in a benchtop centrifuge, and analyze the clear supernatant on SDS-PAGE (soluble fraction).

7. Add 100 μL of 50% Ni-NTA slurry preequilibrated in suspension buffer to the remaining supernatant in each well of the block. Keep on ice for about 30 min with intermittent mixing.

8. Transfer the Ni-NTA and supernatant mix into a 96-well filter block for purification. Place the block on top of a waste block (96-deep-well block) and allow the liquid to drip through the filters.

9. Discard the flow-through and wash filter plate once with 0.4 mL of binding buffer and three times with 0.4 mL washing buffer. After all the liquid has dripped through the filter block, centrifuge the block at 700 g for 1 min to remove the residual liquid.

10. Place the filter plate on top of a 96-well V-bottom microtiter plate. Add 80 μL of elution buffer to each well of the filter plate. Incubate the plate at room temperature for 5 min. Centrifuge at 700 g for 1 min to elute the protein.

11. Transfer 16 μL of the eluted fraction into corresponding positions in a fresh 96-well microtiter plate. Add 4 μL SDS-PAGE loading dye. Pipette up and down to mix the sample.

12. Seal the plate and heat at 98 °C for 10 min. Load 12.5 μL of sample on SDS-PAGE using a multichannel pipette. Run the gels at 150 V for 1 h 30 min. Stain the gel with biosafe Commassie blue and destain to visualize the protein bands.

This study gives a rapid analysis of recombinant baculovirus, which then can direct the expression of target proteins in insect cells and indicate which is the best to scale up (*see* **Note 10**).

3.9   Optimization of Expression Conditions and Functional Titration

Expression levels of a heterologous gene may vary significantly due to different factors. The effect of several parameters can be analyzed by performing the small-scale expression study (see Sect. 3.8). This is very useful as an aid to optimize expression conditions and obtain information regarding scale-up possibilities. We recommend the following:

1. Three volumes of virus from the stock (20 μL, 40 μL, and 80 μL for 2×10^6 Sf9 cells).
2. Two concentrations of Sf9 cells ($\sim 2.0 \times 10^6$ cells per mL and $\sim 3 \times 10^6$ cells per mL).
3. Three harvesting time points (48 h, 72 h, and 96 h).
4. Two cell lines (Sf9 and High Five) (*see* **Note 11**).

Each experimental setup is designed in duplicate, accommodating four 24-deep-well blocks, or multiples of four, to carry out purification steps in 96-well format. The infection procedure, incubation, harvesting, affinity purification, and analysis in small-scale are as described previously. In many cases, the purified protein obtained at this stage is enough to carry out some biochemical analysis and determine the molecular mass using mass spectrometry. With the information obtained by following this method, plaque assays may be replaced, which saves a lot of time-consuming steps. Hence, this method has been presented as a functional titration of virus.

4   Notes

1. Working with insect cell lines for the first time is a challenge. Strict aseptic handling and good laboratory practice must be implemented to keep the cell lines away from contamination. It is advisable to use Pen/Strep solution, if necessary, to avoid bacterial contamination. The Sf9 cells are adapted to a serum-free medium; however, addition of 2% (v/v) FBS helps maintain the cells. It also avoids clumping of the cells and enhances attachment on TC plates. Such cells attach within 15 min. The routine growth and expression study can be carried out in Sf9 cells using Sf-900 II medium either supplemented with FBS or not. However, this medium is not suitable for transfection. Cells grown in Sf-900 II should be washed with Grace's insect medium prior to transfection.
2. Most insect cell culture techniques rely on adherent cultures for the initiation of growth from the stock. The idea behind this is to remove the DMSO present in the storage medium, which hampers cell growth. We developed a different method, as described in Sect. 3.1.1 (steps 2 and 3). This saves the time required for attachment and dislodging of cells in the adherent culture. It is necessary to be gentle and careful while suspending insect cells. Insect cells do not have a cell wall and hence are fragile.

3. In case a large volume of fresh cells are needed at short notice, initiate the culture using the contents of two to three stock vials in 29 mL of Sf-900 II medium and follow the routine method to grow the cells. This helps to reduce the lag period, and high cell density can be achieved more quickly. This also can be adapted as a routine method if sufficient stocks are available. Proper aeration is necessary and must be maintained by keeping the shaker speed as high as 180 rpm while using 250 mL Erlenmeyer flasks. The presence of 2% (v/v) FBS is vital in achieving a high-density cell culture that can be maintained at 3–4×10^6 cells per mL with 100% viability. Such cultures can be diluted to $\sim 2 \times 10^6$ cells per mL and used directly for expression studies.

4. It is worth mentioning here that the recombinant bacmid DNA required for transfection of insect cells should be sterile and free from RNA. We found it difficult to maintain sterility with bacterial lysates purified using commercial kits. This necessitates an extra ethanol precipitation step before using the DNA for transfection. Furthermore, there is the possibility of cross-contamination of bacmids from one well to another. If commercial devices are preferred, make sure systems that have each well of the multi-well block constructed as an individual unit are used, that is, not sharing a large filter unit for a whole plate. We present an easy and economical method of bacmid DNA extraction suitable for transfection. In this simplified method, 96-deep-well blocks are used. The solutions are either ready-to-use commercial reagents or common laboratory chemicals, that is, potassium acetate and isopropanol. Cross-contamination of bacmid DNA and microbial contamination were greatly reduced employing this method. The yield of bacmid DNA also was higher than that obtained from the commercial kit. The user needs to be vigilant in the precise use of multichannel pipettes and tips and concentrate while performing the HT manipulation.

5. From the point of transfection, the frequent use of 24-well plates or blocks and 96-well plates or blocks may increase the risk of contamination and errors. The first problem most novice researchers encounter is the improper orientation of plates and blocks causing confusion in sample transfer from one plate or block to another. Before starting any manipulation, the plates and blocks must be labeled in proper orientation. Another problem arises while loading tips on multichannel pipettes. There might be an air gap, which is a cause of aerosol generation during pipetting. Aerosols can be the cause of cross-contamination within a block as well as spreading contamination across the working area and other materials. Ensure filter tips are used and they are fitted firmly. While dispensing liquid into deep-well blocks, touching the tips on the side wall of the well and releasing the sample gently helps avoid the formation of aerosols. The use of filtered tips is highly recommended.

6. Transfected cells should be observed daily for any signs of infection. Identification of infected cells poses another challenge for the first-time user. Cells from old culture and cells in medium with limited nutrients or insufficient aeration display morphology similar to that shown as a result of infection. The first sign of infection

is the cessation of growth and multiplication. Therefore, each set of experiments should be accompanied by a control cell well for comparison.

The success of transfection depends on the concentration and purity of the bacmid DNA used. When an experiment is set up for HT, finding out the optimum concentration for each target is not possible. Hence, a compromise is made to collect the virus after 72 h of transfection. The amount of virus thus obtained generally is sufficient to use for follow-up amplification. If it does not work for particular targets, repeat the process for these using six-well TC plates. The DNA to cellfectin ratio may need to be optimized for particular targets to achieve successful transfection.

7. During follow-up amplification of recombinant virus, <48 h post infection is the best time to obtain an infective form, that is, a BV form (see Sect. 1). Although the gene responsible for the occlusion formation is replaced by a foreign gene, the viruses produced in the very late phase carry other cellular materials as a result of cell lysis. These might cause problems in subsequent infections. We observed a decrease in efficiency of infection and level of protein production when virus produced after 72 h is used.

8. If the virus stock produces slimy material, centrifuge at 17,500 g, separate the clear supernatant, and titrate it for expression level. Do not filter the viral stock: Filtration results in a nearly complete loss of viral titer.

Even with the best care, virus stocks may lose activity or get contaminated. It is best policy to store the cloned vector and bacmid DNA so that new virus can be easily generated. We recommend storing duplicate sets at 4 °C and −80 °C.

9. Plaque purification of virus stock is highly desirable and gives a clonal isolate. Virus stock generated from such an isolate should give uniform baculovirus clones, which avoids variation in expression products. However, the demand of HT work may lead to a compromise, so that plaque purification and functional titration or functional titration followed by plaque purification is carried out only on selected clones. Functional titration helps select the best clones within 48 h, while plaque purification needs at least 7 d to isolate the virus. We carry out functional titrations on a set of best possible clones directly after P2 virus generation. Any potential clone then is purified by plaque assay before generating the viral stock. This speeds up the entire process of screening by at least 1 week.

10. A small-scale study helps in identifying the best clone for large-scale production of a protein. Large-scale expression study is labor intensive, uses large amounts of consumables, and takes significant time. Moreover, the expression level may differ from time to time and from one batch to another. To monitor such variation, we developed a procedure for rapid extraction, purification, and analysis of small-volume culture samples. The process utilizes (1) the addition of 50 μL of Insect PopCulture Reagent (Novagen 71187) in 1 mL culture samples, (2) vigorous mixing and incubation for 15 min at room temperature followed by (3) test affinity purification (e.g., Ni-NTA purification), and (4) the rapid analysis of samples performed by running them on SDS-PAGE

gels for 50 min and visualizing by Coomassie staining. The overall process can be completed within 2 h 30 min to give an estimate about the level of expression. Close monitoring ensures that any problems in expression of target proteins can be detected early on to avoid unnecessary downstream processing.

Serum-free medium is not suitable for His-affinity purification: A component in the medium may strip-off the Ni^{++} from the column. If heterologous protein is expressed in an extracellular medium, buffer exchange is necessary prior to affinity purification. However, this is not a problem in small-scale expression studies.

11. Once the first expression study is carried out in Sf9 cell lines, we highly recommend analyzing the expression level in HighFive cell lines. The HighFive cell line is generally known to express a higher level of extracellular protein [30]. In our study, most intracellular proteins also were expressed in higher levels than in Sf9 cells. We followed the same protocol to grow and maintain HighFive cells as that used for Sf9 cells. The differences are as follows: (1) the speed of the shaker is maintained at 100 rpm at the beginning and at a maximum of 120 rpm as the cells adapt, (2) the medium used was Express SFM supplemented with L-glutamine, and heparin was added at 10 U/mL to prevent aggregation. The addition of heparin is not necessary once the cells are adapted to suspension culture; (3) suspension cultures should be passaged before they reach a density of 2.0–2.5 × 10^6 cells/mL.

References

1. Harrap KA (1972) The structure of nuclear polyhedrosis viruses. I. The inclusion body. Virology 50:114–123
2. Fraser M (1986) Ultrastructural observations of virion maturation in *Autographa californica* nuclear polyhderosis virus infected *Spodoptera frugiperda* cell cultures. J Ultrastruct Mol Struct Res 95:189–195
3. Summers MD, Anderson DL (1972) Characterization of deoxyribonucleic acid isolated from the granulosis viruses of the cabbage looper, *Trichoplusia ni* and the fall armyworm, *Spodoptera frugiperda*. Virology 50:459–471
4. Vialard JE, Arif BM, Richardson CD (1995) Introduction to the molecular biology of baculoviruses. Methods Mol Biol 39:1–24
5. Blissard GW, Rohrmann GF (1990) Baculovirus diversity and molecular biology. Ann Rev. Entomol 35:127–155
6. O'Reilly D, Miller L, Luckow V (1992) Baculovirus expression vectors: A laboratory manual. W. H. Freeman and Company, San Francisco
7. Ooi BG, Miller LK (1988) Regulation of host RNA levels during baculovirus infection. Virology 166:515–523
8. Knudson D, Harrap K (1976) Replication of a nuclear polyhedrosis virus in a continuous cell culture of *Spodoptera frugiperda*: Microscopy study of the sequence of events of the virus infection. J. Virol 17:254–268
9. Lee HH, Miller LK (1979) Isolation, complementation, and initial characterization of temperature-sensitive mutants of the baculovirus *Autographa californica* nuclear polyhedrosis virus. J Virol 31:240–252

10. Carstens E, Tjia S, Doerfler W (1979) Infection of *Spodoptera frugiperda* cells with *Autographa californica* nuclear polyhedrosis virus I. Synthesis of intracellular proteins after virus infection. Virology 99:386–398

11. Dobos P, Cochran M (1980) Protein synthesis in cells infected by *Autographa californica* nuclear polyhedrosis virus (Ac-NPV): The effect of cytosine arabinoside. Virology 103:446–464

12. Miller L, Trimarchi R, Browne D, Pennock G (1983) A temperature-sensitive mutant of the baculovirus *Autographa californica* Nuclear polyhedrosis virus defective in an early function required for further gene expression. Virology 126:376–380

13. Smith GE, Fraser MJ, Summers MD (1983) Molecular engineering of the *Autographa californica* nuclear polyhedrosis virus genome: Deletion mutations within the polyhedrin gene. J Virol 46:584–593

14. Ooi BG, Rankin C, Miller LK (1989) Downstream sequences augment transcription from the essential initiation site of a baculovirus polyhedrin gene. J Mol Biol 210:721–736

15. Qin JC, Liu AF, Weaver RF (1989) Studies on the control region of the p10 gene of the *Autographa californica* nuclear polyhedrosis virus. J Gen Virol 70, Part 5:1273–1279

16. Weyer U, Possee RD (1989) Analysis of the promoter of the *Autographa californica* nuclear polyhedrosis virus p10 gene. J Gen Virol 70, Part 1:203–208

17. Kost TA, Condreay JP, Jarvis DL (2005) Baculovirus as versatile vectors for protein expression in insect and mammalian cells. Nat Biotechnol 23:567–575

18. Hunt I (2005) From gene to protein: A review of new and enabling technologies for multi-parallel protein expression. Protein Expr Purif 40:1–22

19. Luckow VA, Lee SC, Barry GF, Olins P O (1993) Efficient generation of infectious recombinant baculoviruses by site-specific transposon-mediated insertion of foreign genes into a baculovirus genome propagated in *Escherichia coli*. J Virol 67:4566–4579

20. McCall EJ, Danielsson A, Hardern IM, Dartsch C, Hicks R, Wahlberg JM, Abbott WM (2005) Improvements to the throughput of recombinant protein expression in the baculovirus/insect cell system. Protein Expr Purif 42:29–36

21. Gao M, Brufatto N, Chen T, Murley LL, Thalakada R, Domagala M, Beattie B, Mamelak D, Athanasopoulos V, Johnson D, McFadden G, Burks C, Frappier L (2005) Expression profiling of herpesvirus and vaccinia virus proteins using a high-throughput baculovirus screening system. J Proteome Res 4:2225–2235

22. Philipps B, Rotmann D, Wicki M, Mayr LM, Forstner M (2005) Time reduction and process optimization of the baculovirus expression system for more efficient recombinant protein production in insect cells. Protein Expr Purif 42:211–218

23. Philipps B, Forstner M, Mayr LM (2005) A baculovirus expression vector system for simultaneous protein expression in insect and mammalian cells. Biotechnol Prog 21:708–711

24. Invitrogen (2002) Growth and maintenance of insect cell lines. Invitrogen Life Technologies, version K, Paisley, UK.

25. Aslanidis C, de Jong PJ (1990) Ligation-independent cloning of PCR products (LIC-PCR). Nucleic Acids Res 18:6069–6074

26. Stols L, Gu M, Dieckman L, Raffen R, Collart FR, Donnelly MI (2002) A new vector for high-throughput, ligation-independent cloning encoding a tobacco etch virus protease cleavage site. Protein Expr Purif 25:8–15

27. Je H, Hee Chang J, Young Choi J, Yul Roh J, Rae J, O'Reilly D, Kwon Kang S (2001) A defective viral genome maintained in *Escherichia coli* for the generation of baculovirus expression vectors. Biotechnology Letters 23:575–582

28. Hink W, Vail P (1973) A plaque assay for titration of alfalfa looper nuclear polyhedrosis virus in a cabbage looper (TN-368) cell line. J Invertebrate Pathology 22:168–174

29. Lee HH, Miller LK (1978) Isolation of genotypic variants of *Autographa californica* nuclear polyhedrosis virus. J Virol 27:754–767

30. Davis TR, Trotter KM, Granados RR, Wood HA (1992) Baculovirus expression of alkaline phosphatase as a reporter gene for evaluation of production, glycosylation and secretion. Biotechnology (NY) 10:1148–1150

Chapter 20
Coimmunoprecipitation and Proteomic Analyses

S. Fabio Falsone, Bernd Gesslbauer, and Andreas J. Kungl

Abstract Defining protein-protein interaction networks is a major goal of proteomics. Here, we present a protocol for coimmunoprecipitation, a technique suitable for the isolation of whole protein complexes in vivo and their subsequent identification by either immunoblotting or mass spectrometric sequencing combined to database search.

Keywords coimmunoprecipitation, immunoblotting, protein complexes, protein-protein interaction, mass spectrometry

1   Introduction

In the Italian Commedia dell' Arte, a form of popular theatre of the 18th century, the actors were asked to freely arrange and interpret a conventionally defined plot. The dynamics originating from the continuously changing interplay and attitude of every actor finally generated diversified and exciting performances based on fixed rules of a same story line. Similarly, the proteome is a fascinating, complex world based on the interplay of protein actors continuously associating and dissociating from each other, building and disrupting networks, sometimes undergoing promiscuous partnerships, but all in observance of the "relatively little" information written in the genomic story line (humans have "only" 20,000–25,000 genes; [1]).

Unraveling the complexity of the proteome is a major mission of proteomics [2], wherein one key feature is to deliver clues on the interaction profiles of proteins within their cellular environment [3]. However, this task is complicated by the dynamic lifestyle of a protein. Proteins may be short or long lived; they may be posttranslationally modified and may occur in isoforms, the expression and interaction of a single protein within protein networks can be tissue, age, or disease specific. Therefore, proteomic interaction studies mostly reflect snapshots of "frozen" protein complexes at a given time of cellular life.

From: Methods in Molecular Biology, Vol. 439: *Genomics Protocols: Second Edition*
Edited by Mike Starkey and Ramnath Elaswarapu © Humana Press Inc., Totowa, NJ

Despite these limitations, information inferred from proteomics-based protein interaction profiling is very useful for systems biology to generate global protein interaction networks out of single clusters, as well as to discover disease-related protein interactions patterns. For these reasons, several methods to isolate and characterize protein interaction partners have been developed [4].

The technique presented in this chapter deals with the isolation of protein complexes by coimmunoprecipitation (co-IP), and their identification by immunoblotting or mass spectrometric sequencing (Fig. 20.1). Coimmunoprecipitation is a variant of immunoprecipitation (IP) [5] and relies on the immunodepletion of a protein of interest ("bait") together with all its protein interacting partners ("preys") in solution (e.g., a cell extract). For this purpose, an antibody directed to the antigen acting as a bait is added to the solution. The resulting antibody-bait-prey complex is then captured on a protein A- or protein G-immobilized resin. These cell wall proteins from *Staphylococcus aureus* exhibit unique binding properties for IgG from a variety of mammalian species, binding with the Fc region of immunoglobulins through interaction with the heavy chain. The resin carrying the antibody-bait-prey complex then can be removed from the solution, such as by centrifugation (note that the term immuno*precipitation* is misleading, as the antibody-antigen complex is not insoluble; more precisely, it binds to an insoluble matrix).

The captured protein interaction partners can be resolved with SDS-polyacrylamide gel electrophoresis (SDS-PAGE) and identified by either immunoblotting or in-gel proteolysis followed by nano-HPLC/mass spectrometric (MS) sequencing and database search. Immunoblotting is versatile and useful for the low-throughput identification of one or a few specific interaction partners. Obviously, in this case, the scientist must know which interaction partner he or she is looking for, which is demanding when screening for novel, a-priori unknown interactors. In this case, MS-sequencing is the method of choice, as it requires no a-priori knowledge about possible protein interaction partners.

The protocol presented here was used during the preliminary proteomic investigation of heat shock protein 90 (Hsp90)-interacting proteins [6]. Hsp90 is a molecular chaperone that stabilizes partially folded proteins, preventing their aggregation within the cell. Several of its partner proteins accomplish key oncogenic roles in signal transduction, and inhibition of Hsp90 was shown to induce their degradation [7]. This makes the search for novel interaction partners of Hsp90 particularly intriguing.

Although compiled for Hsp90, the procedure presented here should apply as a general starting protocol for co-IPs involving other proteins. However, remember that every actor has its own attitude, meaning that the optimal protein interaction parameters (buffers, used additives, incubation times) for co-IP must be empirically determined for each interaction system to be investigated (*see* **Note 1**).

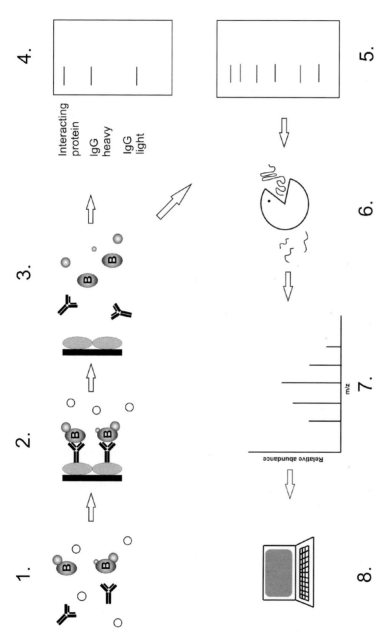

Fig. 20.1 Schematic work through of the co-IP procedure. (1) An antibody directed to an antigen acting as bait (B) is added to a cell extract. (2) The antibody-antigen complex then becomes immobilized on a protein A (or protein G) agarose resin). Physical interactors (prey, small filled gray circles) remain captured to the bait, whereas nonbinding proteins (white circles) can be removed by repeated washing. (3) The complexes then can be eluted from the matrix. (4) Captured interacting proteins can be detected either by immunoblotting using an antibody directed against one particular interacting partner or by mass spectrometric sequencing. For this purpose, the bands of interest are (5) excised from a stained SDS-PAGE gel, (6) cleaved by proteolysis, (7) analyzed by MS, and (8) the spectra are matched to a particular protein by database search

2   Materials

Unless otherwise specified, all reagents are from Sigma-Aldrich and all solutions should be prepared in distilled water.

2.1   Cell Culture

1. Dulbecco's Modified Eagles Medium (DMEM) (*see* **Note 2**), supplemented with 10% (v/v) heat inactivated fetal bovine serum (*see* **Note 3**) (PAA, Pasching, Austria), a penicillin/streptomycin mixture (PAA) (corresponding to a final concentration of 1 u/mL penicillin and 0.1 mg/mL streptomycin), and 2.5 µg/mL amphotericin B (PAA). Store at 4 °C.
2. 1X Sterile Phosphate Buffered Saline (PBS). Prepare 1 L of a 10X stock solution (40 mM KH_2PO_4, 160 mM Na_2HPO_4, 1.15 M NaCl, pH 7.4). Make tenfold dilutions of this stock solution when required and autoclave. Store at 4 °C.
3. Sterile 2 mM EDTA solution in PBS. Prepare a 200 mM EDTA stock solution, dilute 100-fold in 1X PBS and autoclave. Store at 4 °C.
4. 94-mm cell culture dishes without vents (Greiner Bio-One, Kremsmünster, Austria).
5. Cell counting device, such as Blaubrand Neubauer hemocytometer (Brand, Wertheim, Germany).
6. 0.4% (w/v) Trypan blue solution.
7. 25-cm sterile cell scrapers (Greiner Bio-One).

2.2   Cell Lysis

1. Lysis buffer: 20 mM Tris-HCl, 50 mM NaCl, 1% (v/v) Triton X-100, pH 7.4, sterile filtered. Store at 4 °C for one month.
2. Complete® protease inhibitors (Roche Diagnostics, Mannheim, Germany): dissolve one tablet in 2 mL of H_2O and dispense into 50 µL aliquots. Store at −20 °C.

2.3   Coimmunoprecipitation

1. Antibody directed to the antigen that serves as a bait protein. Store as recommended by the manufacturer. If deep-freezing is required, make 10 µL aliquots and store at the requested freezing temperature.
2. Protein A agarose suspension (*see* **Note 4**): suspend 15 mg of protein A agarose in 300 µL of lysis buffer by pipetting up and down. Centrifuge the slurry at 13,000 g for 30 s at room temperature. Discard the supernatant (*see* **Note 5**). Repeat this step

two times. Resuspend the swollen agarose beads in a final volume of 100 μL of lysis buffer. This suspension can be used for at least 1 month if stored at 4 °C.
3. Washing buffer: 20 mM Tris-HCl, 150 mM NaCl, 0.5% (v/v) Triton X-100, pH 7.4, sterile filtered. Store at 4 °C for 1 month.

2.4   SDS-PAGE

1. 4X separating buffer: 1.5 M Tris-HCl, 0.4% (w/v) SDS, pH 8.8. Store at 4 °C.
2. 4X stacking buffer: 0.5 M Tris-HCl, 0.4% (w/v) SDS, pH 6.8. Store at 4 °C.
3. 30% (w/v) Acrylamide:bis-acrylamide 37.5:1 (Rotiphorese 30®, Roth, Karlsruhe, Germany) (see Note 6).
4. 10% (w/v) ammonium persulphate (APS) (see Note 7).
5. TEMED.
6. 2X denaturing, reducing sample loading buffer (Lämmli): 0.5 M Tris-HCl, pH 6.8, 10%. (w/v) SDS, 0.1% (w/v) bromophenol blue, 20% (v/v) glycerol, 5% beta-mercaptoethanol.
7. 10X running buffer: 250 mM Tris base, 2 M glycine, 1% (w/v) SDS.
8. Protein marker (e.g., Mark12® from Invitrogen, Lofer, Germany).
9. Vertical polyacrylamide gel apparatus, such as the Mini-Protean II® protein gel electrophoresis system (Bio-Rad, Hercules, CA), and power supply, such as the PowerPac 300 (Bio-Rad).

2.5   Immunoblotting

1. Primary antibody directed to the protein of interest.
2. Secondary antibody (horseradish peroxidase conjugate).
3. Hybond ECL nitrocellulose membrane (GE Healthcare, Munich, Germany).
4. Blotting buffer: 50 mM Tris base, 35 mM glycine, 20% (v/v) methanol p.a.
5. 5% (w/v) skim milk solution in PBS supplemented with 0.05% (v/v) Tween 20.
6. 0.1% (w/v) Ponceau red in 5% (v/v) acetic acid.
7. PBST membrane washing solution: 0.1% (v/v) Tween 20 in PBS.
8. Western blotting luminol reagent (Santa Cruz, Santa Cruz, CA).
9. Kodak X-OMAT scientific imaging film (13 × 18 cm).
10. Developing cassette.
11. Electro-blotting device, such as the Transblot SD blotting cell (Bio-Rad).

2.6   Mass Spectrometry Compatible Silver Staining

1. Fixing solution: 50% (v/v) methanol, 5% (v/v) acetic acid (see Note 8).
2. Oxidizing solution: 0.02% (w/v) $Na_2S_2O_3$ (see Note 7).
3. Staining solution: 0.1% (w/v) $AgNO_3$ (see Note 7). Store at 4 °C.

4. Developing solution: 2% (w/v) Na$_2$CO$_3$, 0.04% (v/v) formaldehyde (*see* **Note 9**).
5. Stopping solution: 5% (v/v) acetic acid (*see* **Note 8**).
6. Storage solution: 1% (v/v) acetic acid (*see* **Note 8**).

2.7   Mass Spectrometry Compatible Coomassie Blue Staining

1. Coomassie brilliant blue G250 solution: 0.05 % (w/v) Coomassie brilliant blue G250. (Fluka, Buchs, Switzerland) in 50% (v/v) methanol p.a., and 10% (v/v) acetic acid (*see* **Notes 8, 10, and 11**).
2. Destaining solution: 7.5% (v/v) acetic acid, 25% (v/v) methanol p.a. (*see* **Note 8**).

2.8   Preparing the Sample for Mass Spectrometric Sequencing

All the following solutions must be prepared in chromatography-grade H$_2$O (Merck, Darmstadt, Germany). With the exception of solution 8, they should be stored in glass tubes or, for a short time, in plastic tubes flushed three or four times with chromatography-grade H$_2$O (*see* **Note 12**).

1. Clean scalpel.
2. 50 mM NH$_4$HCO$_3$ purum p.a.(Fluka) (*see* **Note 13**).
3. Acetonitrile, gradient grade (Merck).
4. Acetonitrile: NH$_4$HCO$_3$, 1:1 (*see* **Note 7**).
5. 10 mM dithiothreitol (DTT, min. 99%) in 50 mM NH$_4$HCO$_3$ (*see* **Note 14**).
6. 50 mM iodoacetamide (SigmaUltra) in 50 mM NH$_4$HCO$_3$ (*see* **Note 14**). Store in the dark.
7. Trypsin, sequencing grade unmodified (Roche): dilute 25 μg of trypsin in 250 μL of 1 mM HCl (*see* **Note 15**). Make 5–20-μL aliquots and store at −80 °C.
8. 5% (v/v) Formic acid puriss. p.a. for mass spectrometry (Fluka).
9. Eppendorf Thermomixer.

2.9   Nano-HPLC-Ion-Trap-MS/MS

1. Buffer A: 5% (v/v) gradient-grade acetonitrile (Merck), 95% (v/v) chromatography-grade H$_2$O (Merck), 0.1% (v/v) formic acid for mass spectrometry (Fluka).
2. Buffer B: 10% (v/v) chromatography-grade H$_2$O (Merck), 90% (v/v) gradient-grade acetonitrile (Merck).
3. Buffer C: 0.05% (v/v) Trifluoroacetic acid (Fluka) in chromatography-grade H$_2$O (Merck).

4. Nano-HPLC coupled to an ion-trap mass spectrometer, such as the LC Packings UltiMate nano-HPLC system, consisting of a μHPLC pump connected to a UV detection unit, Switchos μ-column-switching device with loading pump, two ten-port valves, and the FAMOS μ-autosampler) coupled online to a Thermo Finnigan LCQ Deca XP plus mass spectrometer via a liquid junction (Proxeon, Odense, Denmark) and Pico Tip emitters (New Objective, Cambridge, MA).

2.10 Data Analysis

Software for the correlation of MS/MS spectra, such as MASCOT (Matrix Sciences, Boston) or SEQUEST (Thermo Finnigan, Waltham, MA).

3 Methods

The experimental scheme first encompasses sample preparation. Samples suitable for co-IPs are biological fluids, such as blood serum, tissue or cell extracts, or generally a protein mixture. Each of these samples requires a specific preparation procedure. Here, we limit our protocol to the preparation of cytosolic extracts from HeLa cells (Sects. 3.1. and 3.2). The next steps describe the co-IP (Sect. 3.3) and the separation of coimmunoprecipitated proteins by SDS-PAGE (Sect. 3.4). The proteins on the gel can then be either transferred to a membrane for immunoblotting (Sect. 3.5) or stained for further mass spectrometric analysis. Staining can be performed by silver staining (Sect. 3.6) or Coomassie blue (Sect. 3.7). The Coomassie blue procedure is less labor intensive but far less sensitive (10–100-fold) than silver staining. Sample preparation for MS-sequencing includes excision and shrinking of the protein band (Sect. 3.8.1), in-gel proteolyis (Sect. 3.8.2), and extraction of the peptides from the gel matrix (Sect. 3.8.3). These are separated on a nano/HPLC online coupled to a ion-trap mass spectrometer (Sect. 3.9). The tandem mass spectra of peptides are correlated with amino acid sequences from protein and nucleotide databases by the appropriate software (e.g., MASCOT or SEQUEST) to identify the proteins from which they are derived (Sect. 3.10).

Before commencing the procedure it is important to appreciate that, due to the particular sensitivity of mass spectrometry, it is essential to work in as clean an environment as possible to avoid sample contamination (*see* **Note 16**).

3.1 Cell Culture

1. Grow HeLa cervical carcinoma cells at 37 °C and 5% CO_2 in 94-mm cell culture dishes without vents in 7–10 mL of DMEM supplemented with 10% (v/v)

fetal bovine serum and antibiotics. Change the cell culture medium every 2 d (*see* **Note 17**).

2. Harvest at 90% confluency. During the following steps, pay attention not to damage or detach the cell monolayer. Carefully aspirate the used DMEM from the dish and add 5–7 mL of cold, sterile PBS by pipetting at the border of the dish. Slightly sway the dish and aspirate the PBS. Repeat this washing step at least one time to completely remove the DMEM (*see* **Note 18**).

3. Add 5 mL of EDTA/PBS and incubate for about 5 min (*see* **Note 19**). Detach the cell monolayer from the surface of the dish by carefully scraping with a disposable plastic scraper.

4. Transfer the cell suspension to a 50-mL sterile Falcon tube and centrifuge at 380 g for 6 min at 4 °C. Aspirate the supernatant with extreme care to avoid loosening the cell pellet. Resuspend the pellet in 5 mL of PBS and centrifuge again.

5. Resuspend the cell pellet in 2 mL of PBS and determine the cell viability using a hemocytometer: dilute 5 μL of the cell suspension with 95 μL of cold PBS. Add an equal volume of 0.4% (w/v) trypan blue solution and incubate for 10 min. Use this solution for cell counting and determination of the cell viability (which should be around 95%). In general, one 94-mm dish yields around 10^7 viable HeLa cells. For this reason, the following steps are all standardized to this number of cells.

6. Centrifuge the cell suspension at 380 g for 5 min at 4 °C and aspirate the supernatant. Use the pellet for cell lysis (Sect. 3.2), or immediately shock freeze it in liquid nitrogen and store at −80 °C.

3.2   Cell Lysis

1. On ice, resuspend the cell pellet in 800 μL of lysis buffer supplemented with 32 μL of 25X Complete® protease inhibitor cocktail (*see* **Note 20**). Homogenize by pipetting up and down for several times and transfer to a 1.5-mL Eppendorf tube.

2. Shake on a rotary shaker for 20 min at 4 °C.

3. Centrifuge at 15,000 g for 10 min at 4 °C. The supernatant should be clear. If not, repeat this centrifugation step (*see* **Note 21**).

3.3   Coimmunoprecipitation

1. If the antibody is frozen, rapidly thaw an aliquot and store on ice.

2. On ice, add 5 μg of antibody to the cell lysate and mix by pipetting up and down several times (*see* **Note 22**).

3. Shake overnight on a rotary shaker at 4 °C.

4. Add 30 μL of protein A agarose suspension (prepared as described in Sect. 2.3, step 2) and shake for further 3 h at 4 °C (*see* **Notes 4, 5, and 22**).

5. Centrifuge at 13,000 *g* for 30 s at 4 °C and carefully remove the supernatant without disturbing the agarose pellet (*see* **Note 5**).

6. Resuspend the agarose in 500 μL of washing buffer and shake on a rotary shaker at 4 °C for 1–2 min. Centrifuge at 13,000 *g* for 30 s at 4 °C and carefully remove the supernatant. Repeat this step four more times.

7. From here on, working on ice is no longer required. Add 60 μL of Lämmli buffer to the agarose pellet and heat at 95 °C for 5 min (*see* **Note 23**). Centrifuge at 13,000 *g* for 30 s and carefully transfer the supernatant (co-IP sample) to a clean 1.5-mL Eppendorf tube. Avoid agarose carryovers (*see* **Note 24**).

3.4 SDS-PAGE

1. Assemble the vertical polyacrylamide gel casting chamber. Use 1 mm spacers between clean glass plates.

2. To cast two 10% (w/v) polyacrylamide separating gels, mix 5 mL of 30% (w/v) acrylamide/bis-acrylamide, 3.75 mL of 4X separating buffer solution and 6.25 mL of H_2O (*see* **Note 25**). Add 50 μL of APS and 10 μL of TEMED and pour immediately. Leave the upper 2 cm of the casting chamber for the stacking gel. Overlay with n-butanol to obtain an even gel surface. Allow the gel to polymerize for approximately 1 h and decant or aspirate the n-butanol phase.

3. To cast two stacking gels, mix 0.65 mL of 30% (w/v) acrylamide/bis-acrylamide, 1.25 mL of 4X stacking buffer solution, and 3.05 mL of H_2O (*see* **Note 25**). Add 25 μL of APS and 5 μL of TEMED. Pour immediately and place in position a ten-tooth (each tooth creating a well easily able to accommodate 30 μL) 1-mm comb. Allow the gel to polymerize for approximately 1 h. After polymerization, carefully pull the comb and rinse the wells with 1X running buffer. Use the gels immediately or store them moistened for 2–3 days at 4 °C.

4. Assemble the gel running chamber. Fill the inner and the outer chamber with 1X running buffer.

5. Load 25 μL of co-IP sample into a well. Into a second well, add 2 μL of protein marker (if you are going to perform silver staining), 5 μL of protein marker for Coomassie blue staining, or 8 μL of protein marker for immunoblotting.

6. Perform electrophoresis at a current of 20 mA per gel until the bromophenol blue front has reached the bottom of the gel.

7. Disassemble the gel running chamber, carefully lift the gel from the glass plate and place it into a plastic dish.

8. For immunoblotting, proceed to Sect. 3.5. For silver staining, go to Sect. 3.6, or for Coomassie blue staining go to Sect. 3.7.

3.5   Immunoblotting (Western Blotting)

1. Cut a piece of nitrocellulose membrane with the dimensions of the polyacrylamide gel.
2. Incubate the membrane and the gel in blotting buffer for 10 min.
3. Place the gel on the membrane between two pieces of blotting paper soaked with blotting buffer and blot for 25 min at 15 V.
4. To monitor the blotting efficiency and mark the protein marker bands, incubate the membrane in Ponceau red for 15 min. Decant the solution and flush the membrane with H_2O. Remark the protein marker bands with a pencil (*see* **Note 26**).
5. Incubate in 5% (w/v) skim milk solution for 2 h at room temperature or overnight at 4 °C.
6. Prepare solutions of primary and secondary antibodies (dilute as recommended by the supplier, usually 1:1,000–1:10,000) in PBST supplemented with 1% (w/v) skim milk powder or 2% (v/v) heat inactivated fetal bovine serum (*see* **Note 27**).
7. Decant the skim milk and add the primary antibody solution. Shake on a laboratory shaker for 1 h at room temperature. Decant the primary antibody solution and wash three times for 10 min each with PBST. Add the secondary antibody and shake on a laboratory shaker for 1 h. Decant the secondary antibody solution and wash three times for 10 min each with PBST.
8. Rinse the membrane with 2 mL of luminol reagent by repeated pipetting for 2 min and pour off the luminol reagent.
9. Place the membrane in a sheet protector and affix it inside a developing cassette. Expose the membrane to an imaging film for an adequate time (usually 30 s–5 min) and develop the film (*see* **Note 23**).

3.6   Silver Staining

1. Agitate (on a laboratory shaker) the gel in 100 mL of fixing solution for 10 min (*see* **Note 8**).
2. Agitate the gel in 50% (v/v) methanol for 20 min and in H_2O for 10–20 min (*see* **Note 8**). During this step, prepare 100 mL of 0.1% (w/v) $AgNO_3$ solution and cool at 4 °C.
3. Agitate the gel in 100 mL of 0.02% (w/v) $Na_2S_2O_3$ oxidizing solution for 1 min.
4. Wash the gel (with agitation) three times with H_2O for 5 min each.
5. Stain (with agitation) the gel with the cold 0.1% (w/v) $AgNO_3$ solution at 4 °C for 20 min.
6. Wash the gel three times with H_2O for 5 min each.
7. Develop the gel with 100 mL of 2% (w/v) $NaCO_3$, 0.04% (v/v) formaldehyde (*see* **Note 9**). It can take several min until the gel becomes fully developed (*see* **Note 28**).
8. Decant the developing solution and add 100 mL of stop solution (*see* **Note 8**). Agitate for 10 min.

9. Store the gel in storing solution at 4 °C (*see* **Note 29**).

3.7   Coomassie Blue Staining

1. Perform the steps 2–5 under a fume hood (*see* **Note 8**).
2. Cover the gel with 100 mL of Coomassie blue G250 stain (*see* **Note 11**).
3. Heat in a microwave oven at 700 W until the solution slightly starts to boil (*see* **Notes 8 and 30**).
4. Cover the gel and agitate for 15 min.
5. Add 100 mL of destaining solution and shake until the background becomes completely colorless (this can take up to several hours). Change the destaining solution frequently.

3.8   Preparing the Gel for Mass Spectrometric Sequencing

3.8.1   Excising and Shrinking the Gel

1. Place the gel on a clean transilluminator.
2. Excise the band of interest around its perimeter with a clean scalpel. Carefully remove unstained borders of the band and chop the gel piece into 1-mm slices. Using the scapel, transfer the slices into a 500 μL snaplock microcentrifuge tube. The tip of the scalpel can be moistened with H_2O to facilitate sticking the small slices on it.
3. Add 100–150 μL of 50 mM NH_4HCO_3 and shake mildly (on an Eppendorf Thermomixer) at room temperature for 10 min.
4. Remove the NH_4HCO_3 and add 100 μL of 1:1 50 mM NH_4HCO_3:acetonitrile. Shake for 10 min. During this and the following steps, be careful not to discard gel slices that might accidentally stick on the pipette tip.
5. Remove the NH_4HCO_3:acetonitrile mixture and add 100 μL of acetonitrile. Incubate for 10 min. During this step, the gel slice should visibly shrink and become opaque. Depending on the intensity of the stained band, the blue color of Coomassie blue G250 should disappear almost completely. Silver stained bands are more difficult to destain. After removing the acetonitrile, the shrunken slice should have a hard consistency when it is slightly pressed with the pipette tip. If the slice is still soft, repeat this step.
6. The slices can be now directly used for step 7 or dried in a vacuum centrifuge and stored at −80 °C for several months (*see* **Notes 31 and 32**).
7. Add 100 μL of 10 mM DTT to the shrunken slice and incubate at 56 °C for 30 min.
8. Remove the DTT and let the slice cool at room temperature for a few min. Add 100 μL of 50 mM iodoacetamide and incubate in the dark for 20 min.
9. Remove the iodoactamide solution, wash, and shrink sequentially with 100 μL of 1:1 50 mM NH_4HCO_3, 50 mM NH_4HCO_3:acetonitrile, and acetonitrile, for

10 min each. Dry the gel slices in a vacuum centrifuge and proceed directly to Sect. 3.8.2, or store at −80 °C.

3.8.2 In-Gel Tryptic Proteolysis

1. Rapidly thaw an aliquot of trypsin/HCl by diluting 1:8 in 50 mM NH_4HCO_3 (e.g., dilute a 5 μL aliquot with 35 μL of 50 mM NH_4HCO_3) (*see* **Notes 13 and 33**).
2. Soak the dry gel slices in 10–20 μL of trypsin solution on ice for 30 min. After this, the slices should be completely rehydrated. Remove excess trypsin solution.
3. Add a volume of 50 mM NH_4HCO_3 to the rehydrated gel slices sufficient to completely cover the slices (around 20 μL). Incubate overnight at 37 °C.

3.8.3 Peptide Extraction (see Note 34)

1. Collect the overnight incubation solution (solution 1) from Sect. 3.8.2, step 3 in a clean 200 μL microcentrifuge tube. Pay attention not to remove the gel slices while removing the solution. Cover the gel slices with 22 μL of 50 mM NH_4HCO_3 (solution 2) and incubate the tube in an ultrasonic bath for 5 min.
2. Remove the solution (solution 2) from the gel slices and add to solution 1 in the 200 μL microcentrifuge tube. Cover the gel slices with 22 μL of 0.1% (v/v) formic acid (solution 3) and incubate in the ultrasonic bath for 5 min.
3. Remove the solution (solution 3) from the gel slices and combine with solutions 1 and 2.
4. Treat the gel slices with a further 22 μL of 0.1% (v/v) formic acid (solution 4), and following sonication, add the supernatant (solution 4) to solutions 1, 2, and 3 (you should now have a 86 μL solution containing extracted peptides).
5. Centrifuge the combined peptide solution at 5,000 g for 10 min to remove any particulate matter that might obstruct the nano-HPLC (see Sect. 3.9).
6. The peptide solution is now ready for mass spectrometric analysis, which should be performed as soon as possible.

3.9 NanoHPLC-Ion-Trap-MS/MS

The protocols that follow summarize parameters adopted in our laboratory for routine peptide sequencing and data analysis via a nano-HPLC coupled to an ion-trap mass spectrometer. For a more comprehensive overview on the manifold application possibilities of this powerful tool, we refer the reader to the available mass spectrometric literature (e.g., [8–11]).

1. Concentrate and desalt the extracted peptides on a precolumn (300 µm ID × 5 mm length, packed with C18 PepMap100, 5 µm, 100Å) using buffer C with a flow rate of 20 µL/min (*see* **Note 35**). After 10 min, switch the precolumn in line with the nano-separation column (75 µm ID × 15 cm length, packed with C18 PepMap100, 3 µm, 100Å). Elute the peptides in the back flush mode using a step gradient separation profile consisting of 0–5% buffer B for 10 min, 5–45% buffer B for 35 min, and 45–90% buffer B for 1 min with a flow rate of 200 nl/min. Monitor the separation by UV at 214 nm (*see* **Note 36**).

2. Introduce the separated peptides online into the MS. Use a 1.5–2 kV spray voltage, 170 °C capillary temperature and 40–50 V capillary voltage (*see* **Note 37**).

3. Collect the data in the centroid mode (*see* **Note 38**) using one MS experiment, followed by three or four MS/MS experiments for three or four of the most intensive ions (intensity at least 0.5–1×10^6) (*see* **Note 37**). Set the collision energy automatically, depending on the mass of the parent ion. For data acquisition, use dynamic exclusion with an exclusion duration of 2 min and an exclusion mass width of ±2 Da.

3.10 Data Analysis

Analyze the raw MS/MS data using the MASCOT or SEQUEST algorithm with the following standard settings: threshold, 10,000; min group count, 1; min ion count, 10; peptide tolerance, ±3 Da; fragment ions, ± 0.8 Da. Set the thresholds for the individual peptide scores to 30 for singly charged and 40 for doubly and triply charged peptides, if using MASCOT, and to 2 for singly charged, 2.5 for doubly charged, and 3.5 for triply charged peptides, if using SEQUEST (*see* **Note 39**). Always check the b- and y-ion series of low score spectra.

4 Notes

1. The best co-IP parameters must be determined individually. First, the antibody must be suitable for IP, which means that it must recognize the native protein and efficiently deplete it from the cellular extract. Second, it must be suitable for co-IP, meaning that it must not disrupt in-vivo complexes between the bait and the prey. If an epitope coincides with a protein interaction site, then the antibody will obstruct this interaction surface or displace other proteins that bind on this interaction surface. To minimize the loss of potentially important interaction partners, it is recommended to repeat the experiments with two or more monoclonal antibodies that recognize separate epitopes on the bait antigen.

   The ideal amount of antibody required to quantitatively precipitate the protein of interest should be determined by immunodepleting the respective protein with increasing antibody concentrations and immunodetection of the residual protein in the depleted extracts.

The buffer composition used for co-IP depends on the nature of the interaction one is going to investigate. To capture a specific interaction between the bait and its protein partners, it is mandatory to add reagents known to be required in vivo for such an interaction, such as cofactors, salts, or small specific ligands. This applies especially to weak interactions, which might not be detectable without an appropriate stabilizing buffer component.

Finally, the work scheme for a co-IP must be optimized to the lifetime of the interaction. Short-lived interactions might disrupt before their detection if the duration of the experiment exceeds their life span.

2. Follow the product data sheet (Sigma-Aldrich, Catalogue number D1152) to prepare a sterile, serum- and antibiotics-free solution of DMEM.

3. To inactivate the fetal bovine serum, thaw the frozen batch to room temperature and heat at 56 °C for 30 min. Finally, cool to room temperature and add to the DMEM.

4. It is indispensable to know the isotype of the antibody used for the immunoprecipitation, as protein A and protein G differ in their affinities for immunoglobulins (see Table 20.1). Although protein G binds to almost every kind of immunoglobulin, when applicable, protein A is the more economic alternative to protein G.

5. Swollen protein A (or protein G) agarose pellets are somewhat difficult to see. The supernatant therefore should be removed with caution, not to unintentionally aspirate the agarose.

6. Monomeric acrylamide is neurotoxic. Wear gloves, safety glasses, and a laboratory coat when handling with it. Dispose of acrylamide by polymerizing to polyacrylamide, which no longer is toxic.

7. Freshly prepare these solutions.

8. Evaporating methanol is toxic and acetic acid is irritating. Discard these solvents into apposite containers.

9. Freshly prepare this solution and add formaldehyde immediately prior to developing.

Table 20.1 Affinity of protein A and protein G for different immunoglobulin isotypes from human and mouse (adapted from [13])

Species	Isotype	Protein A	Protein G
Human	IgG1	+	+
	IgG2	+	+
	IgG3	−	+
	IgG4	+	+
Mouse	IgG1	−	+
	IgG2a	+	+
	IgG2b	+	+
	IgG3	−	+

10. For mass spectrometric compatible staining, we use Brilliant Blue G250 in preference to Brilliant Blue R250, as it yields a lower background and is more easily removed from the gel slices during protein extraction (see Sect. 3.8).

11. Usually, Coomassie blue stain solutions can be reused several times. However, in view of mass spectrometric sequencing, we strongly recommend using a freshly prepared solution.

12. Storing organic solvents in plastic tubes for longer times might dissolve additives, causing contamination of the sample.

13. The NH$_4$HCO$_3$ solution can be stored at room temperature and used for 3–4 days. Control the pH of the solution at times, as a basic pH is crucial for the proteolytic activity of trypsin.

14. Again, these solutions must be freshly prepared.

15. Storing trypsin in an acidic solution suppresses autoproteolysis.

16. The most annoying contamination is keratin (from textile fibers, hair, and skin), which might dramatically complicate, if not make impossible the analysis of proteins, especially those present at low quantities. Therefore, wear gloves, do not wear clothing that might easily shed wool fibers, and try to protect samples from body hair. When possible, work under a fume hood or a laminar flow hood. Store all solutions closed. For SDS-PAGE, wash the glass gel plates thoroughly and avoid contamination of the polyacrylamide gel components. You never get rid again of keratin once your gels have polymerized. Always cover your gels during storage, and do not leave gels exposed for too long during band excision.

17. When handling living cells, work under aseptic conditions.

18. DMEM should be completely removed, since protein contaminations deriving from the bovine serum (in particular the "sticky" serum albumin) might cause significant unspecific background during the coimmunoprecipitation.

19. For HeLa cells, PBS is sufficient for easily detaching the cell layer from the plate, but the addition of EDTA has the advantage of disrupting cell clumps, thus facilitating their counting. For other adherent cell lines (e.g., endothelial cells), the addition of trypsin might be required: add 0.05% (w/v) trypsin to the EDTA solution. Add 5 mL of trypsin/EDTA solution to the cell monolayer and incubate for about 5 min. Under the microscope, you can see the cells rounding up. Slightly shake the plate to release the cells from the surface. Detach the remaining cells by carefully scraping with a disposable plastic scraper. Proceed to step 5.

20. Always add protease inhibitors immediately before cell lysis, as some of the components are not particularly stable in aqueous solutions.

21. Any unsoluble matter will provoke unspecific background during co-IP.

22. In some cases, especially when working with serum samples, a preclearing step is recommended to reduce background derived from unspecific protein binding to the protein A (or protein G) agarose. This step consists of the addition of protein A (or protein G) agarose to the extract before the addition of the antibody. After shaking for approximately 1 h at 4 °C and the removal of the agarose by centrifugation (13,000 g, 30 s), the precleared extract can be handled as

described in Sect. 3.3. step 3. However, remember, that preclearing might deplete some specific protein interaction partners.

23. Although the denaturing treatment of the agarose beads with Lämmli buffer is rapid and straightforward, it causes the antibody to remain in the co-IP sample solution. As a consequence the heavy and light chains of the antibody are visible both on immunoblots (as the secondary antibody recognizes not only the primary antibody directed to the protein of interest added during immunoblotting but also the antibody used for co-IP) and on stained polyacrylamide gels. Given the microgram amounts of antibodies generally added for co-IPs, this might significantly complicate the detection of protein bands around 55 kDa (size of the heavy chain) and 31 kDa (size of the light chain).

To overcome this problem, although possibly more time consuming and/or expensive, we suggest the following alternative protocols:

a. Elute the coimmunoprecipitated proteins from the bait with a high salt buffer. For this purpose, prepare a buffer (high salt buffer) with the same composition as the washing buffer described in Section 20.2.3, but use a NaCl concentration of 600–800 mM instead of 150 mM. This salt concentration should be sufficient to disrupt most of the interactions between the bait protein and the prey proteins without significantly affecting the protein A-antibody-bait complex. Proceed as described in the main protocol until Sect. 3.3, step 7. Add 100 µL of high salt buffer to the agarose beads and shake for 15 min on a rotary shaker at 4 °C. Centrifuge at 13,000 g for 30 s. The supernatant should now contain only coimmunoprecipitated proteins. Transfer the supernatant to a clean 1.5-mL Eppendorf tube and add 2X Lämmli buffer. Proceed to Sect. 3.4.

b. Covalently attach the antibody used for co-IP to the protein A (or protein G) matrix (e.g., by cyanogen bromide coupling; [12]).

c. When performing immunoblots, use secondary antibodies that recognize only native, nonreduced immunoglobulins (e.g., ExactaCruz® from Santa Cruz). These will not recognize the antibody used for co-IP (which is denatured and reduced by the addition of Lämmli buffer) but only the native, nonreduced primary antibody added to the membrane during the immunoblotting procedure (see Sect. 3.5. step 7).

24. Agarose should not be loaded on the polyacrylamide gel, as it causes smearing of the protein bands during PAGE, which is particularly problematic for silver stained gels. It therefore is very important to transfer the co-IP sample supernatant without agarose carryover. To avoid contact with the surface of the agarose pellet, we do not pipette the entire supernatant but rather leave 5–10 µL remaining.

25. You can also prepare a separating and stacking gel stock solution. For 150 mL of a 10% (w/v) separating gel stock solution, mix 50 mL of 30% (w/v) acrylamide/bis-acyrlamide, 37.5 mL of 4X separating buffer, and 62.5 mL of H_2O. For

50 mL of stacking gel solution, mix 6.6 mL of 30% (w/v) acrylamide/bis-acrylamide, 12.6 mL of 4X stacking buffer, and 30.8 mL of H_2O. Store the solutions at 4 °C.

26. As an alternative to Ponceau Red staining, you can use prestained protein markers (e.g., Rainbow markers from GE Healthcare).

27. Usually, these solutions can be reused for several times if stored at 4 °C.

28. Do not expose the gel too long to formaldehyde, as in-gel protein cross-linking can take place, which hampers the mass spectrometric analysis.

29. We recommend that bands of interest (especially weak ones on small gels) are isolated within the first 2 weeks, as longer times significantly reduce the quality of mass spectrometric sequencing.

30. As an alternative to microwave heating, stain the gel by shaking for 1.5 h in the Coomassie blue solution at room temperature.

31. Based on our own experience, deep-frozen slices can be successfully sequenced even after 1 year.

32. Be aware that electrostatically charged, dried gel slices tend to stick on the tube walls or on pipette tips and can be easily lost when handling with the open tube.

33. Keep this solution on ice and do not let the solution stand for too long to minimize autoproteolysis of the now active trypsin.

34. This section deals with the extraction of the peptides generated by overnight trypsinolysis (see Sect. 3.8.2) of the polyacrylamide gel slices. The gel slices are treated sequentially with NH_4HCO_3 (step 1) and then with formic acid (steps 2 and 4). To improve peptide extraction, all these steps are performed in an ultrasonic water bath (we use a Bandelin Sonorex RK255H, working at 35 kHz). The supernatants from each extraction step are collected and combined with the overnight proteolysis solution from Sect. 3.8.2, step 12. It is particularly important to collect this solution, as small peptide fragments may already have diffused out of the gel slices during the overnight proteolysis.

35. The advantage of the preconcentration method over direct injection becomes obvious when the time consumed by the separation is considered.

36. Using a UV detector is not only useful to check the separation efficiency but also to check the system for the appearance of possible void volumes. Insufficiently tight connections result in void volumes, which can act as a "mixing chamber," and the initial focusing becomes lost.

37. Depending on the signal-to-noise ratio, these parameters have to be optimized for every batch.

38. This method is used to improve mass spectral data quality, get better mass assignments, and reduce data file size.

39. These settings fit best with the requirements of our routine measurements.

Acknowledgments This work was supported by the Austrian Genome Network GEN-AU/APP-II of the Austrian Ministry for Education, Science and Culture and by the OeNB (Pr.Nr. 10855).

References

1. Stein LD (2004) Human genome: End of the beginning. Nature 431:915–916
2. Tyers M, Mann M (2003) From genomics to proteomics. Nature 422:193–197
3. Monti M, Orrú S, Pagnozzi D, Pucci P (2005) Functional proteomics. Clin Chim Acta 357:140–150
4. Piehler J (2005) New methodologies for measuring protein interactions in vivo and in vitro. Curr Opin Struct Biol 15:4–14
5. Harlow E, Lane D (eds) (1998) Antibodies: A laboratory manual. Cold Spring Harbor Laboratory Press, Cold Spring Harbor, NY
6. Falsone SF, Gesslbauer B, Tirk F, Piccinini AM, Kungl AJ (2005) A proteomic snapshot of the human heat shock protein 90 interactome. FEBS Lett 579:6350–6354
7. Dai C, Whitesell L (2005) HSP90: A rising star on the horizon of anticancer targets. Future Oncol 4:529–540
8. Mitulovic G, Smoluch M, Chervet JP, Steinmacher I, Kungl AJ, Mechtler K (2003) An improved method for tracking and reducing the void volume in nano HPLC-MS with micro trapping columns. Anal Bioanal Chem 376:946–951
9. De Hoffmann E, Stroobant V. (eds) (2001) Mass spectrometry: Principles and applications. Wiley, New York
10. Aebersold, R, Mann M (2003) Mass spectrometry-based proteomics. Nature 422: 198–207
11. Dass C (ed.) (2000) Principles and practice of biological mass spectrometry. Wiley, New York
12. Pfeiffer NE, Wylie DE, Schuster SM (1987) Immunoaffinity chromatography utilizing monoclonal antibodies. Factors which influence antigen-binding capacity. J Immunol Methods 97:1–9
13. Available at http://www.stressgenbioreagents.com/literature/Protocols-IP.pdf.

Chapter 21
Tandem Affinity Purification Combined with Mass Spectrometry to Identify Components of Protein Complexes

Peter Kaiser, David Meierhofer, Xiaorong Wang, and Lan Huang

Abstract Most biological processes are governed by multiprotein complexes rather than individual proteins. Identification of protein complexes therefore is becoming increasingly important to gain a molecular understanding of cells and organisms. Mass spectrometry–based proteomics combined with affinity-tag-based protein purification is one of the most effective strategies to isolate and identify protein complexes. The development of tandem-affinity purification approaches has revolutionized proteomics experiments. These two-step affinity purification strategies allow rapid, effective purification of protein complexes and, at the same time, minimize background. Identification of even very low-abundant protein complexes with modern sensitive mass spectrometers has become routine. Here, we describe two general strategies for tandem-affinity purification followed by mass spectrometric identification of protein complexes.

Keywords tandem-affinity purification (TAP), His-Bio tag, mass spectrometry, protein complex identification, in-vivo cross-linking, MudPIT

1 Introduction

Tandem-affinity purification combined with mass spectrometric protein identification is a powerful approach to determine the composition of biologically relevant protein complexes. Typically, a protein of interest fused to a tandem-affinity purification tag (TAP-tag) is expressed in cells and used as a bait to purify protein complexes that assemble on the TAP-tagged protein in vivo. Two-step purification based on high-affinity interactions of the TAP-tag with tag-specific binding resins result in highly purified protein complexes that can be analyzed by mass spectrometry. A number of tandem-affinity tags have been developed [1–6]. Here, we describe the procedures for the isolation of protein complexes under native purification conditions using the original TAP-tag developed by the Seraphin group [1, 7] (Fig. 21.1) and a combination of in-vivo cross linking and tandem-affinity purification under denaturing

From: Methods in Molecular Biology, Vol. 439: *Genomics Protocols: Second Edition*
Edited by Mike Starkey and Ramnath Elaswarapu © Humana Press Inc., Totowa, NJ

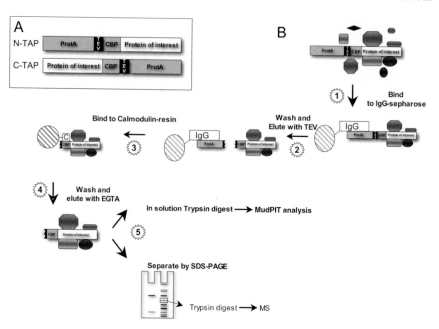

Fig. 21.1 Tandem affinity purification using the protA/CBP tag: (A) N-TAP for N-terminal tagging and C-TAP for C-terminal tagging of proteins. The IgG-interacting domain of protein A (protA) and the calmodulin binding peptide (CBP) are separated by a TEV protease recognition site. Note the different tag configurations of N-TAP and C-TAP. (B) Protein complexes assembled on the tagged bait protein are first purified on IgG sepharose, eluted by site-specific proteolysis with TEV, and further purified on calmodulin beads. The purified protein complex is eluted with EGTA and processed for mass spectrometric analysis (MS). MS analysis (1D-LC-MS/MS) of individual protein bands can be performed after separation by SDS-PAGE and "in-gel" tryptic digest. Alternatively, the entire protein mixture can be analyzed directly by "in-solution" digest followed by 2D-LC-MS/MS (MudPIT)

conditions using the HB tandem affinity tag [6, 8, 9] (Fig. 21.2). Native purification conditions using the Seraphin TAP-tag is the method of choice to identify relatively stable protein complexes, whereas transient or weak interactions can be captured by in-vivo cross-linking followed by tandem affinity purification using HB-tagged bait proteins.

The seraphin TAP tag (protA/CBP tag) consists of the immunoglobulin interacting domain of protein A and a calmodulin-binding peptide (CBP) (Fig. 21.1). The two high-affinity binding domains of the protA/CBP tag are separated by a short recognition sequence for the site-specific tobacco-etch virus protease (TEV protease). The TEV site allows proteolytic elution of the protein complex from IgG-sepharose after the first affinity-purification step, which is based on the protA/IgG-sepharose interaction. The eluted protein complex is further purified by binding to a calmodulin affinity resin, eluted with EGTA and processed for mass spectrometric (MS) analyses (Fig. 21.1).

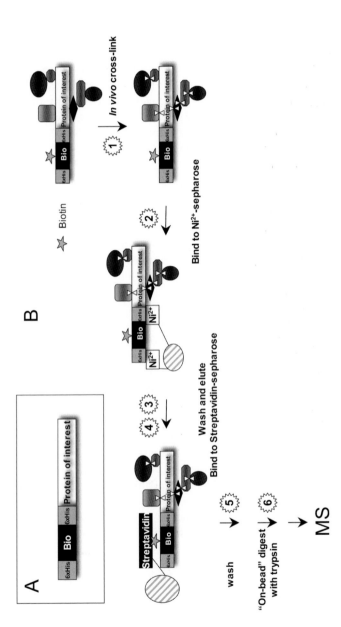

Fig. 21.2 HBH-tag based tandem affinity purification combined with in-vivo cross-linking: (A) The HBH-tag consists of a bacterially derived in-vivo biotinylation signaling peptide (Bio), flanked by hexahistidine motifs (6xHis). The biotinylation signal peptide induces attachment of biotin in vivo and allows purification of HBH-tagged proteins on streptavidin resins. (B) Individual components of protein complexes can be covalently linked to each other by in-vivo cross-linking to preserve their composition during the purification procedure. Complexes are sequentially purified on Ni²⁺ sepharose and streptavidin sepharose. Both purification steps are performed under fully denaturing conditions to minimize background binding. Because of the irreversible nature of the streptavidin-biotin interaction, "on-bead" tryptic digest is used to elute peptides of the bound protein complex from the streptavidin sepharose for mass spectrometric identification

For best sensitivity, MS analysis is done in solution without further separation of the purified protein complexes by SDS-PAGE. As "in-solution" processing typically results in relatively complex peptide mixtures, multidimensional liquid chromatography combined with tandem MS is required for optimal results. This strategy is often referred to as MudPIT (multidimensional protein identification technique) [10, 11]. Alternatively, purified protein complexes can be separated into their subunits by SDS-PAGE and individual, stained protein bands can be excised from the gel and subjected to "in-gel" analyses by MS (Fig. 21.1). The "in-gel" strategy substantially reduces the complexity of the samples, which simplifies MS analyses, but decreases sensitivity due to poor recovery of low-abundant proteins from the gel matrix. Therefore, MudPIT analysis is the method of choice when comprehensive identification of protein complex components is important. The "in-gel" approach is well suited for identification of major subunits of protein complexes.

The protA/CBP tag is very effective for purification of relatively stable associated protein complexes. However, weak or transient interactors often are lost during the purification. Loosely associated components of protein complexes can be covalently attached to each other by in-vivo cross-linking. Typically, cells are treated with cell permeable cross-linking agents, such as formaldehyde, to covalently link protein complexes as they exist in vivo. The challenge is to minimize the background associated with the subsequent tandem-affinity purification as the cross-linking step can amplify nonspecific purification. The use of HBH-tagged bait proteins can reduce this problem, because it tolerates completely denaturing purification conditions that prevent nonspecific interactions [6, 8]. Since protein complexes were covalently linked by in-vivo cross-linking, the denaturing conditions cannot dissociate the components. The HBH-tag is a derivative of the HB-tag series and consists of a bacterially derived biotinylation signal flanked by hexahistidine motifs [6, 8, 9]. The biotinylation signal induces the attachment of biotin to a specific lysine residue in the tag in vivo [12]. HBH-tagged proteins can be sequentially purified on Ni^{2+} chelate resins and streptavidin sepharose (Fig. 21.2). Importantly, both purification steps tolerate highly denaturing conditions, such as 8 M urea or 6 M guanidinium [6, 8]. The high-affinity interaction between the biotin attached to the HBH-tag and streptavidin prevents efficient elution [13]. "On-bead" tryptic digest therefore is used to release peptides from the purified, cross-linked protein complex for MudPIT analysis.

2 Materials

2.1 Yeast Cell Culture and Lysis

1. YEPD medium: 2% (w/v) Bacto peptone, 1% (w/v) yeast extract, 2% (w/v) dextrose, 0.006% (w/v) adenosine and uracil (Sigma-Aldrich, St. Louis, MO).

2. Glass beads, 0.5 mm (Biospec Products, Inc., Bartlesville, OK).
3. Antifoam A (catalogue number A-5758; Sigma-Aldrich).
4. 0.8-μm nitrocellulose membrane (Millipore, Billerica, MA).

2.2 Mammalian Cell Culture and Lysis

1. Dulbeco's Modified Eagle's Medium (DMEM; Mediatech, Herndon, VA) supplemented with 10% (v/v) fetal bovine serum (FBS, GIBCO, Bethesda, MA), 1% penicillin/streptomycin (GIBCO, Bethesda, MA).
2. Formaldehyde (catalogue number F-8775; Sigma).
3. Antibiotics according to selection marker.
4. 10X PBS buffer: 0.58 M Na_2HPO_4, 0.17 M NaH_2PO_4, 0.68 M NaCl, pH 7.4.
5. Cell scraper for cell harvesting.
6. 18G, 21G, and 27G needles to shear DNA (Becton Dickinson, NJ).

2.3 Native Purification Using the ProtA/CBP TAP Tag

1. Lysis buffer: 150 mM of NaCl, 50 mM of molanty? Tris-HCl pH 8.0, 5 mM of EDTA, 10% (v/v) glycerol, 0.2% (v/v) Nonidet P-40, 1 mM of NaF. Immediately prior to use, add (final concentrations): 1 mM phenylmethylsulfonyl fluoride (PMSF) and 1 μg/mL each aprotinin, leupeptin, pepstatin.
2. TEV cleavage buffer: 150 mM of NaCl, 10 mM of Tris-HCl pH 8.0, 0.5 mM of EDTA, 0.1% (v/v) Nonidet P-40. Add dithiothreitol (DTT) to 1 mM just before use.
3. Calmodulin binding buffer: 150 mM NaCl, 10 mM Tris-HCl pH 8.0, 1 mM $MgCl_2$, 1 mM imidazole, 0.1% (v/v) Nonidt P-40. Add $CaCl_2$ to 2 mM and β-mercaptoethanol to 10 mM just before use.
4. Calmodulin elution buffer: 150 mM NaCl, 10 mM Tris-HCl pH 8.0. Add EGTA and β-mercaptoethanol (both to 10 mM) just before use.
5. AcTEV Protease; 10 units/μL (Invitrogen, Carlsbad, CA).
6. IgG Sepharose™ 6 Fast Flow (GE Healthcare, Uppsala, Sweden).
7. Calmodulin affinity resin (Stratagene, La Jolla, CA).
8. Poly-Prep® chromatography columns, 2-mL size (Bio-Rad, Hercules, CA).
9. PMSF: 0.5 M–1 M solution in isopropanol; store at ≤4 °C.

2.4 Purification After In-Vivo Cross-Linking Using the HBH Affinity Tag

Check the pH of all buffers before use and readjust them accordingly.

1. Buffer A-8: 8 M urea, 300 mM NaCl, 50 mM sodium phosphate buffer pH 8.0 (0.68 mL of 0.5 M NaH_2PO_4 and 9.32 mL of 0.5 M Na_2HPO_4 in 100 mL), 0.5% (v/v) Nonidet P-40, pH 8.0.

2. Buffer A-6.3: 8 M urea, 300 mM NaCl, 50 mM sodium phosphate buffer pH 6.3 (8.22 mL of 0.5 M NaH_2PO_4 and 1.78 mL of 0.5 M Na_2HPO_4 in 100 mL), 0.5% (v/v) Nonidet P-40, pH 6.3.

3. Buffer A-6.3-imidazole: same as buffer A-6.3 but also containing 10 mM of imidazole, pH 6.3.

4. Buffer B: 8 M urea, 200 mM NaCl, 50 mM sodium phosphate buffer (9.21 mL of 0.5 M NaH_2PO_4 and 0.79 mL of 0.5 M Na_2HPO_4 in 100 mL), 2% (w/v) SDS, 10 mM of EDTA, 100 mM of Tris, adjust pH of final solution to 4.3.

5. Buffer C: 8 M urea, 0.2 M NaCl, 0.2% (w/v) SDS, 100 mM Tris-HCl, pH 8.0.

6. Buffer D: 8 M urea, 0.2 M NaCl, 100 of mM Tris-HCl, pH 8.0.

7. Ni^{2+} Sepharose™ 6 fast flow beads (GE Healthcare).

8. Immobilized streptavidin beads (Pierce, Rockford, IL).

9. Poly-Prep® chromatography columns (Bio-Rad, Hercules, CA).

10. 50 mM NH_4HCO_3 buffer.

11. Trypsin (Promega, Madison, WI).

12. HPLC-grade H_2O.

13. PMSF, 0.5 M–1 M solution in isopropanol, stored at $\leq 4\,°C$.

2.5   Western Blot Analysis

1. 4X SDS sample buffer: 250 mM Tris-HCl, pH 6.8, 8% (w/v) SDS, 300 mM DTT, 30% (v/v) glycerol, 0.02% (w/v) bromophenol blue.

2. Antibodies: For detection of the protA/CBP tag, peroxidase antiperoxidase complex (catalogue number P1291; Sigma-Aldrich), 1:2000 in blocking buffer and anti-CBP antibody (catalogue number 07-482; Millipore), 1:2000 in blocking buffer. For detection of the HBH-tag, RGS-His antibody (catalogue number 34610; Qiagen, Valencia, CA), 1:2000 in blocking buffer or horse-radish peroxidase conjugated streptavidin (catalogue number PI21126; Fisher, Pittsburgh), 1:10.000 in TBS-T buffer.

3. 10X TBS-T (Tris-buffered saline with Tween), 10X stock: 1.37 M NaCl, 27 mM KCl, 250 mM Tris-HCl, pH 7.4, 1% (v/v) Tween 20.

4. Blocking buffer: 5% (w/v) nonfat dry milk in 1X TBS-T.

2.6   Sample Preparation for "In-Gel" Identification

1. NuPAGE Bis-Tris gel 4-12% with MES running buffer (Invitrogen) or similar.

2. SilverSNAP stain for mass spectrometry (Pierce) or similar.

3. 25 mM NH_4HCO_3 (100 mg/50 mL) in H_2O.

4. 25 mM NH_4HCO_3 in 50% (v/v) acetonitrile (ACN), 50% H_2O.

5. 50% (v/v) ACN, 5% (v/v) formic acid (FA) (Sigma-Aldrich).

6. 10 ng/μL of trypsin (Promega) in 25 mM NH_4HCO_3 in H_2O (*see* **Note 1**).

7. 10 mM DTT in 25 mM NH_4HCO_3 (1.5 mg/mL) (*see* **Note 2**).

8. 55 mM iodoacetamide in 25 mM NH_4HCO_3 (10 mg/mL) (*see* **Note 2**).
9. 0.65-mL siliconized tubes (Fisher).
10. SpeedVac (GMI, Ramsey, MN).

2.7 Sample Preparation for MudPIT Analysis

1. 50 mM of NH_4HCO_3.
2. 0.4 μg/μL trypsin (Promega) in 1 mM trifluoroacetic acid (TFA).
3. Lys-C protease, 1 μg/μL in water (Wako Chemical, Japan).
4. Strong cation exchange PolySULFOETHYL column (The Nest Group, Southborough, MA).
5. Buffer C18-A: 2% (v/v) ACN, 98% (v/v) H_2O, 0.1 % (v/v) FA.
6. Buffer C18-B: 98% (v/v) ACN, 2% (v/v) H_2O, 0.1 % (v/v) FA.
7. Buffer SCX-A: 5 mM of KH_2PO_4, 30% (v/v) ACN, 0.1% (v/v) FA, adjusted to pH 2.7 with FA.
8. Buffer SCX-B: 5 mM of KH_2PO_4, 350 mM of KCl, 30% (v/v) ACN, 0.1% (v/v) FA, adjusted to pH 2.7 with FA.
9. 0.1% (v/v) TFA.
10. Vivapure C-18 microcolumn (Sartorius, Göttingen, Germany).
11. Pepmap C18 capillary column (Dionex, Bannockburn, IL) (length: 15 cm, ID 75 μm), capillary columns from another vender, or self-packed.
12. 500 fmol/μL BSA digest (Michrom Bioresources, Auburn, CA).

3   Methods

3.1 Native Purification Using the ProtA/CBP TAP Tag (Fig. 21.1)

Tandem-affinity purification with the protA/CBP tag very effectively removes the background. However, nonspecific interaction with the purification matrix can never be completely avoided. To distinguish specific interactions from the background, we recommend parallel processing of a sample from cells that do not express the protA/CBP tagged bait protein. Proteins detected in the tagged cell line but not in the untagged cells most likely are in a protein complex with the bait protein.

3.1.1   Growth and Lysis of Yeast Cells Expressing TAP-Tagged Proteins

1. Grow yeast cells in 1 L of YEPD medium to an Absorbance (600 nm) = 1.5 and collect cells by centrifugation (2500 g, 5 min, 4 °C), or by filtration.

2. Wash the cell pellet in 40 mL of ice cold lysis buffer (without added aprotenin, leupeptin, or pepstatin), pellet the cells by centrifugation as previously, remove the supernatant, and quickly freeze the pellet and store it at −80 °C (*see* **Note 3**).

3. Break the cells using glass beads or by an alternative yeast cell lysis method (*see* **Note 4**).

4. Separate the lysate from the glass beads. One easy way is to invert the tube, and with a 21G needle, poke a hole at the bottom of the tube. Insert the tube with the hole into a fresh tube and centrifuge carefully for 30 s at 1,000 g. The lysate will pass through the hole in the upper tube and collect in the lower tube, whereas the glass beads remain in the upper tube.

5. Centrifuge the lysate to remove cell debris (25,000 g, 30 min, 4 °C). Transfer the clarified supernatant to a fresh tube. If the lysate is viscous, use a syringe and pass it once through an 18G needle and twice through a 21G needle to shear the DNA. Save 50 μL of the lysate for analysis. One can expect to obtain between 100 and 250 mg of total protein.

3.1.2 Growth and Lysis of Mammalian Cells Expressing TAP-Tagged Proteins

1. Grow cells expressing the TAP-tagged protein in DMEM to 90% confluence. Use 5–10 150-mm plates. One 150-mm plate of HeLa cells yields about 2–5 mg of total protein.

2. Wash cells on plates twice with 5 mL of ice-cold 1X PBS, pH 7.4.

3. Add 5 mL of ice-cold 1X PBS and detach the cells using a cell scraper. Rinse the plates with 2 mL of ice-cold 1X PBS to increase the quantity of harvested cells. Keep the harvested cells on ice all the time.

4. Transfer the cells into a tube and centrifuge at 400 g for 3 min at 4 °C. Discard the supernatant and freeze the pellet at −80 °C.

5. For a total of 10 150-mm plates of cells harvested, add 20 mL of lysis buffer (*see* **Note 3**) to the frozen pellet. Resuspend immediately and keep on ice for a few minutes.

6. At this point the lysate is very viscous, due to the presence of chromosomal DNA. Shear the DNA by passing the lysates several times through a 27G needle, until the viscosity is similar to that of water.

7. Centrifuge the lysate at 25,000 g for 30 min at 4 °C. Transfer the clarified supernatant to a fresh tube. Save 50 μL of aliquots of the lysate for analysis.

3.1.3 Tandem Affinity Purification Using ProtA/CBP-Tagged Bait Proteins

The optimal amount of IgG beads required to bind most of the protA/CBP-tagged protein complex should be determined. Adding too much IgG beads increases the background, whereas too little IgG resin results in inefficient purification (*see* **Note 5**).

A good starting point is 150 μL of IgG beads for 100 mg of total protein lysate. All steps are at 4 °C unless otherwise stated.

1. Wash the IgG beads twice with 1 mL of lysis buffer. Centrifuge at 100 g for 1 min to pellet the beads.
2. Add the IgG beads to the cell lysates.
3. Incubate for 2 h at 4 °C.
4. Centrifuge at 100 g for 1 min to pellet the IgG beads and carefully remove the supernatant without disturbing the beads. Save 50 μL of the supernatant for analysis (this is the IgG-unbound fraction).
5. Wash the beads three times with 10 mL of ice-cold lysis buffer. Centrifuge and remove supernatant as previously.
6. Wash twice with 10 mL of ice-cold TEV cleavage buffer.
7. TEV protease cleavage: add 1 mL of TEV cleavage buffer and 10 μL of acTEV protease.
8. Incubate for 2 h at room temperature and then overnight at 4 °C. Mix samples gently during the incubation.
9. Centrifuge at 100 g for 1 min and transfer the supernatant into a new 15 mL tube. Save 50 μL for analysis. Store the IgG beads at −20 °C for analysis.
10. Add 3 μL of 1 M CaCl$_2$ to the 1 mL of TEV-eluted protein fraction (final concentration 3 mM of CaCl$_2$).
11. Wash the calmodulin beads three times with 1 mL of calmodulin binding buffer.
12. Resuspend 50 μL of the calmodulin beads in 1 mL of calmodulin binding buffer and add the bead suspension to the TEV-eluted protein sample (total volume is 2 mL) (*see* **Note 6**).
13. Incubate at 4 °C for 2-4 h with gentle mixing.
14. Centrifuge at 100 g for 1 min and carefully remove the supernatant. Retain 100 μL for analysis (this is the calmodulin-unbound fraction).
15. Gently resuspend the beads in 1 mL of calmodulin binding buffer, transfer into a 2 mL Poly-Prep column and allow the column to drain.
16. Wash the beads four times with 1.5 mL of calmodulin binding buffer at 4 °C.
17. Elute proteins from the beads with 1 mL of calmodulin elution buffer and collect in a microcentrifuge tube. Save 50 μL for analysis.
18. Precipitate proteins by adding 335 μL of 100% (v/v) trichloroacetic acid (TCA; final concentration 25%) and incubate on ice for 30 min with occasional vortexing (*see* **Note 7**).
19. Pellet proteins by centrifugation in a microcentifuge at 13,000 g for 30 min at 4 °C.
20. Wash the pellet three times with 0.5 mL of prechilled (−20 °C) acetone. After each wash, pellet the protein by centrifugation at 13,000 g for 10 min at 4 °C.
21. Remove the acetone and air dry the pellet (about 5–10 min).

3.1.4 Western Blot Analysis

To ascertain the efficiency of the purification, the small samples collected from the different purification steps should be analyzed by Western blotting.

1. To evaluate the efficiency of binding to IgG sepharose, analyze 20 µL of the collected lysates and 20 µL of the IgG-unbound fraction (Sect. 3.1.3, step 4.) by Western blotting with peroxidase antiperoxidase complex to detect the protA part of the TAP tag.
2. To analyze the efficiency of TEV elution, boil the IgG beads collected in Sect. 3.1.3, step 9 in 500 µL of 1X SDS sample buffer for 5–10 min and separate 10 µL of this sample, in addition to 20 µL of the TEV-eluted fraction (collected in step 3.1.3.9.) by SDS-PAGE. Analyze by Western blotting using an anti-CBP antibody (*see* **Note 8**).
3. To verify efficient binding to the calmodulin beads and subsequent elution, compare the following fractions by Western blotting using an anti-CBP antibody: 20 µL of the TEV-eluted fraction (Sect. 3.1.3, step 9), 40 µL calmodulin-unbound fraction (Sect. 3.1.3, step 14), and 20 µL of the calmodulin-elution (Sect. 3.1.3, step 17).

3.2 Purification of HBH-Tagged Proteins after In Vivo Cross-Linking (Fig. 21. 2)

3.2.1 Growth, In-Vivo Cross-Linking, and Lysis of Yeast Cells Expressing HB-Tagged Proteins

1. Grow yeast cells expressing the HBH-tagged protein of interest in 1 L of YEPD medium to an Absorbance (600 nm) = 1.5.
2. Add formaldehyde to the medium to a final concentration of 1% (v/v) and incubate at 30 °C for 10 min.
3. Add 2.5 M glycine to a final concentration of 0.125 M and incubate at 30 °C for 10 min to quench the cross-linking reaction.
4. Collect cells by centrifugation (2,500 g, 5 min, 4 °C) or filtration through a 0.8-µm nitrocellulose membrane.
5. Wash the cell pellet in 20 mL of buffer A-8 containing 1 mM PMSF, pellet the cells by centrifugation as previously, remove the supernatant, and quickly freeze the pellet and store it at −80 °C.
6. Break the cells using glass beads or by an alternative yeast cell lysis method (*see* **Note 4**).
7. Separate the lysate from the glass beads. One easy way is to invert the tube, and with a 21G needle, poke a hole at the bottom of the tube. Insert the tube with the hole into a fresh tube and centrifuge carefully for 30 s at 1,000 g. The lysate passes through the hole in the upper tube and collects in the lower tube, whereas the glass beads remain in the upper tube.

8. Centrifuge the lysate to remove cell debris (25,000 g, 30 min, 20 °C). Transfer the clarified supernatant to a fresh tube.

3.2.2   Growth, In-Vivo Cross-Linking, and Lysis of Mammalian Cells Expressing HBH-Tagged Protein

1. Grow cells expressing the HBH-tag in DMEM to 90% confluency (one 150-mm plate of HeLa cells yields about 2 mg of total protein).
2. Add formaldehyde to a final concentration of 1% (v/v) and incubate at 37 °C for 10 min.
3. Quench the cross-linking reaction by adding glycine to a final concentration of 0.125 M and incubate at 37 °C for 10 min.
4. Wash the cells on the plates twice with 5 mL of ice-cold 1X PBS, pH 7.4.
5. Add 5 mL of ice-cold 1X PBS and detach cells from the plate with a cell scraper. Rinse the plate with an additional 2 mL of ice-cold 1X PBS to increase the yield of harvested cells.
6. Transfer the cells into a tube and centrifuge at 100 g for 3 min at 4 °C.
7. Discard the supernatant, for each 150 mm plate of lysed cells add 2 mL of buffer A-8 containing 1 mM PMSF and resuspend the cells.
8. Shear the DNA with a 27G needle, until the viscosity is similar to that of water.
9. Centrifuge the lysate (25,000 g, 30 min, 20 °C). Transfer the clarified supernatant to a fresh tube. Save 50 μL lysates for analysis.

3.2.3   Tandem-Affinity Purification of HBH-Tagged Proteins

To distinguish specific interactions from the background, we recommend parallel processing of a sample from cells that do not express the HBH-tagged bait protein. Proteins detected in the tagged cell line but not in the untagged cells most likely are in a protein complex with the bait protein.

1. Use 70 μL of Ni^{2+} sepharose beads for each 1 mg of protein lysate. Wash the beads three times with at least five bead volumes of buffer A-8 (without PMSF). Pellet the beads by centrifugation at 100 g for 1 min, and remove the supernatant.
2. Add Ni^{2+} sepharose beads to a lysate, followed by imidazole to a final concentration of 10 mM to reduce nonspecific binding.
3. Incubate on a rocking platform at room temperature for 4 h or overnight.
4. Pellet the beads by centrifugation at 100 g for 1 min, and remove the supernatant. Save 50 μL for analysis (Ni-unbound fraction).
5. Wash the Ni^{2+} sepharose beads sequentially with 20 bead volumes of buffer A-8, buffer A-6.3, and buffer A-6.3-imidazole, respectively.

6. Elute HBH- tagged proteins twice with five bead volumes of buffer B. Incubate for at least 10 min at room temperature for each elution step, and pool the eluates. Save 50 μL for analysis (Ni²⁺ eluate).

7. Adjust the pH of the eluate to pH 8.0 (add ~25 μL of 1 M of NaOH to each 1 mL of eluate).

8. Prepare streptavidin sepharose by washing with 2X 3 mL of buffer C. Use 15 μL of beads for each 1 mg protein in the whole cell lysate.

9. Incubate the Ni²⁺ sepharose eluate with streptavidin beads in a Poly-Prep chromatography column on a rocking platform overnight at room temperature.

10. Drain the column. Save 50 μL of the flow-through for analysis (streptavidin unbound fraction).

11. Wash the streptavidin beads (in the Poly-Prep chromatography column) sequentially with 2X 25 bead volumes of buffer-C and buffer-D, respectively.

12. Add 3 mL of 50 mM NH_4HCO_3, pH 8.0 and allow the column to drain.

13. Repeat step 12 once more.

14. Add NH_4HCO_3 buffer (approximately 50% of the bead volume) for trypsinization.

3.2.4 "On-Bead" Digestion for Mass Spectrometric Analysis

1. Dissolve 20 μg of trypsin in 50 μL of 1 mM TFA in the original glass tube.

2. Incubate the sample (from Sect. 3.2.3, step 14) with trypsin at 37 °C for 12–16 h on a rocking platform. Use 1 μg of trypsin for every 2 mg of whole cell lysate used in the first purification step (*see* **Note 9**).

3. Carefully collect the supernatant and add FA to a final concentration of 1% (v/v).

4. Reextract tryptic peptides from the beads two or three times by adding approximately 50% of the bead volume of 25% (v/v) ACN, 0.1% (v/v) FA to the beads.

5. Pool the extracted peptides and concentrate to about 5 μL using a SpeedVac.

6. Add 100 μL of H_2O and concentrate (SpeedVac) to about 5 μL. Repeat this step once more.

7. Acidify by adding 100 μL of 0.1% (v/v) FA.

8. Samples can be stored frozen at this step or used immediately for 1-D- (Sect. 3.3.3) or 2-D-MS analysis (Sect. 3.4.1).

3.2.5 Western Blot Analysis of Purification

To ascertain the efficiency of the purification, the small samples collected from the different purification steps should be analyzed by Western blotting.

1. To evaluate the efficiency of binding to Ni²⁺ sepharose and the extent of cross-linking, analyze 20 μL of the collected whole cell lysates and 20 μL

of the Ni-unbound fraction (Sect. 3.2.3, step 4) by Western blotting with the anti-RGS-His antibody.

2. To analyze the efficiency of binding to streptavidin, use $10\,\mu l$ of the Ni^{2+} eluate (Sect. 3.2.3, step 6) and $10\,\mu L$ of the streptavidin unbound fraction (Sect. 3.2.3, step 10) for Western blot analysis with the anti-RGS-His antibody or horseradish-peroxidase-conjugated streptavidin (*see* **Note 10**).

3.3 Mass Spectrometric Identification of Proteins after Separation by SDS-PAGE

3.3.1 Separation by SDS-PAGE

Components of protein complexes can have a wide range of molecular weights. We therefore recommend the use of a gradient gel for separation of samples by SDS-PAGE. Many manufacturers offer precast gradient gels compatible with subsequent mass spectrometric analyses. The NuPAGE Bis-Tris gel mentioned in this section is only one of many options. To identify protein bands after separation, silver staining offers the best sensitivity. Not all silver staining protocols are compatible with mass spectrometric analysis, but a number of manufacturers offer mass spectrometry-compatible silver staining kits. The SilverSNAP kit is one of several choices.

1. Separate samples from Sect. 3.1.3, step 21 on a 4–12% NuPAGE Bis-Tris gel (or similar) according to the suppliers instructions. Samples should be thoroughly dissolved in the gel-loading buffer supplied with the gels. Load the purified sample from the protA/CBP-tagged cell line and the control sample (untagged cell line).
2. Stain the gel with SilverSNAP stain or similar. Protein bands present in the tagged sample and absent in the untagged control sample can be excised and analyzed.

3.3.2 Excision and "In-Gel" Digestion of Protein Bands

1. Chop each gel slice into small pieces ($1\,mm^2$) and place into a 0.65-mL siliconized tube.
2. Add $100\,\mu L$ (or enough to cover) of 25 mM NH_4HCO_3, 50% (v/v) ACN and vortex for 10 min. Pellet the gel pieces by centrifugation at $16{,}000\,g$ for 30 s.
3. Using a gel-loading micropipette tip, extract the supernatant and discard.
4. Repeat steps 2 and 3 twice more.
5. Dry (SpeedVac) the gel pieces to complete dryness (approximately 20 min).

For low level proteins (<1 pmol), especially those separated by 1D SDS-PAGE, reduction and alkylation (steps 6–11) is recommended. Otherwise, go directly to step 12.

6. Add 25 μL (or enough to cover) of 10 mM DTT in 25 mM NH₄HCO₃ to the dried gel pieces.

7. Vortex and centrifuge at 16,000 *g* for 30 s. Allow the reduction reaction to proceed at 56 °C for 1 h.

8. Remove the supernatant and add 25 μL (or enough to cover) of 55 mM iodoacetamide to the gel pieces. Vortex and centrifuge at 16,000 *g* for 30 s. Allow the alkylation reaction to proceed in the dark for 45 min at room temperature. Occasionally vortex and centrifuge at 16,000 *g* for 30 s.

9. Discard the supernatant. Wash the gel pieces with ~100 μL of NH₄HCO₃, Vortex for 10 min and centrifuge at 16,000 *g* for 30 s.

10. Discard the supernatant and dehydrate the gel pieces with ~100 μL (or enough to cover) of 25 mM NH₄HCO₃, 50% (v/v) ACN, vortex for 5 min and centrifuge at 16,000 *g* for 30 s. Repeat once more.

11. Completely dry the gel pieces in a SpeedVac (approximately 20 min)

12. Add 5–10 μL of trypsin (10 ng/μL) to the dried gel pieces and allow the gel to rehydrate for a few min.

13. Add 25 μL of 25 mM NH₄HCO₃ (or sufficient volume to cover the gel pieces), vortex for 5 min, centrifuge at 16,000 *g* for 30 s and incubate at 37 °C for 4 h, to overnight.

14. Centrifuge at 16,000 *g* for 30 s. Add 10 μL of H₂O, vortex for 10 min and centrifuge at 16,000 *g* for 30 s.

15. Transfer the digest solution (aqueous extraction) into a new 0.65-mL siliconized tube.

16. Add 30 μL (enough to cover) of 50% (v/v) ACN, 5% (v/v) formic acid to the gel pieces, vortex for 10 min and centrifuge at 16,000 *g* for 30 s. Transfer the supernatant to the same extraction tube as in step 15.

17. Repeat step 16 once more.

18. Add 10 μL of 100% (v/v) ACN to the gel pieces, vortex for 5 min and centrifuge at 16,000 *g* for 30 s. Transfer the supernatant to the same extraction tube as in step 15.

19. Centrifuge the extracted digest (extraction tube from step 15) at 16,000 *g* for 30 s and reduce the volume to 5–10 μL in a SpeedVac.

20. Add 5–10 μL of 0.1% (v/v) FA to the tube before mass spectrometric analysis. Samples can be directly analyzed by LC-MS/MS or analyzed by MALDI after desalting.

3.3.3   Mass Spectrometry

1. Condition the Pepmap C18 column sequentially in buffer C18-B for 90 min and buffer C18-A for 30 min. Adjust the column flow rate to 250 nL/min.

2. Inject 1 μL of 500 fmol/μL BSA digest and separate the sample at the following gradient using buffers C18-A and C18-B: 0% buffer C18-B for 5 min, 0–35% buffer C18-B for 30 min, 80% buffer C18-B for 5 min, and 0% buffer C18-B for 30 min. Inject 1 μL of 10–50 ftmol/μL BSA digest using the same gradient. The column is now ready to be used.

3. The protein digest from Sect. 3.3.2, step 20 can be directly injected onto the column for LC-MS/MS analysis.

3.4 Identification of Proteins by MudPIT Analysis

3.4.1 Protease Digestion of Purified Protein Complexes

"On-bead" trypsin digestion of protein complexes purified with HBH-tagged baits, after in-vivo cross-linking, is described in Sect. 3.2.4. Samples prepared in that section (step 8) can be directly analyzed by 1-D-MS/MS, as described in Sect. 3.3.3, step 3, or by 2-D-MS/MS (Sect. 3.4.2). The following steps describe protease digestion of TCA-precipitated affinity-purified proteins as generated in Sect. 3.1.3, step 21.

1. Dissolve the protein pellet from Sect. 3.1.3, step 21 in 100 μL of 8 M urea, 50 mM NH_4HCO_3.
2. Add 2% (enzyme weight/protein weight) Lys-C and digest at 37 °C for 4 h.
3. Add 440 μL of 50 mM NH_4HCO_3 to dilute the urea concentration to ≤1.5 M. Add 1% trypsin (enzyme weight/protein weight) and digest at 37 °C for 8 h or overnight.
4. Reduce the volume to approximately 10 μL in a SpeedVac.
5. Add 100 μL of 0.1% (v/v) FA and proceed with 1-D-MS/MS Sect. 3.3.3, step 3), or 2-D-MS/MS (Sect. 3.4.2) or store the sample frozen at −20 °C.

3.4.2 Peptide Separation by Ion-Exchange Chromatography

1. Separate samples (generated at Sect. 3.2.4, step 8, or at Sect. 3.4.1, step 5) on a strong cation exchange column with the following gradient: 0–5% buffer SCX-B for 2 min, 5–35% buffer SCX-B for 30 min, 35–100% buffer SCX-B for 10 min. Collect the flow-through and each peak fraction.
2. Concentrate each fraction to about 5–10 μL in a SpeedVac.
3. Add 180 μL of 0.1% (v/v) TFA to each fraction.
4. Desalt the samples using Vivapure C-18 microcolumns according to the manufacturer's instructions.
5. After desalting, concentrate samples to 1–2 μL (using a SpeedVac) and add 10 μL of 0.1% (v/v) FA to each sample for LC-MS/MS analysis.

3.4.3 LC-MS/MS Analysis

1. Each desalted fraction from Sect. 3.4.2, step 5 can be analyzed by 1-D-LC-MS/MS (as described in Sect. 3.3.3). As the nongel purified tryptic peptides separated by ion exchange chromatography are more complex than the "in-gel" digestion samples, the gradient should be modified as follows: 0% buffer C18-B

for 5 min, 0–35% buffer C18-B for 80 min, 80% buffer C18-B for 5 min, and 0% buffer C18-B for 30 min.
2. Automatically submit the acquired LC-MS/MS data to commercially available search engines, such as MASCOT, Protein Prospector, or SEQUEST for database searching, protein identification, and characterization of posttranslational modifications.

4 Notes

1. Diluted trypsin solution should be prepared immediately before the digestion.
2. Although this step is optional, if performed, DTT and iodoacetamide solutions need to be freshly prepared for reduction and alkylation of cysteines.
3. It is important to test the extraction of the tagged protein on a small sample first. Some proteins, especially DNA-bound proteins, require higher salt concentrations for efficient extraction. High salt concentrations can disrupt protein complexes, and therefore salt concentrations in the lysis buffer should not exceed that required for extraction of the tagged protein. To determine the best salt concentration, prepare several small aliquots. Extract one sample with lysis buffer containing 2% (w/v) SDS, which serves as a control sample for complete extraction. Extract the other samples with lysis buffer containing increasing amounts of NaCl (e.g., 150, 200, 300, 400, and 500 mM). Analyze all the samples by Western blotting using the peroxidase antiperoxidase complex to compare the extraction efficiencies of each salt concentration. In some cases, increasing the concentration of the detergent in the lysis buffer can improve extraction.
4. A number of methods exist for breaking yeast cells. Very reliable results are achieved when cells are broken with a FastPrep™ FP120 (Qbiogene, Carlsbad, CA) at setting 4.5, for four times for 30 s each at 4 °C, with a 1 min break on ice between the runs. Alternatively, cells can be vortexed with glass beads five times for 1 min at 4 °C, with a 1 min break on ice between the runs. In either case, add two or three pellet volumes of lysis buffer, approximately 1 μL of antifoam A for each 200 μL of lysis buffer and glass beads until the beads reach the top level of the lysis buffer. Cells also can be lysed with a French pressure cell at maximum pressure or by grinding the pellet in liquid nitrogen.
5. Use a small amount of the cell lysate and incubate with different quantities of IgG beads at 4 °C for 2 h. Pellet the beads by centrifugation at 100 g for 1 min. Use 10 μL of the supernatant (lysate after binding to IgG beads) and analyze (along with the equivalent quantity of total lysate) by Western blotting using the peroxidase antiperoxidase complex. Efficient binding of the protA/CBP-tagged protein to the IgG resin is evident by significantly reduced levels of the protA/CBP-tagged protein in the supernatant fraction. Choose the quantity of IgG beads that shows efficient binding, although not necessarily the most efficient binding, as it is important to keep in mind that adding too much IgG resin will increase nonspecific binding.

6. The amount of calmodulin resin required varies depending on the amount of purified tagged protein. Too much resin increases the background and too little resin results in inefficient recovery of the tagged protein. The suggested bead volume should be increased if the analysis of the purification steps suggests inefficient recovery of the tagged protein in this step.

7. If the sample is to be analyzed by MudPIT, in addition to separation by SDS-PAGE and "in-gel" digestion, split the eluted fraction into 2X 500 μL aliquots, and precipitate each with TCA (25% final concentration).

8. TEV cleavage results in a reduction of the molecular weight of the protA/CBP-tagged protein because the protA part of the TAP tag is removed by TEV cleavage (Fig. 21.1).

9. The amount of trypsin suggested is a rough estimate. To determine the amount of trypsin required more accurately, measure the protein concentration in the Ni^{2+} eluate (Sect. 3.2, step 6) and in the streptavidin unbound fraction (Sect. 3.2.3, step 10). The difference inprotein concentration in these two fractions is a good approximation of the amount of protein bound to the streptavidin beads. For every 1 μg of protein bound to the streptavidin beads, use between 0.01 and 0.02 μg of trypsin (the ratio of protein/trypsin is between 1/100 and 1/50).

10. The RGS-His antibody detects HBH-tagged proteins and can be used to measure the efficiency of the purification steps. Horseradish-peroxidase-conjugated streptavidin can be used to analyze the purification after elution from the Ni^{2+} sepharose. However, it is not very useful for analysis of the first purification step because all eukaryotic cells express between four and six endogenous biotinylated proteins. The endogenous biotinylated proteins are relatively abundant and can complicate interpretation of the Western blot results. However, endogenous biotinylated proteins are lost during the first purification step and horseradish-peroxidase-conjugated streptavidin is useful for detection of HBH-tagged proteins after the first purification step, providing important information about the efficiency of biotinylating the HBH-tagged bait protein.

Acknowledgments We thank Christian Tagwerker and Cortnie Guerrero for their contributions in developing the protocols described here and Karin Flick for the figure design. Work in the laboratory of P. Kaiser is supported by grants from NIH (RO1GM-66164, R21CA113823) and the California Breast Cancer Research Program (11NB-0177). L. Huang acknowledges support from NIH (GM074830) and the DOD (PC-041126). D. Meierhofer is supported by an Erwin Schroedinger Fellowship (FWF J2665).

References

1. Rigaut G, Shevchenko A, Rutz B, Wilm M, Mann M, Seraphin B (1999) A generic protein purification method for protein complex characterization and proteome exploration. Nat Biotechnol 17:1030–1032

2. Cheeseman IM, Brew C, Wolyniak M, Desai A, Anderson S, Muster N, Yates JR, Huffaker TC, Drubin DG, Barnes G (2001) Implication of a novel multiprotein Dam1p complex in outer kinetochore function. J Cell Biol 155:1137–1145

3. Denison, C, Rudner AD, Gerber SA, Bakalarski CE, Moazed D, Gygi SP (2004) A proteomic strategy for gaining insights into protein sumoylation in yeast. Mol Cell Proteomics
4. Graumann J, Dunipace LA, Seol JH, McDonald WH, Yates JR III, Wold BJ, Deshaies RJ (2004) Applicability of tandem affinity purification MudPIT to pathway proteomics in yeast. Mol Cell Proteomics 3:226–237
5. Hannich JT, Lewis A, Kroetz MB, Li SJ, Heide H, Emili A, Hochstrasser M (2005) Defining the SUMO-modified proteome by multiple approaches in *Saccharomyces cerevisia*e. J Biol Chem 280:4102–4110
6. Tagwerker C, Flick K, Cu, M, Guerrero C, Dou Y, Auer B, Baldi P, Huang L, Kaiser P (2006) A tandem affinity tag for two-step purification under fully denaturing conditions: Application in ubiquitin profiling and protein complex identification combined with in vivo cross-linking. Mol Cell Proteomics 5:737–748
7. Puig O, Caspary F, Rigaut G, Rutz B, Bouveret E, Bragado-Nilsson E, Wilm M, Seraphin B (2001) The tandem affinity purification (TAP) method: a general procedure of protein complex purification. Methods 24:218–229
8. Guerrero C, Tagwerker C, Kaiser P, Huang L (2006) An integrated mass spectrometry-based proteomic approach: quantitative analysis of tandem affinity-purified in vivo cross-linked protein complexes (QTAX) to decipher the 26S proteasome-interacting network. Mol Cell Proteomics 5:366–378
9. Tagwerker C, Zhang H, Wang X, Larsen LS, Lathrop RH, Hatfield GW, Auer B, Huang L, Kaiser P (2006) HB tag modules for PCR-based gene tagging and tandem affinity purification in *Saccharomyces cerevisia*e. Yeast 23:623–632
10. Link AJ, Eng J, Schieltz DM, Carmack E, Mize GJ, Morris DR, Garvik BM, Yates JR III (1999) Direct analysis of protein complexes using mass spectrometry. Nat Biotechnol 17:676–682
11. Washburn MP, Wolters D, Yates JR III (2001) Large-scale analysis of the yeast proteome by multidimensional protein identification technology. Nat Biotechnol 19:242–247
12. Cronan JE Jr (1990) Biotination of proteins in vivo. A post-translational modification to label, purify, and study proteins. J Biol Chem 265:10327–10333
13. Savage D, Mattson G, Desa, S, Niedlander G, Morgensen S, Conklin E (1994) Avidin-biotin chemistry: A handbook. Pierce Chemical, Rockford, IL.

Chapter 22
Mammalian Two-Hybrid Assay for Detecting Protein-Protein Interactions in Vivo

Runtao He and Xuguang Li

Abstract The mammalian two-hybrid system is a very powerful tool to investigate protein-protein interactions in terms of functional domains and identify potential binding ligands and partners of a protein. Compared with the yeast two-hybrid system, the mammalian two-hybrid system provides the milieu for the bona fide posttranslational modifications and localizations of most eukaryotic proteins and, therefore, should be a better choice to study proteins of mammalian origin. This chapter depicts the detailed experimental procedures adapted by various laboratories. Researchers with experience in molecular biology could modify the procedures according to their own needs, that is, the choice of restriction sites in the cloning process. The reference list could be of use to researchers who wish to understand more of the system and explore its wider applications.

Keywords mammalian two hybrid, secreted alkaline phosphatase (SEAP), protein-protein interaction, PCR, cloning

1 Introduction

Protein-protein interaction is one of the most important mechanisms utilized by mammalian cells to regulate a variety of cellular and molecular activities, such as transcription, translation, signal transduction and enzyme reaction [1, 2]. Proteins of microorganisms also interfere with host cell proliferation and functions by interacting with intracellular proteins [3]. There are various methods to characterize protein-protein interactions including coimmunoprecipitation, mass spectrometry [4], chemical cross-linking [5], and the two-hybrid system [6, 7], which has been used widely in recent years because of its high-throughput screening capability and powerful ability of isolating unknown binding partners of a protein. The first two-hybrid system was yeast based and was introduced by Fields and Song in 1989 [8]. It is composed of two GAL-based transcription activation domains of S*accharomyces cerevisiae*, one is the DNA binding

From: Methods in Molecular Biology, Vol. 439: *Genomics Protocols: Second Edition*
Edited by Mike Starkey and Ramnath Elaswarapu © Humana Press Inc., Totowa, NJ

domain (DBD) domain, while the other is the transcription activation domain (AD). To perform the two-hybrid analysis, an X gene is cloned into a two-hybrid vector fusing to DBD, with a Y gene cloned into another two-hybrid vector fused to the AD. If the X protein interacts with Y protein, the DBD and AD domains come into close proximity and form an active transcription activation complex (Fig. 22.1). The complex binds to and activates its cognate promoter sequence, which is linked to a reporter gene(s), such as *Lac* Z, or an essential nutrient [8]. These gene products generate distinctive phenotypes that can be easily recognized by a change of color of yeast colonies or a gained ability to grow in nutrient-deficient media.

A lot of modifications have been made to improve the sensitivity and specificity of the two-hybrid system: the addition of more reporter genes and genetic alterations of the DBD and AD domains, such as the *LexA* system [9, 10]. However, the yeast-based two-hybrid assay has its limitations. One major issue is that posttranslational modifications such as glycosylation in yeast can be quite different from that in mammalian cells. In addition, the intracellular milieu that supports the protein-protein interaction in yeast is different from that of mammalian cells [11, 12]. Therefore, when considering the study of interactions of proteins derived from mammalian cells, the yeast system may not be able to provide the optimal intracellular environment.

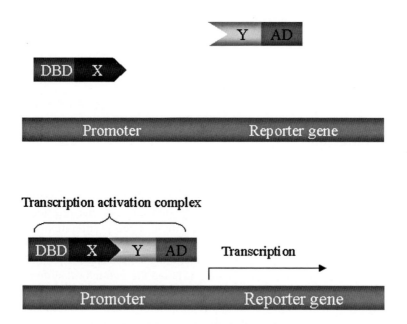

Fig. 22.1 Principles of the-two hybrid system: DBD and AD are DNA binding and transcriptional activation domains, X and Y are the bait and prey genes

To alleviate the pitfalls of the yeast two-hybrid system, the mammalian two-hybrid system was introduced.

The design of the mammalian two-hybrid system is similar to that of the yeast two-hybrid system, which is composed of two vectors carrying DNA binding domain and DNA activation domain, respectively, in two mammalian expression vectors. Two genes of interest (X and Y) are cloned into the vectors and fused with DBD and DA, correspondingly. The interaction of protein X to protein Y brings the DBD and AD to a close proximity and forms the transcriptional activation complex, which binds to the promoter of a reporter gene encoded in a reporter plasmid and starts the gene transcription. The reporter genes can be secreted alkaline phosphatase (SEAP), chloramphenicol acetyl transferaseor (CAT), or luciferase [13–15]. GAL4 DNA binding domain and herpes virus DNA transcription activator VP16 are commonly used in mammalian two-hybrid systems [16]. This chapter depicts the general procedure adopted by various laboratories.

2 Materials

2.1 PCR and Cloning of Genes of Interest into Mammalian Two-Hybrid Vectors

1. High fidelity DNA polymerase (i.e., iProof high fidelity DNA polymerase, BioRad).
2. dNTPs at 25 mM concentration (New England Biolabs).
3. PCR primers (cloning sites are added to the 5′ ends of the primers):

   5′ GTACGAATTCATGTCTGATAATGGACCC 3′.
   5′ GTACGGATCCTTATGCCTGAGTTGAATCAG 3′.

4. PCR template (cDNA reversely transcribed from SARS-CoV genome).
5. Thermal cycler, such as Master Cycler EP (Eppenforf).
6. Thin-wall PCR tubes or 96-well plates.
7. Ampicillin (Sigma-Aldrich).
8. PCR Kleen Spin Columns (BioRad).
9. Restriction enzymes: *Eco* RI and *Bam* HI (New England Biolabs).
10. Mammalian two-hybrid system with pM and pVP16 vectors (Clontech).
11. T4 DNA ligase (New England Biolabs).
12. Taq polymerase for colony screening assay (i.e., HotStar PCR Master Mix Kit, Qiagen).
13. Water bath at 37 °C.
14. Chemical or electro competent cells (Invitrogen).
15. Water bath at 42 °C.
16. Qiaprep Spin Miniprep Kit (Qiagen).

17. 10X TBE buffer: 108 g of Tris-base, 55 g of boric acid, 9.3 g of Na-EDTA, and add ddH$_2$O to 1L.
18. Agarose (Invitrogen).
19. LB (Luria Broth) per liter: bacto-tryptone 10 g, bacto-yeast extract 5 g, NaCl 10 g, pH 7.0; ddH$_2$O to make 1 L.
20. Preparation of LB agarose plates: add 15 g of agarose to 1 L of LB media and autoclave for 20 min, cool to 45 °C, and dispense approximately 25 mL into per 100-mm plate.
21. SOC medium per liter: add 2% bacto-tryptone, 0.5% yeast extract, 10 mM NaCl, 2.5 mM KCl, 10 mM MgCl$_2$, 10 mM MgSO$_4$, 20 mM glucose, pH 7.0, and bring the final volume to 1 L with ddH$_2$O.
22. Shaking incubator (Thermo Electron Corporation).
23. 20% glycerol solution (v/v).

2.2   DNA Purification for Mammalian Cell Transfection

1. Endofree Plasmid Maxi Kit (QIAGEN).
2. 70% ethanol v/v (for the QIAGEN Maxi Plasmid Kit).
3. 99% isopropanol (for the QIAGEN Maxi Pladmid Kit).
4. 50-mL tissue culture conical tubes.
5. Ultracentrifuge tubes.
6. TE buffer: 10 mM Tris-HCl; 1 mM EDTA, pH 8.0.

2.3   Tissue Culture

1. Growth medium: Dulbecco's Modified Eagle Medium (Invitrogen).
2. Fetal calf serum (Invitrogen).
3. 100X penicillin/streptomycin (Invitrogen).
4. Vero E6 cells (ATCC).
5. 0.5% trypsin-EDTA (Invitrogen).
6. Phosphate buffered saline: 8 g NaCl, 0.2 g KCl, 1.44 g Na$_2$HPO$_4$, 0.24 g KH$_2$PO$_4$, add double distilled H$_2$O to 1L. pH 7.4.
7. 24-well tissue culture plates (Nalgene Nunc).
8. Water bath set at 37 °C.
9. Inverted microscope (Zeiss).
10. Hemacytometer (VWR).
11. Tissue culture incubator (Thermo Electron Corporation).

2.4   Transfection and Mammalian Two-Hybrid Analysis

1. Transfection reagent: Effectene (QIAGEN).
2. pM and pVP-based constructs (Clontech, Palo Alto, CA).

3. pG5SEAP reporter vector (included in the mammalian two-hybrid kit).
4. SEAP (secreted alkaline phosphatase) Chemiluminescent Assay Kit (Clontech).
5. 96-well chemiluminescence reader (Tecan) or single-well chemiluminescence reader (Turner Designs).
6. 96-well PCR plates.

3   Methods

3.1   PCR Amplification of Genes of Interest and Preparation of Mammalian Two- Hybrid Vectors

3.1.1   PCR Reaction

1. Add 10 μL of iProof HF buffer, 1 μL of dNTP, two primers at 500 nM each, 0.5 U of DNA polymerase, a PCR template for the genes of interest, and ddH$_2$O to 50 μL.
2. PCR conditions: after 30 s of the initial denaturation at 98 °C, run 25–35 cycles of the following steps—denaturation at 98 °C for 10 s, annealing at 45–72 °C for 30 s, and extension at 72 °C (15–20 s per kb).
3. Clean PCR products with PCR Kleen columns, according to the manufacturer's protocol and elute DNA into a TE buffer.
4. Subject the PCR product to 1.2% agarose gel for electrophoresis, then isolate the band corresponding to the size of the desired gene. Purify the DNA from the gel piece using the Qiagen DNA Gel Purification Kit (optional step).

3.1.2   Vector Preparation

1. Inoculate two single colonies of *E. coli* top 10 strain containing pM and pVP16 vectors into 5 mL of LB respectively, add ampicillin to 100 μg/mL and grow in a shaking incubator at 37 °C and shake at 250 rpm overnight.
2. Harvest the cells by centrifugation at 10,000 g for 1 min.
3. Isolate the constructs using the Qiaprep Spin Miniprep Kit (Qiagen).

3.2   Cloning PCR Genes of Interest into of pM, pVP16 Vectors

3.2.1   Restriction Digestion

In designing the primers, *Eco*RI and *Bam*H I sites were added to the 5′ ends of the two PCR primers not present in the genes to be cloned. Cloning at *Eco*RI site fuses the gene of interest in frame with both DBD and AD domains.

1. Digest pM and pVP vectors (150 ng each) with 3 U of *Eco*RI and *Bam*HI, respectively, in 2 μL *Eco*RI buffer supplied by the manufacturer in 20 μL reaction mixture. Incubate at 37 °C for 1 h.
2. Digest PCR products flanked with *Eco*RI and *Bam*HI sites on the 5′ and 3′ ends respectively, using conditions just described.
3. Clean the digested vectors and PCR products with PCR Kleen columns.

3.2.2   Ligation

1. Add 100 ng of digested pM or pVP16 vector, 500 ng of digested PCR product, 2 μL of T4 Ligase buffer, 1 U of T4 ligase, and add ddH$_2$O to 20 μL.
2. Incubate at 16 °C for overnight.

3.2.3   Transformation

1. Thaw and aliquot 50 mL top 10 chemical competent cells into an Eppendorf tube (1×10^8 cells/mL), and chill on ice for 30 min.
2. Add 10–50 ng of the previously ligated products (see Sect. 3.2.2) into the previous *E coli* cells.
3. Immediately transfer the Eppendorf tube to a 42 °C water bath and incubate for 45 s.
4. Add 1 mL of SOC buffer and incubate in a shaking incubator at 37 °C for 30 min. Transfer the transformed *E coli* cells to a LB-ampicillin plate and incubate at 37 °C for 1 h. Turn the plate upside down and incubate at the same condition overnight.

3.2.4   Screening for Clones That Contain the Gene of Interest

There are many ways to screen for colonies that contain the gene of interest. In this protocol, we introduce a PCR-based screening, which is fast, simple, specific, and economical.

1. Perform PCR reaction with Taq polymerase. For each screening, prepare a master PCR-mix sufficient for 5–10 PCR reactions (48 μL/per PCR). The master mix is prepared by adding appropriate PCR buffer, dNTPs, polymerase, primers used for the previous PCR amplifications of the genes of interest or other primers that can amplify the gene, and ddH$_2$O.
2. Aliquot this PCR mix into 5–10 PCR tubes.
3. Pick a single colony and dip into the PCR mix, then dip the residual of the colony onto a LB-ampicillin master plate with numbers labeled on the bottom. The master plate is incubated at 37 °C overnight.

4. Perform PCR cycling using following program: 1 cycle of 95 °C for 10 min; 40–45 cycles of 95 °C for 1 min, 45–72 °C for 30 s, and 72 °C at 1 kb/min.
5. Subject PCR products to electrophoresis with 1.2% TBE agarose gel prepared in 1X TBE buffer. Samples with a band corresponding to the desired size are the potential positive clones.
6. Inoculate the positive colonies from the master LB-ampicillin plate into 5 mL of culture, incubate at 37 °C overnight, then perform DNA Maxiprep as described next (see Sect. 3.3).
7. Sequence the selected plasmids to verify the correct DNA sequence.
8. The positive *E coli* clone should be stored in LB-ampicillin broth containing 20% glycerol at −80 °C.

3.3 Large-Scale, High-Purity DNA Purification

1. Inoculate from the master plates just described containing pM and pVP16 constructs selected from the previous screenings into two culture tubes with 5 mL LB-ampicillin and incubate overnight at 37 °C.
2. Transfer these two 5 ml culture into two culture flasks containing 100 mL of LB plus ampicillin (100 μg/mL) and incubate until A_{600}=1.0–1.1 (approximately 1 × 10^9cell/mL).
3. Harvest the cells by centrifugation at 4 °C for 10 min. at 5,000 g.
4. Isolate the pM and pVP16-based constructs using a Qiagen Endofree Plasmid Maxi Kit, following the manufacturer's protocol.

3.4 Tissue Culture and Transfection

3.4.1 Culturing and Preparing Vero E6 Cells

1. Supplement DMEM medium with 10% heat-inactivated fetal bovine serum and 1% penicillin/streptomycin.
2. Maintain the Vero E6 cells (African green monkey kidney cells) in culture flasks inside a humidified incubator at 37 °C supplied with 5% CO_2.
3. To set up cell culture plates: Wash the cells with PBS, then treat the cells with 0.5% trypsin-EDTA before incubation at 37 °C for 5–10 min. Add 5 mL of the supplemented DMEM to the culture flask, transfer the supernatant into a screw-top conical tube, and spin at 1,500 g for 5 min. Aspirate out the supernatant and add fresh medium to make the final concentration at 150,000 cells/mL.
4. Aliquot 0.5 mL of these cells into each well of a 24-well plate (75,000 cells/well), then incubate at 37 °C overnight in a humidified chamber supplemented with 5% CO_2.

3.4.2 Transfection

1. Sample arrangement: the following six samples are recommended for each mammalian two-hybrid analysis (see Table 22.1). All samples should be three to four repeat sets.
2. For each transfection: add 150 μL EB buffer, 8 μL enhancer, 1 μg DNA (400 ng pM-X, 400 ng pVP16-Y, and 200 ng pG5SEAP), followed by incubation for 5 min at room temperature. Afterward, add 15 μL effectene (Qiagen) and incubate at room temperature for 10 min.
3. Add 40 μL of this transfection mix dropwise into each well of a 24-well plate, three or four repeats of each sample is needed to generate standard deviation when SEAP assay is performed later.
4. Gently shake the culture plate to mix, then incubate in a tissue culture chamber supplemented with 5% CO_2 at 37 °C for 48 h.

3.5 SEAP Analysis (Chemiluminescence)

3.5.1 SEAP Reaction

1. Aliquot 100 μL of the precedinjg cell culture supernatants from each well, transfer to an Eppdorf tube, and spin for 1 min at 10,000 g.
2. Add 15 μL of 1X dilution buffer from the GreatEscape SEAP Chemiluminescence Kit into each well of a conical bottom 96-well plate.
3. Add 40 μL of the cell culture supernatant into each well of the 96-well plate.
4. Seal the wells with adhesive aluminum foil pads, and incubate at 65 °C for 30 min.
5. Cool the plate on ice for 3 min and equilibrate the plate to room temperature.
6. Add 60 μL of assay buffer to each well and incubate for 5 min at room temperature.
7. Mix 1:20 ratio of chemiluminescent enhancer and CSPD SEAP substrate.
8. Add 60 mL of the mix to each well and incubate the reaction mixture for 15–30 min at room temperature.

Table 22.1 Sample arrangement of mammalian two-hybrid assays

Sample number	pM	pVP16	pG5SEAP vector
1	pM-X	pVP16-Y	pG5SEAP vector
2	pM	PVP16	pG5SEAP vector
3	pM	pVP16-Y	pG5SEAP vector
4	pM-X	PVP16	pG5SEAP vector
5	pM-p53	pVP-(SV40)T	pG5SEAP vector
6	Blank	Blank	Blank
7	Blank	Blank	pG5SEAP vector

Note: In the array of transfection for mammalian two-hybrid analysis, the pM, pVP, and pG5SEAP plasmid constructs are added in 2:2:1 ratio. Each combination should be transfected in triplicates or quadruplicates to obtain the optimal level of standard deviation for SEAP analysis.

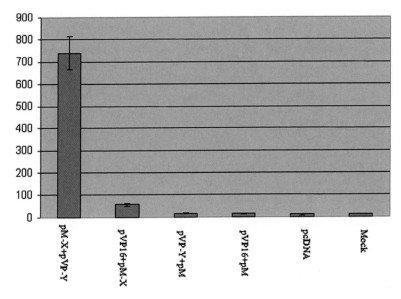

Fig 22.2 SEAP analysis: Chemiluminescence readings of the transfected samples. The depicted constructs were transfected in triplicate, 48 h after transfection, supernatant samples were harvested and subjected to SEAP analysis. The levels of chemiluminescence readings reflect the kinetics of protein-protein interactions

3.5.2 Detection of Chemiluminescence Using Luminometers

1. Detect chemiluminescence using a suitable detector (96-well or a single-tube luminometer can be used for this purpose.
2. Record the readings for phenotype analysis (see Fig. 22.2).

4 Notes

1. The quality of DNA for transfection is very important, we recommend $OD_{260/280}$ to be at least 1.5. The large scale DNA preparation should be aliquoted into several Eppendorf tubes and frozen at −20 °C because frequent freeze-thaw cycles with the same tube may decrease the quality of DNA and affect the analysis.
2. Sometimes the Gal or the VP16 fusion peptides in pM or pVP16 vectors affects the conformation of the targeted protein, it is recommended that pM-X + pVP16-Y, pM-Y, and pVP16-X are analyzed to avoid false negative results.
3. Cells should be 50–70% confluent to allow optimal transcription and translation of the targeted gene.
4. The combination of pM-X + pVP-X is recommended for analysis because many functional forms of proteins are dimers or multimers. In this case, X = Y. The constructs used in this chapter are SARS-CoV nucleocapsid protein in pM and pVP16 vectors (X = Y).

5. The transcription of pM and pVP16 vectors are controlled by a SV40 promoter, which is not as strong as the CMV promoter. Therefore, proteins expressed from these vectors may not be sufficient to perform Western blots or coimmunoprecipitation analyses. Clontech provides another set of expression vectors for the analysis, that is, pCMV-myc and pCM-HA, driven by CMV promoter. Other CMV-based mammalian transcription vectors, such as pcDNA 3.1 can be considered for co-IP or Western blot analysis. Alternatively, there are mammalian two-hybrid vectors carrying CMV promoters, such as pCMX-Gal4-N and pCMX-VP16-N [17].

6. If necessary, other methods should be considered to further confirm the result from the mammalian two-hybrid results. These methods are, but not limited to, GST-pull down assay, fluorescent resonance energy transfer (FRET) or bioluminescence resonance energy transfer (BRET), and mass spectrometry.

7. When pG5SEAP vector alone is used as a negative control, it tends to give high chemiluminescence readings, because no other vector competes for the transfection reagent. When this happens, a "filler" vector (i.e., pcDNA 3.1) that is not related to the mammalian two-hybrid system is recommended (i.e., 800 ng pCDNA 3.1 + 200 ng pG5SEAP).

8. The p53 protein and simian virus large T antigen have very high interaction kinetics. Such positive control included in the mammalian two-hybrid system may yield very high chemiluminescence reading, thereby obscuring the levels of the targeted protein-protein interactions that otherwise would be easily detected as positive reaction. It is recommended that this positive control be used merely to check if the transfection and SEAP assay are successful instead of comparing with the readings from the targeted protein-protein interactions.

9. The chloramphenicol acetyl transferase (CAT) assay is recommended by some manufacturers; however, in some cases, it is not as sensitive as the SEAP assay. The difference of sensitivity level can be more than tenfold.

10. Since mammalian two-hybrid system demonstrates levels of interactions, it can be used for interaction motif mapping. The commonly used mapping methods are sequential deletions or expression of gene fragments. When the sequential deletion method is used, it is recommended to do both amino- and carboxy-terminal sequential deletion so that functional domains can be better defined.

Acknowledgments We thank Alina Radziwon, Melissa Ballantine, and Todd Cutts for their contributions to the mammalian two-hybrid analysis. This work was supported by the Research Funding for Public Health Agency of Canada.

References

1. Fry DC, Vassilev LT (2005) Targeting protein-protein interactions for cancer therapy. J Mol Med 83:955–963
2. Arkin M (2005) Protein-protein interactions and cancer: Small molecules going in for the kill. Curr Opin Chem Biol 9:317–324

3. Tasara T, Hottiger MO, Hubscher U (2001) Functional genomics in HIV-1 virus replication: Protein-protein interactions as a basis for recruiting the host cell machinery for viral propagation. Biol Chem 382:993–999

4. Figeys D, McBroom LD, Moran MF (2001) Mass spectrometry for the study of protein-protein interactions.Methods 24:230–239

5. Trakselis MA, Alley SC, Ishmael FT (2005) Identification and mapping of protein-protein interactions by a combination of cross-linking, cleavage, and proteomics. Bioconjug Chem 16:741–750

6. Causier B, Davies B (2002) Analysing protein-protein interactions with the yeast two-hybrid system. Plant Mol Biol 50:855–870

7. Fields S, Sternglanz R (1994) The two-hybrid system: An assay for protein-protein interactions. Trends Genet 10:286–292

8. Fields S, Song O (1989) A novel genetic system to detect protein-protein interactions. Nature 340:245–246

9. Fields S (2005) High-throughput two-hybrid analysis. The promise and the peril. FEBS J 272:5391–5399

10. Miller J, Stagljar I (2004) Using the yeast two-hybrid system to identify interacting proteins. Methods Mol Biol 261:247–262

11. Wildt S, Gerngross TU (2005) The humanization of N-glycosylation pathways in yeast. Nat Rev Microbiol 3:119–128

12. Lehle L (1992) Protein glycosylation in yeast. Antonie Van Leeuwenhoek 61:133–134

13. Davey MR, Blackhall NW, Power JB (1995) Chloramphenicol acetyl transferase assay. Methods Mol Biol 49:143–148

14. Yang TT, Sinai P, Kitts PA, Kain SR (1997) Quantification of gene expression with a secreted alkaline phosphatase reporter system. Biotechniques 23:1110–1114

15. Gould SJ, Subramani S (1988) Firefly luciferase as a tool in molecular and cell biology. Anal Biochem 175:5–13

16. White J, Brou C, Wu J, Lutz Y, Moncollin V, Chambon P (1992) The acidic transcriptional activator GAL-VP16 acts on preformed template-committed complexes. EMBO J 11:2229–2240

17. Lee, J. W. and Lee, S. K. (2004) Mammalian two-hybrid assay for detecting protein-protein interactions in vivo. Methods Mol Biol 261:327–336

Chapter 23
Detection of Protein-Protein Interactions in Live Cells and Animals with Split Firefly Luciferase Protein Fragment Complementation

Victor Villalobos, Snehal Naik, and David Piwnica-Worms

Abstract Protein fragment complementation has emerged as a powerful tool for measuring protein-protein interactions in the context of live cells. The adaptation of this strategy for use with firefly luciferase now allows for the non-invasive, quantitative, real-time readout of protein interactions in lysates, live cells, and whole animals. Bioluminescence provides a robust imaging modality due to its extremely low background signal and large dynamic range. The split luciferase fusion constructs described here are inducible by addition of ligands, small molecules or drugs, in this example, rapamycin, and have been shown to work in vivo.

Keywords protein fragment complementation, protein-protein interactions, firefly luciferase, split luciferase complementation, bioluminescence, molecular imaging

1   Introduction

Fundamentally, the detection of physical interaction among two or more proteins can be supported if association between the interactive partners leads to the production of a readily observable biological or physical readout [1, 2]. A variety of techniques have been applied to the investigation of protein-protein interactions in cultured cells. Most strategies for detecting protein-protein interactions in intact cells are based on fusion of the pair of interacting molecules to defined protein elements that reconstitute a biological or biochemical function. Examples of reconstituted activities include activation of transcription, repression of transcription, activation of signal transduction pathways, or reconstitution of a disrupted enzymatic activity[1]. Compared with studies of protein interactions in cellulo, strategies for interrogating protein-protein interactions in intact living animals impose even further constraints on reporter systems and mechanisms of detection. To this effect, the development of novel protein fragment complementation assays displaying real-time, quantitative optical readouts may advance the

From: Methods in Molecular Biology, Vol. 439: *Genomics Protocols: Second Edition*
Edited by Mike Starkey and Ramnath Elaswarapu © Humana Press Inc., Totowa, NJ

understanding of how protein interactions function in cellulo and in the context of living animals.

Protein-fragment complementation assays depend on division of a monomeric reporter enzyme into two separate inactive components that can reconstitute function on association. When these reporter fragments are fused to interacting proteins, the reporter is reactivated on association of the interacting proteins. Complementation strategies based on several enzymes, including β-galactosidase, dihydrofolate reductase (DHFR), β-lactamase, green fluorescence protein, and luciferase have been used to monitor protein-protein interactions in mammalian cells [2–12]. A fundamental advantage of complementation assays is that the hybrid proteins directly reconstitute the enzymatic activity of the reporter. In principle, therefore, protein interactions may be detected in any subcellular compartment, and assembly of protein complexes may be monitored in real time. A disadvantage of complementation approaches is that the reassembly of an enzyme may be susceptible to steric constraints imposed by the interacting proteins. Another potential limitation of complementation assays for application in whole animals is that transient interactions between proteins may produce insufficient active enzyme to allow noninvasive detection. Nonetheless, because most complementation strategies are based on reconstituting active enzymes, these systems offer the potential benefits of signal amplification to enhance sensitivity for detecting interacting proteins in living animals.

Several split luciferase variants have been designed based on firefly and *Renilla* luciferases [2, 9, 13–16]. As with the intact luciferases, the primary differences between split *Renilla* and split firefly are its substrates, color, and quality of emission [17]. Intact *Renilla* exhibits more of a short-lived flash type emission, while firefly luciferase emits a more stable output for longer periods of time. *Renilla* luciferase requires calcium and oxygen to react with its substrate coelenterazine, while firefly luciferase requires ATP and oxygen to react with its substrate D-luciferin. Note that coelenterazine has been shown to be transported by the multidrug resistance P-glycoprotein transporter [18] as well as to interact efficiently with superoxide anion and peroxynitrate in light-producing reactions [19], thereby complicating applications of *Renilla* luciferase *in vivo*. Split *Renilla* and firefly luciferase are useful for real-time imaging of inducible and reversible interactions within cells, or animals, and can be efficiently utilized in high-throughput systems with a high degree of sensitivity for protein interaction kinetics. The split luciferase family of reporters also has been shown to be effective in measuring protein interactions in living animals. This paper focuses on the generation of the split firefly luciferase constructs and experiments demonstrating applications of those constructs (Fig. 23.1).

To identify an optimal pair of firefly luciferase fragments that reconstitute bioluminescence activity only on association, we constructed and screened a comprehensive combinatorial incremental truncation library that employed a well-characterized protein interaction system: rapamycin-mediated association of the FRB domain of human mTOR (residues 2024-2113) with FKBP-12 [9, 20]). Initial fusions of FRB and FKBP with inactive overlapping N- and C-terminal fragments of luciferase, respectively, were designed to minimize

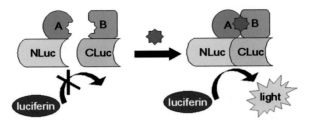

Fig. 23.1 Protein fragment complementation shows a drug induced interaction of split luciferase fusion constructs. Prior to drug administration, proteins of interest (A and B) are not associated, hence the NLuc and CLuc fragments are not in close enough proximity to be enzymatically active. On addition of the drug, A and B interact, bringing the two split luciferase fragments close enough to induce final folding, thereby regenerating luciferase activity and emitting light in the presence of ATP, O_2, and the chemical substrate D-luciferin

enzymatic activity of the individual starting fragments. The initial fragments of *P. pyralis* luciferase (derived from pGL3; Promega) also contained a linker composed of a flexible Gly/Ser region to minimize steric constraints between the luciferase fragments and target proteins. From these constructs, N- and C-terminal incremental truncation libraries were generated by unidirectional exonuclease digestion and validated as described [9, 21]. The libraries were coexpressed in *Escherichia coli* and screened for bioluminescence in the presence of rapamycin. From this screen, we identified an optimal pair of overlapping amino acid sequences for the NLuc and CLuc fragments that produced a very low signal in an uninduced state and abundant biolumines-cence when induced by the addition of rapamycin [9]. The final constructs maximized the induced complementation to yield the highest fold-inducibility and total photon output possible.

2   Materials

2.1   Cloning and DNA Production

1. PGL3 firefly luciferase plasmid (Promega, Madison, WI).
2. RLucN1 *Renilla* luciferase plasmid (Biosignal Packard, Quebec, Canada).
3. pTriEx3Neo triple expression vector (EMD Biosciences, San Diego, CA).
4. Primers (Integrated DNA Technologies, Coralville, IA).
5. DNA restriction enzymes (New England Biosciences, Ipswich, MA).
6. T4 DNA ligase (Invitrogen, Carlsbad, CA).
7. Calf intestinal phosphatase (CIP) (New England Biosciences).
8. QIAquick Gel Extraction Kit (Qiagen, Valencia, CA, USA).
9. Top 10 bacterial cells (Invitrogen).

10. LB agar (Sigma-Aldrich, St. Louis, MO).
11. LB broth (Sigma-Alrich).
12. Selection media: LB agar plates supplemented with 100 μg/mL ampicillin

2.2 Tissue Culture Materials

1. Dulbecco's Modified Eagle's Medium (DMEM) (Gibco/BRL, Bethesda, MD), supplemented with 10% (v/v) fetal bovine serum (FBS, HyClone, Ogden, UT) and 1% (w/v) glutamine (Sigma-Aldrich).
2. 10X stock solution phosphate buffered saline (PBS): 1.37 M NaCl, 27 mM KCl, 100 mM Na_2HPO_4, 18 mM KH_2PO_4, adjusted to pH 7.4.
3. Trypsin (Gibco/BRL).
4. Black plastic 24-well plate for bioluminescence imaging (Corning Life Sciences, Kennebunkport, ME).
5. Fugene6 transfection reagent (Roche Applied Sciences, Indianapolis, IN).

2.3 Bioluminescence Assay Materials

1. Modified Earle's balanced salt solution (MEBSS): 144 mM NaCl, 5.4 mM KCl, 800 μM $MgSO_4$, 800 μM NaH_2PO_4, 1.2 mM $CaCl_2$, 5.6 mM glucose, 4 mM HEPES, pH 7.4
2. Reporter lysis buffer (Promega).
3. Reporter assay buffer: MEBSS with 10% (v/v) heat-inactivated fetal bovine serum and 150 μg/ml D-luciferin.
4. D-luciferin (Biosynth, Naperville, IL) (100X stock:15 mg/mL in distilled water).
5. Coelenterazine (Biotium, Hayward, CA) (1,000X stock: 1 mg/mL in ethanol).
6. Lysis buffer for Western blots: 1% (w/v) SDS, 1.0 mM of sodium ortho-vanadate, 10 mM of Tris, pH 7.4.
7. Solution A: 2.7 mM KCl, 1.47 mM KH_2PO_4, 139 mM NaCl, 8.1 mM Na_2HPO_4, pH 7.4.
8. Matrigel (BD Biosciences, San Jose, CA).
9. Rapamycin (LC Laboratories, Woburn, MA).
10. BCA Assay Kit (Pierce Biotechnology, , Rockford, IL).

2.4 Western Blot Materials

1. Tris-buffered saline with Tween (TBS-T): 10 mM Tris base, 100 mM NaCl, pH to 7.5, then add 0.1% (v/v) Tween 20.
2. Running buffer: 25 mM Tris base, 192 mM glycine, 0.1% (w/v) SDS, pH 8.3.
3. Transfer buffer: 25 mM Tris base, 190 mM glycine, pH to 8.3.

4. 30% acrylamide (Bio-Rad Laboratories, Hercules, CA).
5. N,N,N,N′-Tetramethyl-ethylenediamine (TEMED) (Sigma-Aldrich).
6. 10% ammonium persulphate (APS) solution (Sigma-Aldrich).
7. Radioimmunoassay bovine serum albumin (RIA-BSA) (Tago, Burlingame, CA).
8. Goat anti-PGL3 luciferase antibody (Promega).
9. Antigoat antibody conjugated to horseradish peroxidase (EMD Biosciences, San Diego, CA).
10. ECL Western Assay Kit (Amersham Biosciences, Piscataway, NJ).
11. Western blot molecular weight standards (Bio-Rad Laboratories).
12. Western blot membrane. 0.45-μm pore size Immobulin-P (Millipore, Billerica, MA).

2.5   Instrumentation

1. Xenogen IVIS 100 (Alameda, CA) (*see* **Note 1**).
2. UV/Vis spectrophotometer (NanoDrop Technologies, Wilmington, DE).
3. Gene Pulser XCell electroporator (Bio-Rad Laboratories).
4. Sonicator: Sonifier 450 (Branson, Danbury, CT).

2.6   Analysis Software

1. Igor Living Image 2.50 (Xenogen, Alameda, CA).
2. Microsoft Excel.
3. Graphpad Prism (GraphPad Software, San Diego, CA).

3   Methods

3.1   DNA Construction and Manipulation into TriEx3Neo Expression Cassette

The split luciferase initially was designed for bacterial expression to screen for optimal pairs of firefly luciferase fragments that reconstitute bioluminescence activity only on association [9]. Excision and ligation into a more versatile plasmid backbone was performed to facilitate both bacterial and mammalian experimentation without having to use different constructs (*see* **Note 2**). For this purpose, the TriEx3Neo cassette was chosen, offering sites for expression in bacteria, mammalian cells, and lentivirus. As the split luciferase constructs are reasonably small (~400 bases and 1,200 bases for C and N terminal fragments, respectively), PCR cloning is recommended to introduce appropriate restriction digest sites for insertion of sequences encoding proteins of interest. While the present method focuses on the use of rapamycin-induced protein interactions of FK506 binding protein (FKBP) and FKBP-12 rapamycin binding domain (FRB), this technique also has been applied successfully to create split luciferase fusion constructs with STAT-1 proteins, Cdc25C, and 14-3-3 for quantitative measurement of protein interactions [9].

For DNA manipulation,

1. Digest 1 µg of split luciferase plasmids with 5 units of *Bam*HI and 5 units of *Xho*I in *Bam*HI buffer at 37 °C for 1 h.
2. Double digest 2 µg of TriEx3Neo with 5 units of *Bam*HI and 5 units of *Xho*I in *Bam*HI buffer at 37 °C for 1 h.
3. Treat the double-digested TriEx3Neo with 5 units of CIP at 37 °C for 30 min (to remove phosphates from the vector to reduce nonrecombinant colony growth).
4. Resolve the digested fragments by 1% agarose gel electrophoresis and purify the required fragments using the QIAquick gel extraction kit as per the manufacturer's instructions.
5. Measure the concentrations of the double-digested DNAs by UV spectrophotometry (optical density at 260 nm).
6. Set up a split luciferase plasmid: TriEx3Neo ligation reaction using T4 DNA ligase (as per manufacturer's recommendations) and incubate at 17 °C overnight.
7. Electroporate 1 µL of the ligation reaction into top 10 cells, add 250 µL of LB broth, and incubate at 37 °C for 1 h prior to plating on selective media.
8. Select individual colonies and grow 5-ml overnight cultures in LB broth containing an appropriate antibiotic.
9. Purify plasmid DNA (using the Qiagen Miniprep Kit) and perform diagnostic digest with *Apa*I (although the choice of restriction enzyme will be based on which plasmid is used) to confirm the identity of recombinant expression plasmids.
10. Further confirm that the required sequence has been created sequence by PCR sequencing the plasmid clone inserts using the TriExUp and TriExDown sequencing primers.

3.2 In Vitro Mammalian Cell Lysate

This method describes the optimal conditions for imaging split firefly luciferase in vitro. Cotransfection into the same mammalian cell prior to lysis yields better inducibility and signal than combining cell lysates after single fragment transfections. This also can be carried out using recombinant protein from bacteria (*see* **Note 3**). The readout for this system can be modified for use in fluorometers, or luminometers, to obtain spectra, or total photon output/inducibility, respectively.

3.2.1 Preparation of Mammalian Cell Lysates

1. To prepare DNA for transfection into a 35-mm dish, combine 500 ng of each construct (FRB-NLuc and CLuc-FKBP-12) in one tube.

2. Separately add 3 µL of Fugene6 to 500 µL of DMEM (prewarmed to 37 °C).
3. Add 503 µL of Fugene6/DMEM mixture to each DNA mixture and incubate at room temperature for at least 20 min.
4. Meanwhile, plate 1×10^6 HEK 293T cells out in a 35-mm dish.
5. After the requisite incubation of the DNA mixture, add the DNA:Fugene6/ DMEM solution to the cells, ensuring that the solution is spread over the whole 35-mm dish.
6. Incubate at 37 °C for 24 h.
7. After 24 h, remove the media from the cells and wash with 2 mL of prewarmed PBS or solution A.
8. Remove the PBS/solution A and add 1 mL of 1X reporter lysis buffer, and incubate on a rocking platform for 15 min.
9. Scrape cell fragments from the dish and transfer into a microcentrifuge tube.
10. Centrifuge the cells at 18,000 g for 15 min at 4 °C, collect the supernatant, and discard the pellet.
11. Store the lysate at −80 °C, or use immediately.

3.2.2   Imaging of Mammalian Cell Lysates

1. Add reporter assay buffer (1.0 mL) to each of 2×2 mL Eppendorf tubes. Add rapamycin, to a final concentration of 100 nM, to one tube and preheat the tubes to 37 °C.
2. Add 5–10 µL per well of paired lysates to a preheated black plastic 96-well plate.
3. Add 100 µL of prewarmed assay buffer, with or without rapamycin, to each appropriate well and measure the luminescence using the IVIS CCD camera.
4. Acquire images as 1 min acquisitions, binning 4, field of view (FOV) C, 20 × 20 cm.
5. Obtain an aliquot from each lysate and measure the concentration of protein (using a BCA assay) to be able to normalize luciferase data to protein levels.

3.3   Western Blot Analysis

3.3.1   Preparation of Mammalian Cell Lysates for Western Blot Assay

1. Wash previously transfected cells with 2 mL of 1X PBS at room temperature.
2. Add 250 µL of lysis buffer to each well. Scrape the cells from each plate well and immediately incubate the lysates at 95 °C for 10 min.
3. Maintain the lysates at 4 °C and sonicate using a micro tip limit of 6 and 50% duty cycle for 20–25 pulses.
4. Centrifuge at 18,000 g for 10 min at 4 °C.
5. Use samples immediately for SDS-polyacrylamide gel electrophoresis, or store at −20 °C.

3.3.2 SDS-Polyacrylamide Gel Electrophoresis

1. Separate mammalian cell lysates by electrophoresis through a 10% separating gel (with a 5% stacking gel). Include appropriate protein markers to track desired bands at 30 and 60 kDa, representing CLuc and NLuc fusion fragments, respectively.
2. Prepare a Western blot of the gel.

3.3.3 Immunoblotting

1. Immerse the Western blot in TBS-T, supplemented with 1% (w/v) RIA-BSA, for 1 h at room temperature to ensure minimal background activity.
2. Immerse the blot in a 1:200 dilution of primary goat antiluciferase polyclonal antibody in TBS-T, supplemented with 1% (w/v) RIA-BSA, for 1 h at room temperature.
3. Rinse the blot three times in TBS-T for 10 min each, at room temperature.
4. Incubate the blot in a 1:1000 dilution of antigoat antibody conjugated to HRP in TBS-T, supplemented with 1% (w/v) RIA-BSA, for 1 h at room temperature.
5. Using the ECL assay kit, develop the blot.

3.4 Live Mammalian Cell Bioluminescence Assays

Cell lines with high transfection efficiencies, such as HEK293 or 293T, are preferred for transient transfection experiments to ensure adequate transfection of both constructs into the same cell. If cell lines stably expressing the constructs are needed, such as for tumor xenografts, HeLa or a variety of other relevant tumorigenic cell lines can be used. High efficiency transfection reagents such as Fugene6 for HEK293T cells, or similar cell line-appropriate reagents, should be used. Figure 23.2 shows an example of live cell imaging.

3.4.1 Mammalian Cell Transfection

1. Combine 75 ng of each split luciferase fusion construct with 5 ng of RLuc DNA (155 ng total) per well of a 96-well plate.
2. Combine sufficient Fugene6 (0.45 µL per well) with serum-free media (DMEM), such that 10 µL of Fugene/media mixture can be added to each desired well, and incubate at room temperature for 20 min. A DNA:Fugene6 ratio of 1:3 is commonly used (*see* **Note 4**).
3. Plate 1×10^4 cells per well in 140 µL of media with serum. Add 10 µL of DNA: Fugene6/media mixture to each 1×10^4 aliquot of cells.
4. Allow transfection to proceed for 48 h. Proceed directly to imaging.

Fig. 23.2 Protein fragment complementation in live cells. This experiment illustrates wells from a black plastic 96-well plate, cotransfected with either FN+FC (FRB-NLuc, CLuc-FKBP, and RLuc), RLuc (RLuc alone), or ΔFN+FC (S2035I mutant FRB-NLuc, CLuc-FKBP, and RLuc). The mutant FRB precludes rapamycin association with the FKBP:rapamycin complex and functions as a control negative. The top four rows were treated with 100 nM rapamycin for 6 h prior to imaging and the bottom four rows were treated with vehicle (DMSO). The left panel shows cells exposed to D-luciferin (150 μg/mL) and imaged 10 min later. The cells then were washed and reimaged in the presence of coelenterazine (0.1 μg/mL), a *Renilla* luciferase substrate not recognized by firefly luciferase, to document uniform transfection efficiency. The left and right panels represent the identical cells imaged with the different bioluminescent substrates

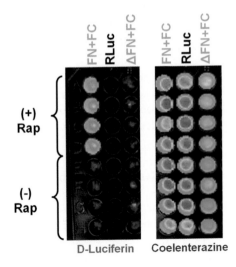

5. For a drug-induced interaction, add an appropriate drug or vehicle control, such as 100 nM rapamycin to the treatment group to induce FRB/FKBP interactions.
6. Incubate for an appropriate length of time, such as 6–8 h for rapamycin induction.

3.4.2 Imaging Live Mammalian Cells Using the Xenogen IVIS (In Vivo Imaging System)

1. Precool the camera to the appropriate temperature as required.
2. Prepare the warming tray to the desired temperature. Most cellular assays are performed at 37 °C.
3. Load the LivingImage 2.5 program and log onto the system.
4. Initialize the system as per the manufacturer's instructions.
5. Replace the media with prewarmed (37 °C) MEBSS containing 1% (v/v) heat-inactivated FBS and 150 μg/mL D-luciferin.
6. Load the plate containing the cells onto the heated imaging platform and allow to equilibrate for 5–10 min.
7. Depending on the signal output, perform several test images to optimize the readout. The camera saturates at ~80,000 counts per pixel, so the binning and acquisition time should be adjusted accordingly to obtain high-quality images with significant signal over background (at least 1 log-fold), while taking care not to saturate the camera. (*see* **Note 5**).
8. To analyze data, create a requisite region of interest (ROI) and click on Measure, then export the data to a software of choice for analysis.

3.5 In Vivo Mouse Imaging Experiments Using Pseudotumors

Pseudotumors are useful for monitoring the availability of drugs to a nonvascularized subcutaneous compartment in mice, while eliminating the waiting period for standard tumor xenografts to implant and grow. They yield high bioluminescence signals and tolerate transient transfection conditions for imaging the reporter.

3.5.1 Transfection of Mammalian Cells for Implantation

Perform transient transfection of cells as per the protocol described in Sect. 3.4.1 but with the following modifications:

1. Use 2 million cells per tumor per mouse.
2. Use 500 ng of each split luciferase construct, combined with 500 ng of RLuc DNA, per tumor per mouse.
3. Perform the transfection in a 35-mm tissue culture-treated dish, or a six-well plate. For example, for an experimental cohort consisting of five mice, where two tumors will be implanted into each mouse, use 2.5 μg of NLuc, 2.5 μg of CLuc, and 0.5 μg of RLuc, 15 μL of Fugene6 and 10 million cells for each of the two tumors.
4. Prechill all material required for matrigel usage, including pipette tips, needles, and syringes.

3.5.2 Preparation of Mammalian Cells For Implantation

1. Add 1 mL of Trypsin solution to washed cells from Sect. 3.4.1.
2. Add 1 mL of DMEM with 10% serum to the trypsinized cells and transfer 2 mL of the cell suspension to an Eppendorf tube.
3. Centrifuge at 18,000 g for 5 min.
4. Remove the supernatant. Resuspend in 2 mL of media. Centrifuge at 18,000 g for 5 min.
5. Remove the supernatant. Resuspend in 2 mL of PBS. Centrifuge at 18,000 g for 5 min.
6. Remove the supernatant. Place the pellet on ice.

3.5.3 Pseudotumor Implantation

The following procedures should be performed on ice.

1. Add an appropriate amount of matrigel (*see* **Note 6**) to each cell pellet for $n+1$ tumors; for example, if four tumors are to be implanted with a given construct, add 500 μL of matrigel.

2. Very gently resuspend the cells by slowly drawing the solution in and out of the pipette tip. The mixture will be viscous.
3. Very slowly draw up the matrigel-cell mix into a 1 mL syringe using a 27.5 gauge needle. Ensure that minimal air is drawn up.
4. Prepare the syringe for injection by removing all air bubbles and priming the needle.
5. Replace the filled syringe on ice, ensuring that the syringe shaft is covered with ice.
6. Anaesthetize a mouse (*see* **Note 7**) and position the animal prone with paws spread out away from the body. Keep the mouse under anesthesia for the duration of this procedure.
7. Gently shake the syringe to ensure proper mixing of its contents.
8. Insert the needle into the mouse at a very shallow angle, just under the skin.
9. Inject 100 μL of matrigel-cell mix. The injected area can be identified as an elevated wheel under the skin.
10. Repeat as required for each of the tumors to be placed on the mouse.

3.5.4 Imaging of Pseudotumors

1. Weigh mice 16–24 h after implantation.
2. Inject coelenterazine at 1 mg/kg of body weight (BW) i.v. and image mice in the IVIS under anesthesia (*see* **Note 8**).
3. Allow the signal to decay (for approximately 2 h). Confirm negative signal by imaging.
4. Inject D-luciferin intraperitoneally (i.p.) at 150 mg/kg BW.
5. Using the IVIS, image the mice, while under anesthesia after thermoequilibrating them in the IVIS for 30 min.
6. For inducible interactions, inject intraveneously (i.v.) drug or vehicle, as appropriate; for example, 1 mg/kg BW rapamycin in dimethyl sulphoxide (DMSO) or DMSO alone prior to imaging.

3.6 In-Vivo Mouse Experiments Using Hydrodynamic Injections for Hepatocellular Somatic Gene Transfer

Hydrodynamic injections require a bolus injection of 10% body weight of a solution containing DNA to induce extravasation into the liver and promote the uptake of large amounts of DNA [22, 23]. Hepatocytes begin expressing protein from injected DNA within hours, with a signal first measurable within 4 h and peaking typically at 24 h. The entire bolus injection (*see* **Note 9**) via tail vein is required to get full extravasation of fluid into the mouse liver, temporarily inducing congestive heart failure. Once it has hemodynamically recovered, the mouse will behave normally again. Figure 23.3 shows an example of imaging drug-inducible protein-protein interactions in vivo following hydrodynamic somatic gene transfer.

Fig. 23.3 Imaging hydrodynamic hepatocellular somatic gene transfer and protein fragment complementation in vivo. A mouse was injected with a high volume DNA mixture of FRB-NLuc, CLuc-FKBP and RLuc constructs as described in Sect. 3.6.1 and allowed to recover overnight. D-luciferin (150 mg/kg BW) was injected i.p. and the first image (left) was obtained 10 min later. Rapamycin (1 mg/kg BW) then was injected i.p., and 6 h later, the mouse was reinjected with D-luciferin i.p. and reimaged. The second image (right) reflects a 53-fold induction in signal with total photon flux of 2.6×10^7 photons/s postrapamycin, approximately 10^4-fold over the background

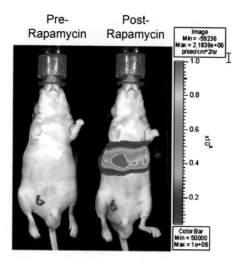

3.6.1 Injection of DNA

1. Combine 15 μg of each split luciferase fusion DNA with 2 μg of RLuc per mouse.
2. Weigh the mice.
3. Mix the DNA with solution A to bring the total volume up to 10% of the body weight of the mouse; for example, 2.5 mL total volume for a mouse weighing 25 g.
4. Use a warming light to increase the diameter of the tail veins during injection; this yields more successful transfection events.
5. Rapidly inject the entire DNA solution into the tail vein of the mouse without using anesthesia (a restraining device can be used to facilitate the injection). The full solution must be injected within approximately 5 s to yield adequate transfections.

3.6.2 Mouse Imaging

1. Weigh the mice 16–24 h after the hydrodynamic injections.
2. Inject coelenterazine at 1 μg/g BW i.v. and image mice in the IVIS under anesthesia.
3. Allow the signal to decay (1–2 h). Confirm by IVIS imaging.
4. Inject D-luciferin i.p. at 150 mg/kg BW. While under anesthesia, image mice in the IVIS.
5. For inducible interactions, inject drug or vehicle, as appropriate; for example, 1 mg/kg BW rapamycin in DMSO or DMSO alone, i.p. Inject D-luciferin i.p. repetitively (allow at least 4 h intervals to enable signal to decay to background between imaging sessions) and reimage throughout appropriate drug delivery and induction period.
6. To quantify data, create requisite ROI and click on Measure, then export the data to analysis software. For analysis, normalize the total photon flux (photons/s) from drug-treatment FLuc images to RLuc images, or to pretreatment FLuc images, to determine the fold-induction of FLuc signal by drug.

4 Notes

1. Due to the low photon output of these constructs, a cooled CCD camera is essential for acquiring accurate data. Standard luminometers with photomultiplier tubes can be used for in vitro experiments but require longer acquisition times and appropriate optimization.
2. TriEx3Neo contains a Neo cassette following an internal ribosomal entry site, which potentially reduces expression of the desired constructs. In addition, this is a plasmid with a low copy number, requiring special attention during DNA purification.
3. Cotransfection in bacteria yields better inducibility. We were unable to reconstitute activity by combining lysates from individual transfections. The same was observed in mammalian cell lysates.
4. Fugene6 should always be added into a tube containing DMEM. It is recommended that premixed Fugene6-DMEM solution is added to DNA.
5. Binning is a more appropriate setting to change, since it has a greater effect on the amount of light captured than changing acquisition time. For example, a binning of 8 captures 16 times more light than a binning of 2 but gives a lower resolution.
6. Matrigel is a liquid only at 4 °C and solidifies at both −20 °C and 37 °C.
7. Anesthesia for the mice is typically set at 2% (v/v) isofluorane with a flow of 1 L/min. Different mice respond differently, therefore, their breathing must be monitored to maintain a proper level of anesthesia.
8. Bolus i.v. injection of coelenterazine in 10% (v/v) ethanol can cause cardiac arrest in mice. If this occurs, perform chest compressions (gently massage the area above the heart with one finger) and monitor the mice closely to ensure recovery before placing under anesthesia again. It is recommended that a 1% (v/v) ethanol solution be used as the coelenterazine carrier.
9. The entire bolus must be injected within 3–5 s to ensure high transfection efficiency in the liver. Partial injections (for example, two consecutive 1 mL injections within 10 s) will not give adequate expression. The solution carrying the DNA must be at isotonic salt concentrations and pH 7.4.

Acknowledgments Special thanks to colleagues at the Molecular Imaging Center for valuable discussions. This educational project was supported by NIH grant P50 CA94056.

References

1. Toby G, Golemis E (2001) Using the yeast interaction trap and other two-hybrid-based approaches to study protein-protein interactions. Methods 24:201–217
2. Luker K, Piwnica-Worms D (2004) Optimizing luciferase protein fragment complementation for bioluminescent imaging of protein-protein interactions in live cells and animals. Methods Enzymol 385:349–360
3. Rossi F, Charlton C, Blau H (1997) Monitoring protein-protein interactions in intact eukaryotic cells by beta-galactosidase complementation. Proc Natl Acad Sci USA 94:8405–8410

4. Wehrman T, Kleaveland B, Her JH, Balint RF, Blau HM (2002) Protein-protein interactions monitored in mammalian cells via complementation of beta-lactamase enzyme fragments. Proc Natl Acad Sci USA 99:3469–3474

5. Remy I, Michnick S (1999) Clonal selection and *in vivo* quantitation of protein interactions with protein-fragment complementation assays. Proc Natl Acad Sci USA 96:5394–5399

6. Remy I, Wilson I, Michnick S (1999) Erythropoietin receptor activation by a ligand-induced conformation change. Science 283:990–993

7. Galarneau A, Primeau M, Trudeau L-E, Michnick S (2002) b-Lactamase protein fragment complementation assays as *in vivo* and *in vitro* sensors of protein-protein interactions. Nat Biotechnol 20:619–622

8. Ozawa T, Kaihara A, Sato M, Tachihara K, Umezawa Y (2001) Split luciferase as an optical probe for detecting protein-protein interactions in mammalian cells based on protein splicing. Anal Chem 73:2516–2521

9. Luker KE, Smith MC, Luker GD, Gammon ST, Piwnica-Worms H, Piwnica-Worms D (2004) Kinetics of regulated protein-protein interactions revealed with firefly luciferase complementation imaging in cells and living animals. Proc Natl Acad Sci USA 101:12288–12293

10. Ozawa T, Umezawa Y (2001) Detection of protein-protein interactions in vivo based on protein splicing. Curr Opin Chem Biol 5:578–583

11. Ozawa T, Nogami S, Sato M, Ohya Y, Umezawa Y (2000) A fluorescent indicator for detecting protein-protein interactions in vivo based on protein splicing. Anal Chem 72:5151–5157

12. Hu CD, Kerppola TK (2003) Simultaneous visualization of multiple protein interactions in living cells using multicolor fluorescence complementation analysis. Nat Biotechnol 21:539–545

13. Kaihara A, Kawai Y, Sato M, Ozawa T, Umezawa Y (2003) Locating a protein-protein interaction in living cells via split Renilla luciferase complementation. Anal Chem 75:4176–4181

14. Kim SB, Ozawa T, Watanabe S, Umezawa Y (2004) High-throughput sensing and noninvasive imaging of protein nuclear transport by using reconstitution of split Renilla luciferase. Proc Natl Acad Sci USA 101:11542–11547

15. Paulmurugan R, Gambhir SS (2005) Firefly luciferase enzyme fragment complementation for imaging in cells and living animals. Anal Chem 77:1295–1302

16. Paulmurugan R, Gambhir S (2003) Monitoring protein-protein interactions using split synthetic Renilla luciferase protein-fragment-assisted complementation. Anal Chem 75:1584–1589

17. Wilson T, Hastings JW (1998) Bioluminescence. Annu Rev Cell Dev Biol 14:197–230

18. Pichler A, Prior J, Piwnica-Worms D (2004) Imaging reversal of multidrug resistance in living mice with bioluminescence: *MDR1* P-glycoprotein transports coelenterazine. Proc Natl Acad Sci USA 101:1702–1707

19. Tarpey M, White C, Suarez E, Richardson G, Radi R, Freeman B (1999) Chemiluminescent detection of oxidants in vascular tissue. Lucigenin but not coelenterazine enhances superoxide formation. Circ Res 84:1203–1211

20. Chen J, Zheng X, Brown E, Schreiber S (1995) Identification of an 11-kDa FKBP12-rapamycin-binding domain within the 289-kDa FKBP12-rapamycin-associated protein and characterization of a critical serine residue. Proc Natl Acad Sci USA 92:4947–4951

21. Ostermeier M, Nixon A, Shim J, Benkovic S (1999) Combinatorial protein engineering by incremental truncation. Proc Natl Acad Sci USA 96:3562–3567

22. Wolff JA, and Budker V (2005) The mechanism of naked DNA uptake and expression. Adv Genet 54:3–20

23. Hagstrom JE (2003) Plasmid-based gene delivery to target tissues in vivo: The intravascular approach. Curr Opin Mol Ther 5:338–344

Chapter 24
Subcellular Localization of Intracellular Human Proteins by Construction of Tagged Fusion Proteins and Transient Expression in COS-7 Cells

John E. Collins

Abstract Identifying the subcellular compartment of a protein is an important step toward assigning protein function. Starting with a clone containing the open reading frame (ORF) of interest, it is possible to attach a variety of short amino acid tags or fluorescent proteins and detect the location of the protein, after transfection into a cell line, using fluorescent microscopy. By collecting data from various expression clone constructs, using a range of cell lines and double labeling with cellular compartment markers, a picture of the localization of a gene can be built up. This chapter describes how to obtain the ORF clone for your gene of interest, clone it into your choice of mammalian expression vector or vectors, transiently transfect for visualization, and where to get started when interpreting the results.

Keywords subcellular localization, open reading frame, mammalian expression clone, fluorescent microscopy

1 Introduction

The study of mammalian protein function gained momentum as more genomic sequences have been completed. Building up a picture of how a protein works requires asking a variety of questions and investigating many avenues [1]. One important piece of information, which gives a clue to function, is the identification of the protein's subcellular compartment [2]. The subcellular localization of a protein can be identified directly in a cell using fluorescence microscopy. A gene-specific antibody can be used to detect a protein. However, many proteins do not currently have antibodies available, and a suitable cell line expressing the protein is not necessarily available. In this chapter, an alternative method is described, using an open reading frame (ORF) cloned into a mammalian expression vector, which fuses the subsequent protein to a fluorescent protein molecule or short amino acid tag detectable by a generic antibody. After transfection into a suitable cell line, the protein's subcellular compartments can be identified by fluorescence microscopy.

From a practical point of view, there are numerous options and variations on the basic subcellular localization protocol. Commonly used methods for protein detection are the fusion of a complete fluorescent protein to the ORF, such as green fluorescent protein (GFP) or fusion of a short epitope tag such as V5, an epitope found in the P and V proteins of simian virus 5 [3]. Such epitopes can be detected with a specific antibody and then visualized using a secondary antibody with a fluorescent molecule attached. Different colored fluorescent fusion proteins or a variety of tags associated with different colored fluorescent molecules allow multiple proteins to be observed in a single cell. For example, a protein with a known localization fluorescing in red can be compared to a protein of unknown localization labeled in green. Another consideration is determining a suitable cell type. The cells need to attach to glass slides, transfect easily, and have a morphology suitable for accurate determination of localization. Both COS-7 African green monkey kidney cells and HeLa human cervical carcinoma cells are used regularly. However, it is wise to use a number of cell lines to confirm results. The strength of the promoter used in the expression construct also may vary among cell lines. With a strong promoter like cytomegalovirus (CMV) enhancer-promoter, it is possible to get very high quantities of fusion or tagged protein expressed, which can affect interpretation of the results. Alternative promoters, such as simian virus 40 (SV40) early promoter-enhancer, can be used to lower the protein expression in some cell lines. The final practical point to consider is the method of cloning the ORF sequence into the expression vector. A traditional restriction digest and ligation system works well, but is difficult to scale up for high-throughput applications. However, the Gateway® recombinational cloning system [4, 5] is well adapted to high-throughput projects. Currently, a multinational group of laboratories, referred to as the ORFeome Collaboration, is producing a publicly available, fully sequence-verified, Gateway compatible set of ORF clones representing all the currently defined genes in the human genome (www.orfeomecollaboration.org). This will allow a cloned ORF to be transferred into any Gateway-compatible expression vector using a simple recombination reaction, thus simplifying the production of a transfection-ready clone.

Several projects have used high-throughput techniques to determine subcellular protein localizations for substantial numbers of genes and have produced databases of these localizations. The LIFEdb database, (www.LIFEdb.de) [6], contains many human GFP fusion protein localization images [7]. In the mouse, the LOCATE database (http://locate.imb.uq.edu.au) used data from immunofluorescence experiments and literature review to compile a dataset representing approximately 40% of the genome [8].

This chapter details methods to clone a protein into a tagged expression vector, transiently express it in COS-7 cells, detect it by fluorescence microscopy, and analyze the results. Protocols are presented to enable low-throughput subcellular localization analysis, as well as a description of the future availability of Gateway-compatible resources that could lead to high-throughput analysis.

2 Materials

2.1 Making an Expression Vector Construct

1. Clone containing chosen open reading frame (see Sect. 3.1).
2. Expression vector of choice (see Sect. 3.2.1).
3. KOD Hot Start DNA polymerase (Novagen 71086-4).
4. Agarose gel and electrophoresis equipment.
5. Mach1 T1 phage-resistant cells (Invitrogen C8620-03).
6. QIAquick gel extraction column (QIAgen 28706).
7. Rapid DNA Ligation Kit (Roche 11 635 379 001).
8. Shrimp alkaline phosphatase (UBS 70092Y).
9. Exonuclease I (UBS 70073Z).
10. QIAprep Spin Miniprep Kit (QIAgen 27104), QIAgen Plasmid Mini Kit (QIAgen 12123), or Wizard Plus Minipreps DNA Purification Systems (Promega A7100).
11. LB agar. Prepare 10 g of tryptone (Fisher BPE1421-500), 5 g of yeast extract (Oxiod LP0021), 5 g of sodium chloride (BDH 3012375), and 20 g of bacto agar (BD 214030), make up to 1 L with water and autoclave.
12. Agarose (Invitrogen 15510-027).
13. Ethidium bromide (Sigma E1510).
14. Thermocycler (supplier MJ Reasearch).
15. NotI restriction enzyme (NEB R0189S).
16. XbaI restriction enzyme (NEB R0145S).
17. HindIII restriction enzyme (NEB R0104S).
18. NheI restriction enzyme (NEB R0131S).

2.2 Cell Culture and Transfection

1. COS-7 cells (available from the European Collection of Cell Cultures).
2. Dulbecco's Modified Eagle Medium (DMEM; Invitrogen 41965-039).
3. Fetal bovine serum (Invitrogen 10108-165).
4. 100X penicillin-streptomycin-glutamine (Invitrogen 10378-016).
5. Gene Juice (Novagen 70967-3).
6. PBS. Prepare 36.65 g of soduim chloride (BDH 3012375), 11.8 g of disodium hydrogen phosphate (BDH 104363A), and 6.6 g of sodium dihydrogen phosphate (BDH 442444H), make up to 1 L with water and autoclave.
7. Cell culture facility.
8. Trypsin (Gibco 25300).

2.3   Preparing Slides for Fluorescence Microscopy

1. Chamber slides, eight chamber polystyrene-vessel tissue-culture-treated glass slides (BD Falcon Ref 354118).
2. Fume hood.
3. Gelatin from cold water fish (Sigma G7765). Prepare 2% gelatine in PBS.
4. Saponin from quillaja bark (Sigma S4521). Prepare 0.5% saponin in PBS.
5. Paraformaldehyde (BDH 294474L). Prepare 3.7% paraformaldehyde, 4% sucrose in PBS, heat to 80 °C in a water bath to dissolve and add a few drops 1 M NaOH until clear. Cool before use. Note that paraformaldehyde is hazardous and should be handled in an appropriate chemical fume hood.
6. Ammonium chloride (Sigma A0171). Prepare 50 mM in PBS.
7. Vectashield with DAPI (Vector Laboratories H-1200).
8. Primary antibody that recognizes epitope tags; for example, T7 tag monoclonal antibody (Novagen 69522-3) and anti-V5 antibody (Invitrogen R960-25).
9. Antimouse Alexa Fluor® 488 (Molecular Probes A11070; if the primary antibody was raised in a mouse).
10. Alexa Fluor® 546 phalloidin (optional) (Molecular Probes A22283).
11. Fluorescent microscope.

2.4   Western Blot Analysis (Optional)

1. Western blotting gels and equipment (*see* **Note 1**).
2. Fume hood.
3. Protein sample buffer: 2% SDS (BDH 442444H), 10% glycerol (BDH 101184K), 60 mM Tris-HCl, pH 6.8, 0.01% bromophenol blue (BDH 20015 2E), and 5% β-mercaptoethanol (Sigma M3148).
4. Milk powder for antibody blocking (for example, Marvel).
5. Tween20 SigmaUltra (Sigma P7949). Prepare 0.1% Tween diluted in PBS.
6. Western Lightning chemiluminescence reagent (Perkin Elmer NEL102).
7. Primary antibody (see Sect. 2.3, step 8).
8. Secondary peroxidase conjugate antibody which binds to the primary antibody.
9. Magic Marker XP Western protein standards (Invitrogen LC5602).

3   Methods

3.1   Obtaining an Open Reading Frame Clone

A number of projects produced collections of human full-length cDNA clones comprising the 5′ untranslated region, the coding region, and the 3′ untranslated region, while others made or converted full-length cDNA clones into expression-ready clones comprising just the open reading frame (the ATG start codon to the stop or penultimate codon) [9–14] (www.hip.harvard.edu). The Mammalian Gene Collection (MGC; http://mgc.nci.nih.gov) [15] contains many sequence-verified full-length cDNAs for human genes, as well as from other animals. These are available via the distributors of the IMAGE consortium (http://image.llnl.gov). The human MGC clones subsequently have been used in a large-scale project to produce Gateway-compatible clones of just the in-frame ORF sequence (referred to as pENTR clones), usually without the stop codon, which are ready to recombine into expression vectors [14]. These clones, along with clones from other laboratories, currently are undergoing full-length sequence verification to produce a Gateway-compatible pENTR ORFeome collection in an international multilaboratory project, the ORFeome Collaboration (www.orfeomecollaboration.org). As with the MGC clones, the ORFeome Collaboration clones will be available via the distributors of the IMAGE consortium. Gateway pENTR ORF clones also are available from Invitrogen.

If you are unable to find the gene or particular splice variant you require in existing collections, it is possible to PCR amplify and clone it from cDNA by a relatively straightforward procedure (see Collins et al. 2004 for a full protocol [16]). Briefly, design two pairs of nested PCR primers that will amplify the ORF plus part of the 5′ and 3′ untranslated region, picking the sequence to give optimal primers. Then amplify the ORF fragment with two rounds of nested PCR from a suitable cDNA source using proofreading KOD Hot Start DNA polymerase [17] and clone the fragment into any vector, such as pGEM. It is advisable to fully sequence the insert of the clone to verify it is as expected. The ORF then can be transferred to an expression vector using the protocol described later (see Sect. 3.2.2) or converted to a pENTR clone using the Gateway Technology PCR Cloning System for subsequent recombinational cloning (Invitrogen).

3.2 Preparing the Expression Vector

3.2.1 Choosing an Expression Vector

A number of companies offer ready-made expression vectors, including various color fluorescent fusion proteins and short amino acid epitope tags; for example Invitrogen (www.invitrogen.com), Sigma (www.sigmaaldrich.com), BD Clontech (www.clontech.com). and Stratagene (www.stratagene.com). The larger fusion proteins are easier to use, as they can be viewed directly in live or fixed and mounted cells, whereas the smaller epitope tags require antibody development for visualization but potentially have less artifactual effect on the localization of the test protein. Most expression systems use the strong CMV promoter, but others are available. Invitrogen supplies expression clones that are Gateway compatible. However, by using the Invitrogen Vector Conversion System (Invitrogen 11828-019), it is possible to convert any expression vector into a Gateway-compatible version.

3.2.2   Transferring an ORF into an Expression Vector

The method for transferring a gene into an expression vector depends on the ORF clone. If the clone is a Gateway-compatible in-frame pENTR clone, from the ORFeome Collaboration, for example, it can be transferred into a destination or pDEST clone using LR clonase (Invitrogen 11791-020), following the protocol that comes with the enzyme. If not, the ORF can be extracted from a full-length cDNA clone and converted to a pENTR clone or cloned using the restriction enzyme cloning method described next. A pENTR clone also can be used in the restriction enzyme cloning method if preferred. The following is an example of how to clone an ORF into an expression vector that has been altered to allow expression with an epitope tag fused in-frame at either the N- or C-terminus of the protein (Fig. 24.1). The principle is the same for the commercially available tagged vectors, the most important consideration being to maintain the reading frame.

1. Design primers that contain the correct restriction enzyme site at either end of the PCR product followed by at least 18 bases of the 5′ or 3′ end of the ORF making sure that the reading frame is maintained between the ORF and the tag (see Fig. 24.1). The diagram shows how to make two constructs for each gene: the ORF-tag (which is the tag at the C-terminal end of the protein) and the tag-ORF (which is the tag at the N-terminal end of the protein). The primer for an ORF-tag construct should also contain a Kozak consensus sequence [18] (shown here as ACC) before the start codon. The 3′ primer for the tag-ORF construct should include the stop codon but the ORF-tag 3′ primer should not. It is a good idea to check the primers for regions of self complementarity and either add or remove bases to reduce the effect.

2. Dilute a single streaked colony or a small number of ice crystals from a glycerol stock of the parent ORF clone in 100 μL of sterile water.

3. Amplify the ORF in a 25 μL KOD Hot Start DNA polymerase reaction using the supplied buffer, $MgSO_4$, and nucleotides (following the manufacturer's instructions), and using a cycling profile of 94 °C,prewarm, 2 min at 94 °C, 30 cycles of 15 s at 94 °C, 30 s at 60 °C, and 3 min at 68 °C, finishing with 5 min at 68 °C.

4. Separate by electrophoresis on a 1% agarose gel, excise the fragment of the expected size and purify using a QIAquick gel extraction column.

5. Digest the PCR fragment with the appropriate enzymes, for example, for the situation illustrated in Fig. 24.1 use *Not*I and *Xba*I for the tag-ORF construct and *Hind*III and *Nhe*I for the ORF-tag construct. For best results, clean the digested product using a column purification system or ethanol precipitation.

6. Prepare the expression vector by restriction digest with the appropriate enzymes.

7. Separate digested expression vector on a 1% gel, excise the fragment of the expected size, and purify using a QIAquick gel extraction column.

8. Ligate approximately 100 ng of ORF fragment to digested expression vector at a 3:1 molar ratio using the Rapid DNA Ligation Kit according to the manufacturer's instructions.

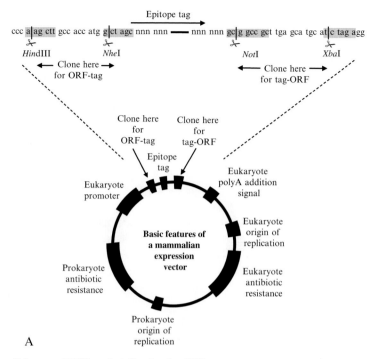

Primers and PCR product digestion for ORF-tag construct

5' primer ggccaagctttagaccatgnnnnnnnnnnnnnnnnnnnn
3' primer ggccgctagcnnnnnnnnnnnnnnnnnnnn

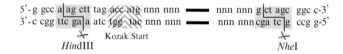

Primers and PCR product digestion for tag-ORF construct

5' primer ggccgcggccgcnatgnnnnnnnnnnnnnnnnnnnn
3' primer ggcctctagaNNNnnnnnnnnnnnnnnnnn
(the capital Ns are the stop codon)

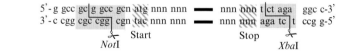

B

Fig. 24.1 Expression vector with N and C-terminal cloning sites. Panel A shows the main features required for a mammalian expression vector with the ORF cloning region detail above. The epitope tag is depicted as a series of *n* lying between the cloning sites. The ORF-tag construct consists of a gene followed by a tag (i.e., the tag at the C-terminal end of the protein), where as the tag-ORF consists of a tag followed by a gene (the tag at the N-terminal end of the protein). Panel B **shows** examples of primers and the resulting amplified PCR product with the cloning restriction digest site indicated. Note that the reading frame is shown by sets of three bases, this makes it simpler to check that the frame is retained between the tag and the ORF in the final construct

9. Transform into competent cells with a transformation control to check the efficiency. It is a good idea to use cells resistant to the T1 phage, we used the fast-growing Mach1™ T1 phage-resistant cells (Invitrogen).
10. Plate on LB agar plates with appropriate antibiotic and incubate overnight at 37 °C.

3.2.3   Confirming Expression Clone

3.2.3.1   Identifying Colonies Containing the ORF

1. PCR amplify with KOD Hot Start DNA polymerase, as described in Sect. 3.2.2, step 3, using a colony picked into 100 µL of sterile water as a template and the primers used to amplify the ORF (see Sect. 3.2.2, step 2).
2. Visualize by gel electrophoresis on a 1% agarose gel.

3.2.3.2   Confirming the Integrity of Primer Sequences

1. PCR amplify the template from Sect. 3.2.3.1, step 1, with vector primers, approximately 150 bases from either side of the cloning site, in a 25 µL KOD Hot Start DNA polymerase reaction (it is worth amplifying Gateway-recombined clones with vector primers to confirm that the recombination has worked correctly, but sequencing normally is unnecessary).
2. Confirm amplification by running 3 µL on a 1% agarose gel.
3. Prepare 7.9 µL of fragment for sequencing by inactivating the unincorporated nucleotides with 1 unit of shrimp alkaline phosphatase and digesting the PCR primers with 0.1 U of exonuclease I in a 10 µL reaction, buffered with 20 mM of Tris-HCl, pH 8.0, and 10 mM of MgCl$_2$ at 37 °C for 30 min, followed by 15 min at 80 °C to inactivate the enzymes.
4. Sequence PCR fragments with the vector primers and check they are correct (*see* **Note 2**).
5. If desired, fully sequence the clone to check for polymerase errors, although these errors should be rare. Base changes causing synonymous amino acid changes may be acceptable.

3.3   *Transfection of Expression Vector*

3.3.1   Preparing DNA for Transfection

Many commercially available DNA preparation kits claim to provide "transfection quality" DNA. We used QIAprep Spin Miniprep Kit, QIAgen Plasmid Mini Kit, and

Wizard Plus Minipreps DNA Purification Systems to prepare DNA for transfection with success. For each transfection, 250 ng of DNA is required with an optical density 260/280 ratio between 1.8 and 2.0. Before transfecting cells on microscope slides, it is useful to check that the protein is expressed in the chosen cell line and it compares to the predicted molecular weight by performing a Western blot (see Sect. 3.5). When looking at protein expression, it is useful to transfect two constructs, one tagged at the N-terminal and one tagged at the C-terminal.

3.3.2 Transfecting for Fluorescence Microscopy

1. Grow COS-7 cells to at least three passages after defrosting; healthy rapidly growing cells give optimum results.
2. In a tissue culture flow hood, trypsinize the cells, count, and dilute to a concentration of 2×10^4 cells/mL with fresh DMEM containing 10% FBS and 1X penicilin-streptomycin-glutamine (*see* **Note 3**).
3. Add 0.5 mL of cell suspension to each well of an eight-well tissue-culture slide. Select one of the eight wells as a negative control, which receives no DNA, and a second well as a vector only (no ORF insert) control (*see* **Note 4**). It is also useful to use a well to calculate the transfection efficiency to monitor the levels of transfection between experiments (*see* **Note 5**).
4. In a premix, for each transfection, add 20 μL prewarmed DMEM (serum-free media) and 0.75 μL Gene Juice into a tube, mix, and incubate at room temperature for 5 min.
5. Place 250 ng of each sample DNA (desirably between 1–10 μL) in a tube and add 20.75 μL of the DMEM/gene juice solution to each, pipetting gently up and down to mix, and incubate for 5 min at room temperature.
6. Add each mix dropwise to the cells in the eight-well tissue culture slides.
7. Place the slides in a plastic box on a tray above two reservoirs of sterile water to maintain humidity.
8. Leave on the bench for 30–60 min to allow the cells to settle (*see* **Note 6**) before transferring to the incubator at 37 °C, 5% CO_2 with humidity for 2 d (*see* **Note 7**), ensuring box is closed but not completely sealed.

3.4 Visualising Protein Expression

3.4.1 Fixing and Mounting Cells

1. Due to the paraformaldehyde, use an external venting tissue culture flow hood to fix the cells.
2. Carefully aspirate the media from each well and wash the cells three times in 0.5 mL PBS.
3. Fix the cells in 0.5 mL 3.7% paraformaldehyde solution for 20 min.

4. Remove the paraformaldehyde and wash the cells three times in 0.5 mL PBS (collect the paraformaldehyde and washes for hazardous waste deposal).

5. Remove the well formers with the adaptors provided. If using several primary or secondary antibodies or performing a titration experiment, leave the well former to keep the specific antibody incubations separate.

6. Quench the cells with 50 mM of NH$_4$Cl in PBS for 15 min. The slide now can be removed from the flow hood.

7. In a Coplin jar, wash cells three times in PBS.

8. If using an epitope tag, complete Sect. 3.4.2 before mounting; if using a fluorescent fusion protein, mount immediately.

9. Place a coverslip on blotting paper and dispense Vectashield onto the coverslip. Carefully lower the slide face down on to the coverslip and apply gentle pressure, removing any excess Vectashield with blotting paper.

10. Seal the slides with clear nail varnish and leave to dry in the dark for 1–2 h.

11. View using a fluorescence or confocal microscope.

12. Store the slidse in the dark at 4 °C.

3.4.2   Detecting Gene Epitope Tags with Antibody

1. Block and permeablilize the cells by incubating in a Coplin jar of 0.2% gelatin and 0.05% saponin solution for 15 min.

2. For a whole slide, dilute the primary antibody (which detects the epitope tag) in 500 µL of 0.2% gelatin and 0.05% saponin (*see* **Note 8**). If using several different primary antibodies or performing a titration experiment without removing the well former, use 200 µL of solution per well. If two constructs expressing different epitopes are being used, incubate both antibodies together, but these antibodies must be raised in different animals to allow separate detection by secondary fluorescently conjugated antibodies.

3. Pipette the antibody mix onto parafilm, lower the slide face down onto the solution ensuring no air bubbles and incubate for 45 min at room temperature.

4. Rinse cells twice in 0.05% saponin solution.

5. Wash cells three times for 10 min in 0.05% saponin solution with very gentle shaking.

6. Incubate with 0.2% gelatin and 0.05% saponin solution for 15 min.

7. Prepare 500 µL of the fluorescent antibody conjugate diluted in 0.2% gelatin and 0.05% saponin solution (*see* **Note 8**). Multiple secondary fluorescence antibody conjugates may be used to bind the different primary antibodies if used in step 2 (*see* **Note 9**). If the well former remains, use 200 µL of antibody solution per well.

8. Incubate on parafilm as in step 3 for 30–45 min at room temperature protected from light.

9. Rinse cells twice in 0.05% saponin solution.

10. Wash cells three times for 10 min in 0.05% saponin solution protected from light.

11. Wash cells for 60 min in PBS then replace the PBS and leave overnight, all protected from light.
12. Mount the slide as described in Sect. 3.4.1.

3.5 Performing a Western Blot

1. Place 40,000 COS-7 cells in 0.5 mL of fresh DMEM containing 10% FBS, 1X penicilin-streptomycin-glutamine into each well of a 24-well tissue-culture dish.
2. Transfect 250 ng of DNA into as described in Sect. 3.3.2.
3. As controls, transfect one well with vector only (no ORF insert) DNA and leave another with cells but no transfected DNA.
4. Incubate the cells for 2 d at 37 °C, 5% CO_2 with humidity.
5. Remove the media from each well and add 100 μL of protein sample buffer to extract the protein.
6. Detach the cells from the bottom of each well of the plate by scraping with a pipette tip and transfer the lysate to separate microfuge tubes.
7. Heat to 100 °C for 5 min (a dry heating block is best) and cool before opening the tubes. You may need to make a hole in the top of the tube to prevent the lid popping open. Perform this step in a fume hood due to the β-mercaptoethanol.
8. Electrophorese the protein on a denaturing PAGE gel and Western blot according to the instructions of your chosen system (see **Note 1**). Include a marker lane, we used Magic Marker XP Western protein standards (Invitrogen LC5602).
9. Develop the blot by first blocking in 10% milk powder, 0.1% Tween in PBS for 30 min with gentle shaking.
10. Take an antibody that detects the epitope tag being used, dilute in 10% milk powder, 0.1% Tween in PBS and incubate at room temperature for 60 min with gentle shaking (the antibody may need to be titrated to get an optimal signal to background ratio).
11. Rinse twice in 0.1% Tween in PBS and wash three times in 0.1% Tween in PBS for 10 min with gentle shaking.
12. Take a peroxidase-conjugated antibody that binds to the primary antibody, dilute in 10% milk powder, 0.1% Tween in PBS, and incubate for 60 min with gentle shaking (again the antibody may need to be titrated).
13. Rinse twice in 0.1% Tween in PBS and wash three times in 0.1% Tween in PBS for 10 min with gentle shaking.
14. Mix 5 mL of each Western Lightning chemiluminescence solution and develop the blot for 1 min.
15. Visualize the blot by autoradiography and compare the size of the fragments to the expected size (see **Note 10**). The fragments may not be the expected size due to posttranslational processing, such as cleavage or addition of carbohydrate modifications.

3.6   Analysing the Results

Images can be captured using a fluorescence or confocal microscope. When interpreting the results, it is important to consider all the evidence before deciding in which subcellular compartment or compartments the protein is located. Look at many cells from each transfection to get an overall picture of the fluorescent signal. A good place to start the analysis is by studying the collection of images showing representative localizations of ORF-GFP fusions to a number of cellular compartments in Vero cells (African green monkey kidney) [7]. The LIFEdb database (www.LIFEdb.de) [6] contains numerous examples of localizations and is useful for comparison with your images.

When initially looking at the slide, assess the level of fluorescence, which is likely to vary from cell to cell. Ideally, representative cells should have enough signal to detect localization but should not be saturated with fluorescence, as the accumulation of large amounts of artificially overexpressed protein could result in mislocalization (*see* **Note 7**). Changing the cell line or promoter may improve the image. Also, cells may have more than one localizsation within a single well. This could be due to variation in the physiology of the cells, such as the position in the cell cycle or apoptosis, for example. There are ways to detect processes like the cell cycle including antibodies to cycle-stage-specific proteins. By selecting an antibody raised in a different animal to the antibody that detects the epitope, it is possible to detect both antibodies with different colored secondary fluorescent conjugates in the same cell (see www.abcam.com for examples of marker antibodies you may wish to test).

To assist with localization, software to predict the cellular compartment of an ORF from the amino acid sequence and programs to help with image analysis are available. A summary of localization tools hosted by the University of Colorado Health Science Center can be found at (http://compbio.uchsc.edu/Hunter_lab/Zhiyong/subcellular). A good example is the signal peptide prediction program, called SignalP, at (www.cbs.dtu.dk/services/SignalP) [19]. This program searches the amino acid sequence and gives the likelihood of the N-terminal region being a signal peptide. These are signals for protein translocation and are removed as the protein passes through the plasma membrane. The presence of a signal peptide suggests that the protein may be found in one of the compartments in the secretary pathway. A fluorescent protein or amino acid tag at the N-terminal end of a protein could disrupt this process by obscuring the signal peptide or could be removed during cleavage, resulting in no signal. Running SignalP on the protein with the fusion or tag gives an indication of whether this is likely. If the N-terminal fusion or tag does disrupt the signal peptide, the protein might localize to the wrong cellular compartment, but in such cases, the C-terminal fusion or tag can help with interpretation. A general purpose localization tool is PSORT (http://psort.hgc.jp/form.html), which gives an overall subcellular compartment prediction using a number of different programs [20]. It is important to remember that this prediction tool uses just the primary amino acid sequence and does not take into account other factors, such as posttranslational modification, and therefore may not always agree with your observed localization. Another place to explore is the Murphy Lab website from Carnegie Mellon University (http://murphylab.web.cmu.edu), which has useful tools and information [21].

To aid interpretation of results, consider using a second fluorescent colour to define the morphology of the cell or stain a specific compartment. Using DAPI in the mounting media (see Sect. 3.4.1) will define the nucleus, but be aware of bleed-through into the green spectrum. A good method of defining the whole cell is phalloidin staining. Phalloidin is a phallotoxin stain of F-actin that binds at nanomolar concentrations to give a background cellular stain. Alexa Fluor® 546 phalloidin used in conjunction with a green fluorophore such as GFP or Alexa Fluor® 488 makes a good background for cell morphology. Alternatively, a gene-specific antibody can be used to define a cellular compartment. This antibody needs to be raised in a different animal than the antibody detecting the epitope tag, allowing the two secondary antibody fluorophore conjugates to be excited at different wavelengths. A further option is to cotransfect the ORF construct being tested with a second construct containing a sequence that localizes a fluorescent protein to a subcellular compartment. Various companies produce sets of ready-made constructs, such as the Living Colors series (BD Clontech). Here, a range of constructs available in different colors localize to compartments, again with constructs in the red wavelengths, making a good contrast to green test proteins. These vectors are easy to use, as they require no developing step, but it is important to consider that a second artificially overexpressed protein is being added into the cell, which may lead to artifact. A comparison of signals from cells transfected with just the test protein and cells cotransfected with the marker construct as well is advisable.

4   Notes

1. We used precast 12% gels plus PAGE and Western blot apparatus from Biorad and 4–12% gels in the NuPAGE and Western blot system from Invitrogen, blotting onto Hybond ECL filters (Amersham Biosciences RPN68D).
2. Due to errors during primer synthesis, there is a potential problem with loss, gain, or substitution of bases in cloned primers. This can alter the amino acid sequence or interfere with the reading frame between the fusion or tag and the ORF. To retain the frame, it is particularly important to look for insertions and deletions of bases. Base substitutions in the primers (or any of the ORF sequence) may be acceptable if they are synonymous amino acid changes.
3. When using a new cell line or if few cells are on the slide after processing, titrate the cell number used to seed the wells. Ideally, after the posttransfection growth period, there should be plenty of cells on the slides, but they should not be so dense as to preclude a good image of a single cell.
4. Transfection and antibody developing controls help interpret the results and identify when something has gone wrong. Assess the level of background by

developing a well that has not been transfected and a well that has been transfected with the expression vector containing no insert. A construct containing an ORF of known localization tagged with the epitope being used makes a good positive control for the antibody developing process. A construct containing an ORF of known localization attached to a fluorescent fusion protein, such as GFP, makes a good positive control for transfection.

5. To measure the efficiency of transfection, stain the nuclei with DAPI (this already is in the Vectashield, listed in the materials) and compare the number of cells fluorescing to the total number of cells.

6. If the eight-well slide is returned to the incubator immediately after transfection, the cells sometimes settle around the edge of the wells. To reduce this effect, leave the cells outside the incubator for 30–60 min.

7. To alter the amount of protein produced and therefore fluorescent signal, try increasing the posttransfection incubation to 3 d to increase the signal or incubate for 1 d to decrease the signal.

8. It is wise to titrate the primary and secondary antibodies to achieve the best possible signal with the lowest background fluorescence. To titrate the primary antibody, take an eight-well slide and seed all the wells with cells. Then, using a construct containing an ORF of known localization, transfect it into four wells of the eight-well slide. Taking pairs of transfected and untransfected wells, develop with one of four primary antibody concentrations. Develop all the wells with the same concentration of secondary antibody and compare the level of the signal in transfected wells to the background fluorescence in the untransfected wells in each pair. The secondary antibody could be titrated in a similar way, using a primary antibody against a protein found in all cells, such as Pan Cadherin, followed by developing with the various secondary antibody concentrations as previously, compare the signal to background in each pair.

9. We generally use antimouse Alexa Fluor® 488 in the green spectrum, Alexa Fluor® 546 for a contrasting red (Molecular Probes), and have used antimouse FITC in the green spectrum (Jackson Immuno Research).

10. The European Bioinformatics Institute has a web tool to calculate the expected molecular weight from the amino acid sequence (see www.ebi.ac.uk/emboss/pepinfo). Select the Pepstats option and paste in the amino acid sequence.

Acknowledgments We thank Begoña Aguado, Carol Edwards, Catherine Taylor, Charmain Wright, Frida Andersson, Gözde Akdeniz, James Grinham, Matthew Davis, and Meera Mallya for assistance with developing the protocols. We also thank Catherine Taylor and Ian Dunham for critically reading the manuscript. This work was supported by the Wellcome Trust.

References

1. Vidal M (2001) A biological atlas of functional maps. Cell 104:333–339
2. Simpson JC, Pepperkok R (2006) The subcellular localization of the mammalian proteome comes a fraction closer. Genome Biol 7:222

3. Southern JA, Young DF, Heaney F, Baumgartner WK, Randall RE (1991) Identification of an epitope on the P and V proteins of simian virus 5 that distinguishes between two isolates with different biological characteristics. J Gen Virol. 72, Part 7:1551–1557

4. Hartley JL, Temple GF, Brasch MA (2000) DNA cloning using in vitro site-specific recombination. Genome Res 10:1788–1795

5. Walhout AJ, Temple GF, Brasch MA, Hartley, JL, Lorson MA, van den Heuvel S, Vidal M (2000) GATEWAY recombinational cloning: Application to the cloning of large numbers of open reading frames or ORFeomes. Methods Enzymol 328:575–592

6. Mehrle A, Rosenfelder H, Schupp I, del Val C, Arlt D, Hahne F, Bechtel S, Simpson J, Hofmann O, Hide W et al (2006) The LIFEdb database in 2006. Nucleic Acids Res 34: D415–418

7. Simpson JC, Wellenreuther R, Poustka A, Pepperkok R, Wiemann S (2000) Systematic subcellular localization of novel proteins identified by large-scale cDNA sequencing. EMBO Rep 1:287–292

8. Fink JL, Aturaliya RN, Davis MJ, Zhang F, Hanson K, Teasdale MS, Kai C, Kawai J, Carninci P, Hayashizaki Y et al (2006) LOCATE: A mouse protein subcellular localization database. Nucleic Acids Res 34:D213–D217

9. Temple G, Lamesch P, Milstein S, Hill DE, Wagner L, Moore T, Vidal M (2006) From genome to proteome: Developing expression clone resources for the human genome. Hum Mol Genet 15, Spec No 1:R31–R43

10. Ota T, Suzuki Y, Nishikawa T, Otsuki T, Sugiyama T, Irie R, Wakamatsu A, Hayashi K, Sato H, Nagai K et al (2004) Complete sequencing and characterization of 21,243 full-length human cDNAs. Nat Genet 36:40–45

11. Strausberg, RL, Feingold EA, Grouse LH, Derge JG, Klausner RD, Collins FS, Wagner L, Shenmen CM, Schuler GD, Altschul SF et al (2002) Generation and initial analysis of more than 15,000 full-length human and mouse cDNA sequences. Proc Natl Acad Sci USA 99:16899–16903

12. Wiemann S, Arlt D, Huber W, Wellenreuther R, Schleeger S, Mehrle A, Bechtel S, Sauermann M, Korf U, Pepperkok R et al (2004) From ORFeome to biology: A functional genomics pipeline. Genome Res 14:2136–2144

13. Nakajima D, Saito K, Yamakawa H, Kikuno RF, Nakayama M, Ohara R, Okazaki N, Koga H, Nagase T, Ohara O (2005) Preparation of a set of expression-ready clones of mammalian long cDNAs encoding large proteins by the ORF trap cloning method. DNA Res 12:257–267

14. Rual JF, Hirozane-Kishikawa T, Hao T, Bertin N, Li S, Dricot A, Li N, Rosenberg J, Lamesch P, Vidalain PO et al (2004) Human ORFeome version 1.1: A platform for reverse proteomics. Genome Res 14:2128–2135

15. Gerhard DS, Wagner L, Feingold EA, Shenmen CM, Grouse LH, Schuler G, Klein SL, Old S, Rasooly R, Good P et al (2004) The status, quality, and expansion of the NIH full-length cDNA project: The Mammalian Gene Collection (MGC). Genome Res 14:2121–2127

16. Collins JE, Wright CL, Edwards CA, Davis MP, Grinham JA, Cole CG, Goward ME, Aguado B, Mallya M, Mokrab Y et al (2004) A genome annotation-driven approach to cloning the human ORFeome. Genome Biol 5:R84

17. Takagi M, Nishiok, M, Kakihara H, Kitabayashi M, Inoue H, Kawakami B, Oka M, Imanaka T (1997) Characterization of DNA polymerase from Pyrococcus sp. strain KOD1 and its application to PCR. Appl Environ. Microbiol 63:4504–4510

18. Kozak M (1999) Initiation of translation in prokaryotes and eukaryotes. Gene 234:187–208

19. Bendtsen JD, Nielsen H, von Heijne G, Brunak,S (2004) Improved prediction of signal peptides: SignalP 3.0. J Mol Biol 340:783–795

20. Nakai K, Horton P (1999) PSORT: a program for detecting sorting signals in proteins and predicting their subcellular localization. Trends Biochem Sci 24:34–36

21. Glory E, Murphy R. (2007) Automated subcellular location determination and high-throughput microscopy. Dev Cell 12:7–16

Chapter 25
GeneFAS: A Tool for Prediction of Gene Function Using Multiple Sources of Data

Trupti Joshi, Chao Zhang, Guan Ning Lin, Zhao Song, and Dong Xu

Abstract Characterizing gene function is one of the major challenging tasks in the postgenomic era. To address this challenge, we developed GeneFAS (gene function annotation system), a computer system with a graphical user interface for cellular function prediction by integrating information from protein-protein interactions, protein complexes, microarray gene expression profiles, and annotations of known proteins. GeneFAS can provide biologists a workspace for their organism of interest, to integrate different types of experimental data and annotation information, and facilitate biological discovery and hypothesis generation using all the information. It also provides testing and training capabilities for users to utilize and integrate their data more efficiently. GeneFAS is freely available for download at http://digbio.missouri.edu/genefas.

Keywords function prediction, microarray data, protein-protein interaction, high-throughput data, meta-analysis

1 Introduction

Determination of gene function is one of the most challenging problems in the postgenomic era. The traditional wet laboratory experiments for this purpose are accurate, but the process is time consuming and costly. Various high-throughput biological data provide a source for genome-scale function prediction. However, high-throughput data have their own advantages and disadvantages. The yeast two-hybrid assays may not detect some protein-protein interactions due to post-translational modifications, while mass spectrometry may fail to identify some transient and weak interactions. In a microarray clustering analysis, the genes with similar functions may not be clustered together due to lack of similar expression profiles. Clearly, different types of high-throughput data indicate different aspects of the internal relationships between the same set of genes, and no single experimental method can obtain all the relevant information. Each type of

From: Methods in Molecular Biology, Vol. 439: *Genomics Protocols: Second Edition*   369
Edited by Mike Starkey and Ramnath Elaswarapu © Humana Press Inc., Totowa, NJ

high-throughput data has its strengths and weaknesses in revealing certain relationships. Therefore, different types of high-throughput data complement each other and offer more information than a single source. The combination of high-throughput data from various sources also provides a basis for cross-validating the data [1]. While most current methods use a single source of high-throughput data for function prediction, it is evident that integrating various types of high-throughput data will help handle the data quality issue and better retrieve the underlying information from the data for function prediction. The key to new biological discovery lies in the successful integration and hypothesis generation from these data.

To address data integration for gene function prediction, we developed a computer system, GeneFAS, based on an integrated probabilistic approach, which combines high-throughput data of protein-protein interactions, protein complexes, microarray gene-expression profiles, and genomic sequences [2, 3]. We quantified the relationship between functional similarity in the Gene Ontology (GO) biological process annotation [4] and high-throughput data (including protein-protein interactions, protein complexes, microarray gene expression profiles), and coded the relationship into a "functional linkage graph," where each node represents one gene and the weight of each edge is characterized by the Bayesian probability [5, 6] of function similarity between the two connected genes. We subsequently used the Boltzmann machine [7] and simulated annealing [8] to perform optimization for assigning gene functions based on the global information of the functional linkage graph.

GeneFAS provides a prediction system and a graphical user-interface targeted toward biologists to provide them with a workspace to integrate different experimental data for their organism of interest. GeneFAS allows a user to generate hypotheses and predict functions for their genes of interest and to get a global view of the relationship among genes. Users can retrieve information based on a search of an open reading frame (ORF) name, gene name, or annotation keyword. It also facilitates sequence-based searches. GeneFAS provides users with the option to select different data types and upload both public and private datasets for integrating them in function prediction. Users also can get a global understanding of the relationships among different gene products by viewing the "neighboring genes," defined to be neighbors on the basis of the distance calculated from high-throughput data and functional classification pie charts. It provides biologists with testing and training capabilities based on different datasets and evaluates the performance based on sensitivity and specificity plots.

The subsequent sections describe the different components of the tool, the software usage, and step-by-step instructions with examples from the yeast, *Saccharomyces cerevisiae*, featuring experimental microarray and protein interactions data.

2   Materials

2.1   GeneFAS Components

GeneFAS has four components—database, engine, graphical user interface (GUI), and façade—as shown in Fig. 25.1. The overall architecture of the software is as shown in Fig. 25.2. The database component stores the public data provided with the application, the engine component runs function prediction, while the façade component integrates all the other data. Public data provided with GeneFAS are stored in the database folder. Private data uploaded by the user along with the results are stored in the workspace folder.

2.1.1   Database Component

The database component is the main source for storage and retrieval of data. It stores various public data for the GeneFAS application. The data are stored primarily in the form of XML and plain text files. The private data uploaded by the user and the intermediate, or final results, saved by the user are stored in the workspace directories under user-defined projects.

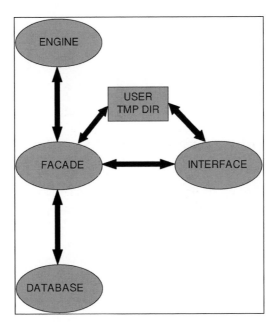

Fig. 25.1 The four components of GeneFAS

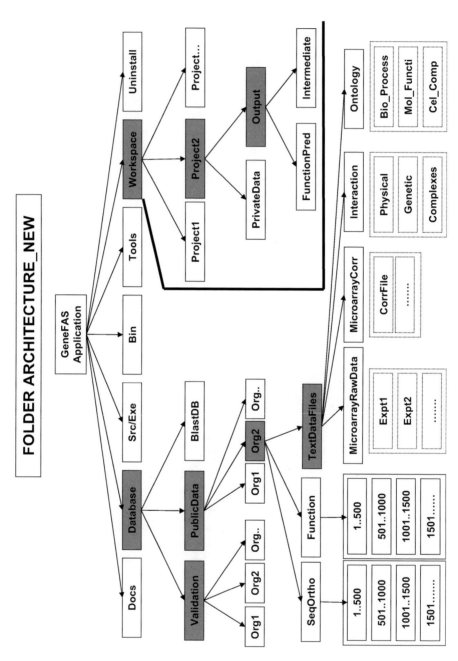

Fig. 25.2 GeneFAS folder architecture

GeneFAS currently includes public data for *Saccharomyces cerevisiae* (yeast), *Arabidopsis thaliana*, and *Mus musculus* (mouse). In the future, we plan to incorporate public data from *Drosophila melanogaster* (fruit fly), *Caenorrhabditis elegans* (worm), *Homo sapiens* (human), *Rattus norvegicus* (rat), and *Oryza sativa* (rice) as part of the database.

The database is organized in such a way that all organism-specific data are stored in a directory named after the organism. Each organism has two directories to store the XML modules and text data files for microarray raw data, microarray correlation coefficient data, binary and complex interactions data, and gene ontology files. The directory also stores the validation text file for each organism, with information about all valid ORF names, gene names, and primary annotation, along with indexing information.

The schema for XML modules is as follows:

1. Pub_ Species_ORFName_SeqOrtho.xml: This module stores the details for database type, species, organism, ORF name, gene name, sequences including protein, mRNA, genomic including those upstream of transcriptional start sites, and ortholog ORF names from other species (see Fig. 25.3).
2. Pub_ Species_ORFName_Function.xml: This module stores the function annotations from GO [4] biological process, molecular function, and cellular component for known cases genes as well as genes for which we have predicted functions using our engine (see Fig. 25.4). Each GO annotation has a GO ID and a description. Fields highlighted by the black box in Fig. 25.4 indicate that the same XML structure is followed for molecular function and cellular component ontologies.

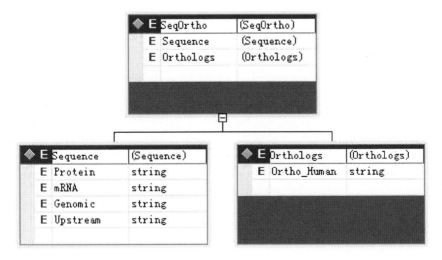

Fig. 25.3 The SeqOrtho module of the database

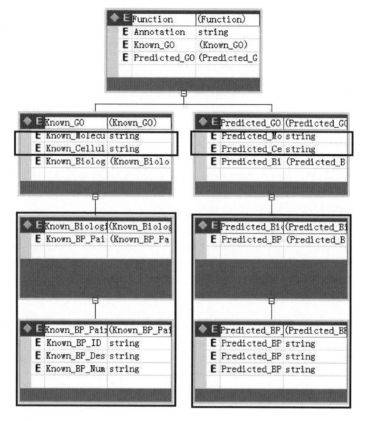

Fig. 25.4 The function module of the database

2.1.2 Engine Component

The engine component is the core of the GeneFAS application. It takes user-defined inputs, including different types of public and private data, and uses the information derived from it for a global function prediction of unannotated genes using the Boltzmann machine.

2.1.3 GUI Component

The GUI component is designed to be user friendly. It provides users a complete array of functionalities including but not limited to

- Creating and naming projects.
- Searching information based on ORF name, gene name, or annotation keywords.
- Viewing functional annotation pie charts and neighboring gene networks.
- Uploading private experimental data for analysis.

- Conducting testing and training on available experimental data.
- Generating and viewing sensitivity and specificity plots.
- Running function predictions using public and private experimental data.
- Viewing function prediction results.
- User documentation for help.

2.1.4   Façade Component

The façade component is an integral functional part of the application. All the other components are connected to each other and share information via the functionalities of the façade. It performs functions such as automatic database updates, parsing GO files into numerical indices, calculating correlation coefficients for microarray data, displaying probability plots based on a user-defined cutoff, testing and training with experimental data, displaying sensitivity and specificity plots, and displaying function prediction results.

2.2   Data Sources

The ORF names, gene names, and sequences (protein, mRNA, genomic, upstream of transcriptional start sites) were acquired from ftp://genome-ftp.stanford.edu/pub/yeast/data_download/ and ftp://ftpmips.gsf.de/yeast/ for *Saccharomyces cerevisiae* (yeast) and from ftp://ftp.arabidopsis.org/home/tair/home/tair/Sequences/blast_datasets/ and ftp://ftp.arabidopsis.org/home/tair/home/tair/seq_analysis_updates/ for *Arabidopsis thaliana*. The GO function annotations were acquired from http://geneontology.org . The experimental data for physical interactions, genetic interactions, and protein complexes interaction were acquired from BIOGRID at www.thebiogrid.org and ftp://ftpmips.gsf.de/yeast/catalogues/complexcat. Microarray data was acquired from the GEO website at www.ncbi.nlm.nih.gov/geo.

3   Methods

3.1   Search by ORF Name, Gene Name, or Keyword

Figure 25.5 shows the flowchart for a user-defined search based on ORF name, gene name, or annotation keyword. When searching by ORF name or gene name, the interface passes the search parameters to the façade. The façade validates the name against the validation document to see if the name belongs to the selected organism and entry in the database. After the validation file confirms the gene existence, it displays the results from corresponding files. The interface

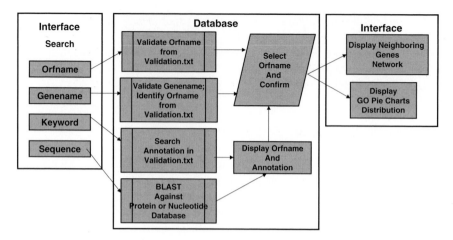

Fig. 25.5 Flowchart for search based on ORF name, gene name, or annotation keyword

also generates a list of all data available for that gene, including microarray, binary protein interaction, and complexes. The user then has the option of selecting the data to view, and accordingly, the interface will generate the graph of the neighboring genes based on the distance calculated from high-throughput data and color the edges based on type of data. In addition, it will generate a GO pie chart for neighboring genes and present the distribution of GO functional categories observed in this set of neighboring genes.

When searching by keyword the interface passes the search parameters to the façade. The façade searches through the primary annotations in the validation files for all genes in the selected organism. It generates a list of all identified ORF names and their annotations.

To search by ORF name, gene name, or annotation keyword,

1. Start GeneFAS by clicking on the GeneFAS icon.
2. Select the check box for organism on the left panel to start the organism-specific search.
3. Enter the complete ORF name with which to search and click the Submit button.
4. The search will return a list of hits based on the query (see Fig. 25.6).
5. Select the check box next to the hit of interest, and click the Confirm button at the bottom to view the information on the gene.
6. Select the Search Result tab to view the general information about the gene (see Fig. 25.7), Neighbouring Tree and Pie Chart tabs to view the neighboring gene tree (see Fig. 25.8) and pie charts (see Fig. 25.9) based on experimental evidence, respectively.
7. View pie chart distributions for the three GO annotations by clicking on the tabs in the pie chart window.

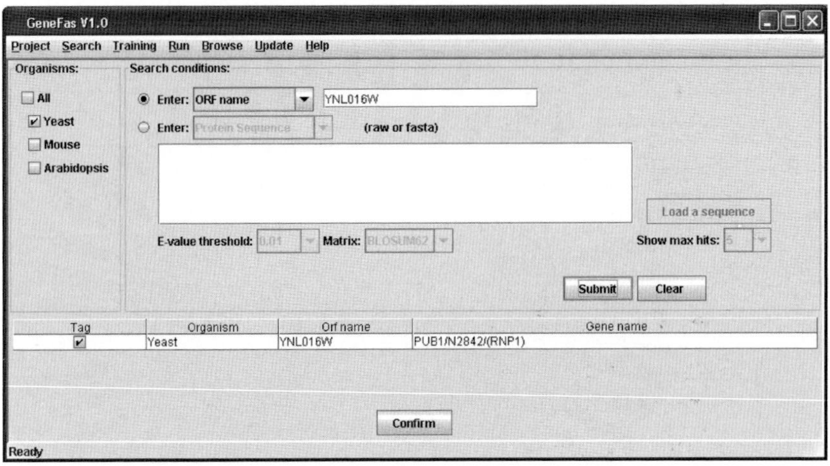

Fig. 25.6 Search by ORF name YOR260W and the result

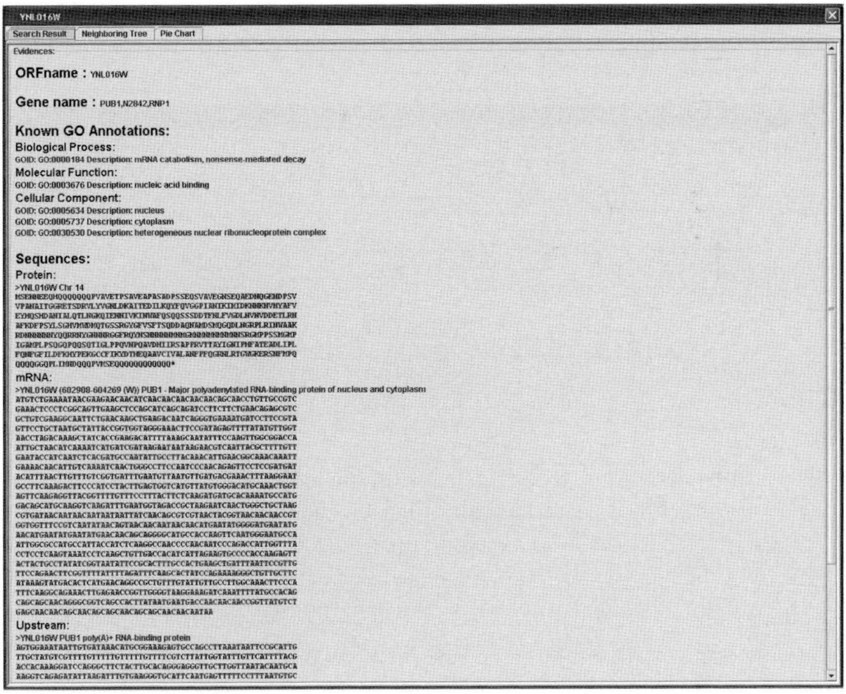

Fig. 25.7 Results of search by ORF name YNL016W

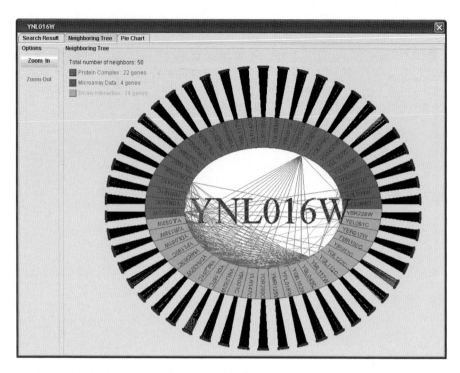

Fig. 25.8 Neighboring gene tree for ORF YNL016W

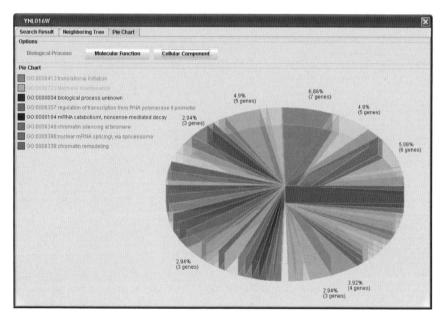

Fig. 25.9 Pie chart distribution of GO biological process functions of the neighboring genes of ORF YNL016W

The search can also be conducted using a partial gene name or keyword (see Fig. 25.10) or by submitting a protein or nucleotide sequence. The sequence is searched for similarity (using BLAST; [9]) against the selected organism databases and the results are displayed as a list of top hits from the organisms (see Fig. 25.11).

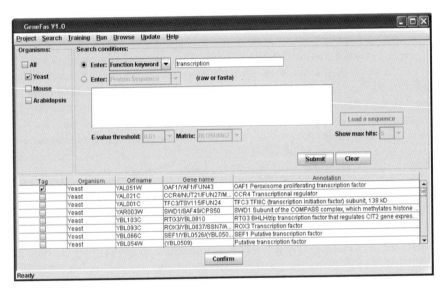

Fig. 25.10 Results of search by keyword "transcription"

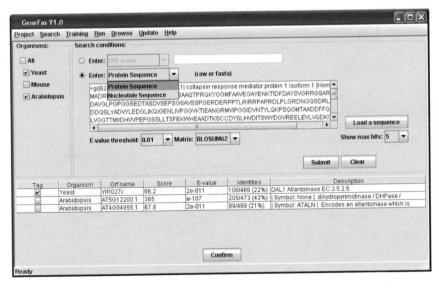

Fig. 25.11 Results of query sequence BLAST against yeast and *Arabidopsis* protein sequence databases

To view information about the top hits, select the check box next to it and click on the Confirm button. Displayed is general information about the gene, including gene names, known GO annotations, and sequences.

3.2   Prediction of Function of the Gene of Interest

Figure 25.12 shows the flowchart for running a function prediction using GeneFAS. To run function prediction, users may choose to utilize the public experimental data provided with GeneFAS or to upload their own experimental data (microarray data, interactions data, etc.), which will be stored as private data in their projects workspace after validation. Users also can upload a list of genes of interest to predict functions. The GUI provides a list of available public experimental datasets as well as private datasets uploaded by the user. The user can select a combination of these datasets and the type of probability function to use for function prediction. The façade component calculates the microarray correlation coefficient and generates the probability curves based on selected datasets.

Based on the probability curves (shown later, in Fig. 25.15) generated for two genes to share the same function in terms of different GO indices, the user can specify the correlation coefficient cutoff and interval to be utilized for the function prediction. At this point, the interface calls the engine and the engine component runs the function prediction with the provided data. The results of the prediction can be viewed using the GUI and stored as files in a user-defined output directory within the user-defined project.

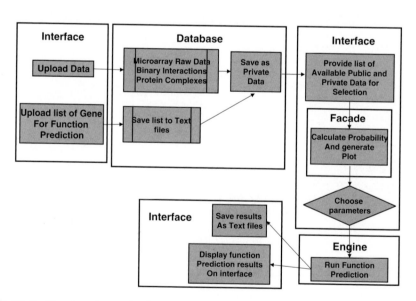

Fig. 25.12 Flowchart for running function prediction using GeneFAS

To predict the function of a gene of interest,

1. Upload a file with ORF names using the Upload Data option from the Run menu then click the Submit button (see Fig. 25.13). Users have the option to upload their own raw microarray raw data or protein interactions data in the specified format and choose to use the data in the function prediction. Users also can select the data to be used in function prediction by selecting the check box next to them and choose the type of probability to be calculated from the pull down menu (see Fig. 25.14).
2. Select the Run option to generate the probability curves based on selected data (see Fig. 25.15). The higher the correlation coefficient, the higher is the probability value. The lower the GO similarity index, the higher is the probability value.

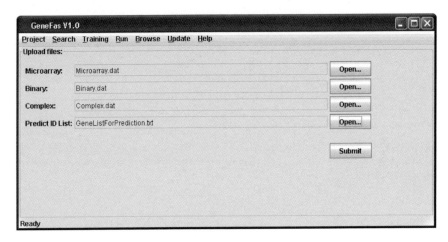

Fig. 25.13 Upload experimental data and a list of genes to predict function

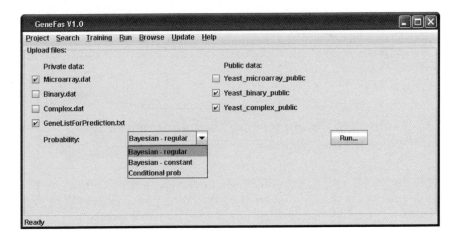

Fig. 25.14 Public and private data selection for function prediction

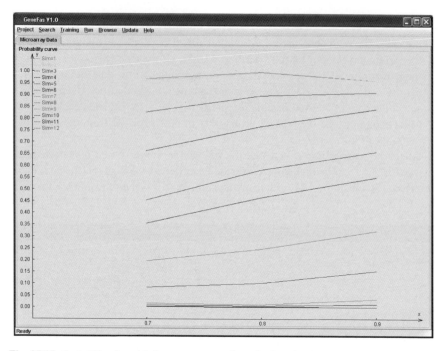

Fig. 25.15 Probability (y-axis) for two genes to share the function in terms of the same level of GO indices versus the correlation coefficient between the microarray expression profiles of the two genes (x-axis)

3. On the basis of the probability plot trend, select the cutoff for microarray correlation coefficient to be used in function prediction.
4. Select the Predict option from the Run menu to run the function prediction.
5. View the function prediction results using the Results option accessed via the Run menu (see Fig. 25.16). The results show the predicted GO ID, description, and reliability score. It also lists the evidence from experimental data with correlation coefficient values and annotations.

3.3 Testing and Training Using Different Probability Functions and Datasets

Figure 25.17 shows the flowchart for the testing and training functionality. It gives the user the option of testing and training on a user-defined percentage of genes with known functions using different probability functions and datasets. This functionality helps the user evaluate which probability function or datasets give the best prediction results.

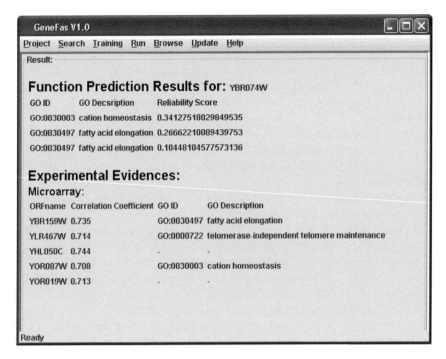

Fig. 25.16 Function prediction results for gene YBR074W

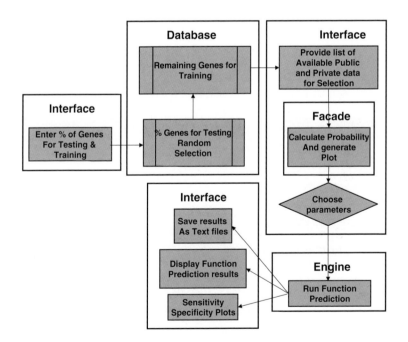

Fig. 25.17 Flowchart for running testing and training using GeneFAS

1. Select Options from the Training menu to specify the percentage of genes to use in testing, the correlation coefficient interval, and the cutoff for the microarray data. Click on the Train button to randomly select the user-defined percentage of genes as the testing set and the remaining as training set (see Fig. 25.18).
2. On completion of training, view the probability curves (equivalent to Fig. 25.15), based on the training dataset, using the Prob Curve option under the Training menu.
3. On the basis of the training dataset probability plot trend, select the cutoff for microarray correlation coefficient to be used in testing.
4. Select the Test option from the Training menu to run the function prediction on the testing dataset.
5. View the function prediction results for the testing set by using the Results option accessible via the Training menu.
6. View the sensitivity and specificity plot based on the training and testing using the S-S curve option under the Training menu (see Fig. 25.19). This plot compares the predicted functions for the testing set with their actual known functions and generates sensitivity-specificity plots based on the different probability functions and datasets used for prediction. For a given set of proteins K, let n_i be the number of the known functions for protein P_i. Let m_i be the number of functions predicted for the protein P_i by the method. Let k_i be the number of predicted functions that are correct (the same as the known function). Thus, sensitivity (SN) and specificity (SP) are defined as

$$SN = \frac{\sum_{1}^{K} k_i}{\sum_{1}^{K} n_i} \qquad (25.1)$$

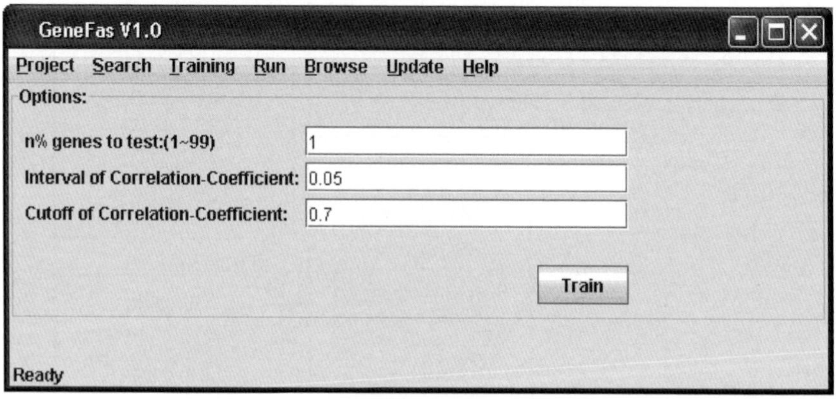

Fig. 25.18 Defining testing and training set

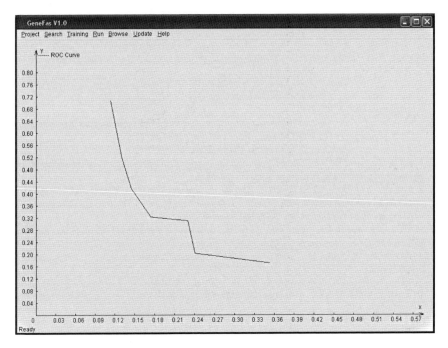

Fig. 25.19 Sensitivity and Specificity plot for testing and training

$$SP = \frac{\sum\limits_{1}^{K} k_i}{\sum\limits_{1}^{K} m_i} \qquad (25.2)$$

The sensitivity (y-axis) is plotted against specificity (x-axis) as per Eqs. 25.1 and 25.2.

4   Notes

1. Information about the GeneFAS development, team, and release can be obtained by clicking on the About option under the Help menu, as well as at the website http://digbio.missouri.edu/genefas.

Acknowledgments This work is supported by a startup fund provided to Dong Xu at the University of Missouri-Columbia. We thank Yu Chen for helpful discussions and Gyan Prakash Srivastava for technical assistance.

References

1. Chen Y, Xu D (2003) Computation analysis of high-throughput protein-protein interaction data. Current Peptide Protein Sci 4:159–181
2. Joshi T, Chen Y, Becker J, Alexandrov N, Xu D (2004) Genome-scale gene function prediction using multiple sources of high-throughput data in yeast *Saccharomyces cerevisiae*. OMICS: J Integrative Biol 8:322–333
3. Chen Y, Xu D (2004) Global protein function annotation through mining genome-scale data in yeast *Saccharomyces cerevisiae*. Nucl. Acids Res 32:6414–6424
4. The Gene Ontology Consortium (2000) Nature Genetics 25:25–29
5. Letovsky S, Kasif S (2003) Predicting protein function from protein/protein interaction data: a probabilistic approach. Bioinformatics 19:I197–I204
6. Troyanskaya O, Dolinski K, Owen A, Altman R, Botstein D (2003) A Bayesian framework for combining heterogeneous data sources for gene function prediction (in *Saccharomyces cerevisiae*). Proc Natl Acad Sci USA 100:8348–8353
7. Ackley DH, Hinton GE, Sejnowski TJ (1985) A learning algorithm for Boltzmann machines. Cognit Sci 9:147–169
8. Kirkpatrick S, Gelatt CD, Vecchi MP (1983) Optimization by simulated annealing. Science 220:621–680
9. Altschul S, Madden T, Schaffer A, Zhang J, Zhang Z, Miller W, Lipman D (1997) Gapped BLAST and PSI-BLAST: A new generation of protein database search programs. Nucl Acids Res 25:3389–3402

Chapter 26
Comparative Genomics-Based Prediction of Protein Function

Toni Gabaldón

Abstract The era of genomics has opened new possibilities for the computational prediction of protein function. In particular, the comparison of fully
sequenced genomes allows us to investigate the so-called genomic context of a
gene, which includes its chromosomal positioning relative to other genes as well
as its evolutionary record among the genomes considered. This information can
be exploited to find functionally interacting partners for a protein of unknown
function and thus obtain information on the biological process in which it is play-
ing a role. Such comparative genomics-based techniques are increasingly being
used in the process of genome annotation and in the development of testable
working hypothesis.

Keywords Genomic context, phylogenetic profile, orthology, gene fusion, gene
neighborhood, gene order, coevolution

1 Introduction

As I write this chapter, there are 409 published complete genomes, with at least
1,662 additional genomes in the process of being sequenced (Genomes Online
Database; www.genomesonline.org). It would be no surprise that, when this
book becomes available, the number of sequenced genomes will have almost
doubled. Assigning of functions to the proteins encoded in these genomes
constitutes a major challenge of the genome era. Ideally, one would like to have
the function of every protein characterized through experimental analysis.
However, despite recent developments in high-throughput techniques, the
experimental characterization of proteins lags far behind the availability of
new sequences; therefore, the annotation of genomes relies mostly on computational methods.

The most common and straightforward approach for computationally assigning
a function to a protein sequence consists in the detection of homologous proteins
with a known function to subsequently transfer its functional annotation. This

methodology, known as *homology-based function prediction*, is based on the fact that proteins that have a certain level of sequence similarity have similar structures and, therefore, are likely to perform related functions. However, this established approach has several important limitations, mainly because there is no simple relationship between sequence similarity and function. For instance, it has been shown that enzyme classification numbers tend to be conserved completely only among proteins with more than 80% sequence identity [1]. Below this level, changes in substrate specificity or enzyme activity might lead to incorrect annotations. Moreover, this technique is of little help when there are no experimentally characterized homologues. The combined effect of these drawbacks leads to a situation in which there is no clear functional prediction for about 40% of the genes in most genomes [2].

Complementary to homology-based methods, alternative approaches for the computational characterization of proteins have been proposed. In particular, the so-called genomic context methods [3] exploit relations between genes on the genome, such as gene fusions, chromosomal proximity, their distribution across species, and the like (Fig. 26.1). The kind of information provided by such techniques is qualitatively different from that of the classical, homology-based one: While the detection of characterized homologues of a sequence gives us information on its molecular function, a genome context analysis is useful to

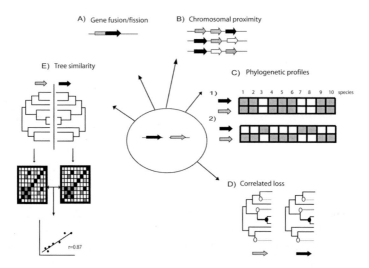

Fig. 26.1 Overview of the comparative genomic-based methods described in the chapter. A functional link between a pair of genes (center) can be inferred from: (A) gene fusion or fission, both genes form a single gene in other genomes; (B) chromosomal proximity, both genes are close to each other in a significant number of genomes; (C) phylogenetic profiling, both genes have similar distributions across diverse organisms; (D) correlation of gene loss events, the history of losses of both families is similar; (E) mirror tree, similarity of the phylogenetic trees of both protein families can be measured by the correlation coefficient of the distance matrices from both trees

discover the interacting partners of a given protein and, thus, the biological process in which it is participating.

The aim of this chapter is to provide the essential theoretical background and the necessary practical tools to allow the reader to implement comparative genomic-based function prediction approaches to his or her own data. The protocol assumes a set of sequences coding for proteins of unknown function as the starting point. This set may consist of just a single protein sequence, a large number of them coming from a fully sequenced genome (*see* **Note 1**), or any number in between. By comparing this set of sequences with other sequences encoded in complete genomes, we search for evolutionary conserved traits that might give us hints about their functional role.

Working in a UNIX environment, such as GNU/Linux (www.gnu.org), and having some skill in a computer scripting language, such as python (www.python.org), greatly facilitates the comparative genomics operations needed for applying the techniques described here. Alternatively, no computer literacy is needed to use integrated servers such as STRING (see below), in which by pasting the sequence of interest we can obtain precomputed results from comparative genomics analyses. This strategy, however, is limited to cases in which the number of proteins to analyze is reduced, since little automation is allowed.

2   Materials

2.1   *Databases*

Repertoires of proteins encoded in fully sequenced genomes (*see* **Note 2**) can be downloaded from their respective genome sequencing projects' website. However, a number of repositories that store data from different completed genomes greatly facilitate the task of gathering the sequences of interest. Here, I list some of the most complete and commonly used repositories (*see* **Note 3**):

1. Genome resources at NCBI. The National Center for Biotechnology Information (NCBI) provides complete genomic sequences for both eukaryotic and prokaryotic organisms as well as several resources for genomic analyses (www.ncbi.nlm.nih. gov/Genomes).
2. Ensembl database: The Ensembl site is one of the leading sources of genome sequence annotation for human and other model eukaryotic species (www. ensembl.org)
3. Integr8 database: Through the combination of information from databases maintained at major bioinformatics centers in Europe, this database provides comprehensive information on completed genomes and proteomes (www.ebi.ac.uk/integr8).

4. GOLD: The Genomes Online Database (GOLD) provides the largest available and most detailed monitoring of genome sequencing projects, both completed and ongoing. Therefore GOLD is the site at which to check for what genomes are, or soon will be, available to plan a comparative-genomics project (www.genomesonline.org).

These databases can be searched online through their Web interfaces or, alternatively, stored locally. In the latter case, the use and repetition of automated searches is favored.

2.2   Sequence Comparison

Sequence comparison is inherent to comparative genomics, since it allows the establishment of evolutionary relationships among the protein sequences encoded in complete genomes. A large number of algorithms to search for similar sequences in a database exist and are thoroughly described elsewhere [4]. Here, I provide just some indication about two of the most commonly used tools:

1. BLAST. The Basic Local Alignment Search Tool (BLAST) finds regions of local similarity between sequences. In addition to a BLAST score, the program provides an expect value (e-value) that indicates the statistical significance of the hit by taking into account the size of the database and the length and composition of the hit. To put it simple, it gives a measure of the chance of observing a hit of that score or higher just by chance. Therefore, the lower the e-value, the higher is the statistical significance of the hit. The use of filters to mask regions of "low complexity" is recommended (*see* **Note 4**).
2. Psi-Blast. The Position-Specific Iterated (PSI)-BLAST is a very sensitive implementation of the BLAST program and, therefore, appropriate for finding distantly related proteins. PSI-BLAST starts with a standard BLAST search and use the hits with e-values lower than a given threshold to derive a position-specific scoring matrix (PSSM). This PSSM takes into account, specifically, the type of residues found at a given position in the multiple sequence alignment of the hits. The next iteration continues by using the PSSM from the previous search and updating it with the new hits. The process is repeated until convergence (no new sequences are found) or a specified number of rounds is completed.

For genome-context function prediction, we use either BLAST or Psi-Blast to define homologous groups of proteins, to subsequently build groups of orthologues, or perform a phylogenetic analysis. In doing so, and to avoid hits restricted to a specific protein domain, it is useful to request not only that the homologous proteins share a region of similarity under a given e-value threshold but that this region covers a significant fraction of the full length of the sequences considered. BLAST and Psi-Blast can be used through the NCBI website (www.ncbi.nlm.nih.gov/blast) or can be downloaded and installed locally (www.ncbi.nlm.nih.gov/BLAST/download.shtml).

2.3 Multiple-Sequence Alignment and Phylogenetic Reconstruction

Phylogenetic reconstruction is needed for a number of genomic-context function prediction approaches, in particular for the establishment of phylogeny-based orthology relationships and the detection of coevolving protein families. The process of phylogenetic reconstruction involves two main steps that use specific programs: the alignment of the homologous proteins considered and the phylogenetic inference itself.

1. Multiple sequence alignment. An alignment of protein sequences basically aims at placing homologous residues of different proteins on top of each other. Programs such as Clustal W [5] and MUSCLE [6] implement fast, efficient multiple sequence algorithms and are easy to use. They basically require as an input a multiple sequence file in the appropriate format (usually FASTA or PHYLIP) (*see* **Note 5**). A manual curation of the alignment or the removal of unreliably aligned regions of the alignment (e.g., regions with many gaps) is recommended before the phylogenetic analysis. However, this is not always possible when working at a large scale (*see* **Note 6**).

2. Phylogenetic reconstruction. Phylogenetic reconstruction aims at explaining the differences observed in a protein alignment by means of the evolutionary distances among the sequences involved, given a *subyacent* evolutionary model. Several strategies exist for phylogenetic reconstruction [7], but in this chapter we will focus on just two of them, neighbor joining (NJ) and maximum likelihood (ML). Neighbor joining is a fast heuristic algorithm that estimates a "minimal evolution" tree, from a matrix of pairwise distances of the sequences involved. Since it is very fast and fairly accurate, it usually is the method of choice for large-scale analyses. Clustal W is a fast, easy to use program that can perform both the multiple sequence alignment and the neighbor joining analysis (*see* **Note 7**).

   Maximum likelihood (ML) methods try to find the tree with the maximum likelihood to have produced the variation observed in the alignment. To produce results in reasonable time, all practical ML methods rely on certain heuristics. I recommend PhyML [8], which has gained popularity because of its speed and ease of use (*see* **Note 8**).

2.4 Integrated Servers

The local installation of data and software allows the researcher to perform tailored comparative genome analyses, that is, to use the preferred algorithms and programs on the set of genomes of choice. However, this is not always simple, and some researchers might prefer to use standard analyses on standard datasets that, in return, have faster results and user-friendly interfaces. Fortunately, several platforms on the Internet allow access to precomputed comparative genomics analyses or enable doing such analyses online. The choice is huge, and here I focus on three complementary ones: a general comparative genomics server (NCBI), a server that integrates phylogenetic tools (Phylemon), and a genome-context-based function prediction server (STRING). Most of the techniques described here can be performed through these servers, although, as I pointed out before, the analysis cannot be applied to large datasets.

1. NCBI. In its section "Genomic Biology" (www.ncbi.nlm.nih.gov/Genomes/), the NCBI offers a number of tools for comparative genomics. These include, organism-specific BLAST searches, sequence repositories, and predefined clusters of orthologous groups (see Sect. 3.1).
2. Phylemon. Phylemon is a recently developed Web server that integrates a complete suite of phylogenetic analysis tools. Applications range from sequence alignment to very sophisticated phylogenetic analyses and evolutionary tests. A useful property of the server is that all applications can be easily interconnected in a user-friendly pipeline (http://phylemon.bioinfo.cipf.es).
3. STRING. The Search Tool for the Retrieval of Interacting Genes/Proteins (STRING) is a database that integrates known and predicted interactions between proteins of fully sequenced genomes. By establishing orthology relationships among genes in different species, the predicted interactions are transferred across organisms. STRING implements the most important types of genomic context, and therefore, it can be used directly to predict the function of a query protein (http://string.embl.de).

3 Methods

3.1 Detection of Orthology

A basic requirement of comparative genomics is to identify which genes are common in different genomes. This ideally is done by establishing orthology, rather than just homology, relationships (*see* **Note 9**). The detection of orthology is not straightforward when comparing distantly related genomes, and several operational definitions do exist.

The most widely used method for automatically establishing orthology relationships is based on the detection of best reciprocal hits (BRHs), that is, pairs of sequences from different species that, reciprocally, are the best hit of each other in a sequence search [9]. Extensions of the BRH approach include the definition of "tri-angular" BRH relationships across a minimum of three species, such as COG [10]. More recently, some methods approximate the classical, phylogeny-based approach for detecting orthology. Such methods generally rely on the detection of duplication and speciation events by comparing the gene tree with the species tree [11]. Here, we selected some of the most widely used techniques to define orthology relationships, which range from less (BRH) to more (phylogeny based) time consuming; therefore, the choice of the methods depends on the size and characteristics of the set to analyze. Alternatively, one can use predefined orthologous sets, if these are available for the species of interests. Websites like COG [12], KOG [13], Ensembl [14], or multiparanoid [15]) provide such predefined orthologous sets.

3.1.1 Best Reciprocal Hits

1. Perform a BLAST search for every protein encoded in genome A against a database consisting of all proteins encoded in genome B.
2. Select the hit with the highest score for each protein.
3. Perform the reciprocal search (a BLAST search for every protein in genome B against the database of proteins in genome A) and select the best hit for each protein.
4. Compare the lists of best hits from both searches and define BRHs as those pairs of proteins from different genomes that, reciprocally, are the best hits of each other.

3.1.2 COG-like Procedure

1. Proceed as in Sect. 3.1.1 with all pairs of genomes involved.
2. Detect "triangles" of BRHs; for example, if a protein pA from genome A reciprocally is a best hit of protein pB in genome B and of protein pC in genome C, and similarly pB and pC also are BRHs, then pA, pB and pC constitute a cluster of orthologues.
3. Merge triangles of BRHs into a single COG if they share one side.
4. Repeat step 3 until no more COGs are fused.

3.1.3 Phylogeny-Based Orthology Inference

1. Retrieve homologues in the genomes of interest by using BLAST or Psi-Blast.
2. Perform a multiple sequence alignment.
3. Remove gap-rich regions in the alignment (e.g., those columns having a gap in more that 20% of the sequences).
4. Perform a phylogenetic reconstruction based on that alignment.
5. Compare the resulting tree with the phylogeny of the species involved and map duplication and speciation events. Do this automaticaly by using the RAP algorithm implemented by Dufayard and colleagues (http://pbil.univ-lyon1.fr/software/RAP/RAP.htm) [16].
6. Orthologues then are defined as genes from different species that derive from a common ancestor only through events of speciation.

3.2   *Gene Fusion and Fission*

A gene fusion (or fission) event is revealed when two or more proteins encoded by separate genes are found to be encoded by a single gene in a different species (Fig. 26.1A). This genomic event constitutes the most direct type of genomic context association, and it usually is indicative of a very close functional relationship, often involving physical interactions. At the molecular level, the fusion of two proteins can enhance their functional interaction, such as by facilitating the channeling of a substrate. The discovery of functional relationships between fused proteins on a small scale has been intuitively exploited in the past. The most widely known example perhaps is the fusion of the alpha and beta subunits of tryptophan synthase in yeast and other fungi [17]. An extension of such strategy at a genomic scale was proposed in the late 1990s [18, 19]. Based on the fact that a significant fraction of the observed fusions involved proteins known to interact functionally (*see* **Note 10**), the authors proposed the use of this technique on a large-scale to predict other interactions. In agreement with the aforementioned substrate channeling effect, most of the observed fusion events involve metabolic enzymes, although not necessarily from consecutive steps in a given pathway.

Although gene fusion is a rare event, the number of proteins for which a fusion can be found increases exponentially with the number of genomes compared. For instance, when comparing three bacterial genomes, Enright et al. found 88 gene fusions events [19], while just two years later and using 30 bacterial genomes, 10,073 fusions were detected [20].

3.2.1   Detection of Fusion Events

1. Perform an all-to-all BLAST comparison of the genomes involved. For each protein, store the best hit in each genome and information regarding the region of similarity e.g., residue x to residue y in the query protein aligns with residues a to b in its best hit.
2. Fusion (or fission) events are evidenced by those proteins that are the best hit of two proteins in another genome and their regions of similarity with the query protein do not overlap.

3.2.2   Conservation of Gene Order

Comparisons of bacterial genomes have shown that there generally is a lack of gene order conservation, even among closely related species [9]. This indicates that recombination events rapidly shuffle the genomes during evolution. Despite this overall trend, some gene clusters, mainly in prokaryotes, are conserved in terms of their relative order over large evolutionary distances. A closer inspection of such genes reveals that they usually constitute operons [21] and are functionally related [22, 23]. The use of gene order conservation to predict functional interactions was first proposed in 1998 by two independent studies [22, 23] that measured, respectively, the conservation of genes in runs (gene sets separated by less than 300 bp) and the conservation of pairs of neighboring genes (Fig. 26.1B). In eukaryotes, the chromosomal proximity of the genes rarely is indicative of functional association, although several remarkable examples do exist [24, 25]. In any case, one can use chromosomal proximity to predict the function of eukaryotic proteins if they have orthologues in prokaryotic species, assuming that the function has been conserved. Next, I describe the use of chromosomal proximity using the two aforementioned strategies.

3.2.3 Detection of Chromosomal Proximity by Conservation of Genes in Runs

1. Establish orthology relationships among the genes encoded in the genomes involved. Discard very closely related species (e.g., belonging to the same genus), since gene order might be too conserved in these cases.
2. For each genome, define runs as sets of genes localized in the same strand of the chromosome and separated by less than 300 nucleotides (*see* **Note 11**).
3. A cooccurrence event between two genes is defined as these genes being part of the same run in a given genome.
4. The chromosomal proximity score between two genes is defined as the number of cooccurrence events observed divided by the total number of genomes analyzed that contain at least one of the genes. Scores higher than 0.3 usually are considered statistically significant [26].

3.2.4 Detection of Chromosomal Proximity by Conservation of Gene Neighborhood

Proceed as in Sect. 3.3.1, but define a cooccurrence event between two genes when these are localized next to each other in the chromosome.

3.3 Phylogenetic Profiling

The phylogenetic profiling techniques [9, 27], compare distributions across complete genomes of proteins to subsequently predict functional links between those proteins that exhibit similar distributions (Fig. 26.1C). The technique is based on the observation that proteins having a similar species distribution often are found to be involved in the same biological process. In addition, the finding of proteins with complementary distributions might result from a nonorthologous gene displacement and thus suggests that both protein families perform a similar function [28].

The distribution of a gene across different organisms can be formalized by means of a phylogenetic profile, a binary series that codes the presence or absence of a protein in a set of genomes, then detects proteins with similar profiles. Distances between profiles can be computed by simply counting the differences (Hamming distance) or using more sophisticated scores, like mutual information [29] or Pearson correlation coefficient [30].

3.3.1   Detection of Genes with Similar Profiles

1. Derive orthology relationships among the genes encoded in the genomes of interest by using any of the approaches described in Sect. 2.1.
2. Code the information of presence or absence in the form of a phylogenetic profile, indicating absence as a 0 and presence as 1. For example, the profile 00010010000111, indicates that the gene is absent from all genomes except those coded in the positions 4, 7, 12, 13, and 14.
3. Compare profiles using any of these distance measures:

   a. Hamming distance. The Hamming distance between two strings of equal length is the number of positions with different numbers. It measures the number of substitutions required to transform one profile into the other. The hamming distance between the profiles 001000111001 and 000100011001 is 3.
   b. Mutual information. Mutual information is a measure of the independence of two distributions of discrete variables. Mathematically, this measure is equivalent to the log-odds ratio of the expected cooccurrence of pairs of genes, based on their individual frequencies, to the oserved cooccurrence. Thus, the mutual information between the profiles i and j (M_{ij}) is

$$M_{ij} = \Sigma_{ij}(P_{i,j}/P_i P_j)$$

where $P_{i,j}$ is the frequency of observed cooccurrence events (both genes are in the same genome) and P_i and P_j are the individual frequencies of each gene.

3.3.2   Detection of Complementary Phylogenetic Profiles

Proceed as in Sect. 3.3.1 but search for genes having dissimilar phylogenetic profiles (i.e., greater Hamming distances). Perfect complementarity is found between the profiles of two genes if the Hamming distance equals the length of the profile.

3.4   Lineage-Specific Gene Loss

A variant of the aforementioned phylogenetic profiling technique incorporates information on the evolutionary relationship of the species to identify genes that

Fig. 26.2 How to expand a phylogenetic profile to include ancestral nodes. Given a phylogenetic profile and the evolutionary relationships among the species considered (a), the expanded profile can be derived by assigning to every ancestral node (numbered in bold) a new position in the profile that will describe the inferred presence (1) or absence (0) of that gene in the ancestor (digits in bold)

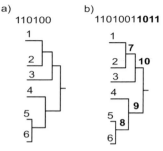

have been lost concomitantly during evolution (Fig. 26.1D). This method originally was used for the prediction of genes involved in pathogenicity by searching for those genes present in a pathogenic organisms but absent in closely related non-pathogenic species or strains [31, 32]. More recently, the method was extended to the detection of genes that have been concomitantly lost along the phylogeny of a group of species [33]. A prerequisite of the method is that the differences in the observed phylogenetic profiles are the result of differential loss across the lineages rather than differential gain of genes (*see* **Note 12**). Another requirement is that the evolutionary relationships among the species considered are known (*see* **Note 13**).

For the detection of similar gene-loss patterns,

1. Proceed as in the previous section to derive phylogenetic profiles of the genes considered.
2. Use the species phylogeny and a parsimony method to infer the presence or absence of a gene on every ancestral node. The presence of a gene is inferred when any of the descendants have that gene (*see* **Note 12**). In this manner, the original phylogenetic profile is expanded to a new, bigger one that includes the presence or absence of the genes in the ancestral nodes (Fig. 26.2).
3. Hamming distances and mutual information can be applied to these extended profiles. However, Gabaldón and Huynen [33] found that the following coevolution score provided the best results: concordances of the type 1,1 (both genes are present) minus discordances of the types 0,1 and 1,0 (only one of the genes is present). Note that species or ancestral nodes in which neither of the two genes is present are not taken into account in the score.

3.5   *Other Forms of Coevolution*

The coevolution of two proteins can be followed more closely by using the evolutionary information contained in the protein alignments of their respective families. For specific cases of interacting protein families, such as insulin and its receptors [34], it has been shown that their phylogenetic trees are more similar to each other than expected, based on the species divergence. This suggested the existence of correlated evolution resulting from similar evolutionary constraints. Pazos and Valencia [35] used this property to search for interacting partners within the

Escherichia coli proteome by measuring the correlation between the distance matrices used to build the phylogenetic trees, a technique they call *mirror tree* (Fig. 26.1E).

1. Perform a phylogenetic reconstruction of each protein family as described in Sect. 3.1.3 (steps 1–4), using the neighbor joining approach.
2. For each pair of trees, compute the correlation coefficient of the distance matrices used to make the tree.
3. Proteins with high correlation coefficients in their distance matrices have similar phylogenetic trees and therefore show coevolution.

4   Notes

1. When the techniques presented here are applied to sequences coming from genomes for which no complete sequence is available, the prediction of orthologues might be compromised, since the "true" orthologue can be missing in the incomplete genome.
2. The complete set of proteins encoded in a single genome is also known as the *proteome*, although this term can be applied to the complement of proteins expressed in the cell during a given period or under specific conditions.
3. Since the Internet is a dynamic entity, the websites and links provided in the chapter might be relocated or disappear. Usually the authors will indicate the new address. If this is not the case, one can try searching with Google (www. google.com) or a similar search tool or writing to the authors of the site of interest. It is important to check for the date of the last update of the site, since it might get obsolete. In these cases and those in which the site has apparently disappeared, look for equivalent sites.
4. Low-complexity sequences are stretches of residues with a very limited variability in terms of amino acid composition. For example, the protein sequence QQQKKQQQQQQQLLKGQQQQ has low complexity. Regions of low complexity are problematic because they can produce artifactual hits with high scores. If the low-complexity filter is used, BLAST will substitute the low complexity regions by a string of Xs and not use this regions in the computation of the alignment scores.
5. Different programs produce or require different formats, and one often needs to convert the results from one program into the format required for the next program in the pipeline. The program readseq from EBI (www.ebi.ac.uk/cgi-bin/readseq.cgi) is a very useful tool that allows the conversion across many different formats.
6. For this purpose, a multiple sequence alignment viewer such as jalview (www.jalview.org) and the program Gblocks (http://molevol.ibmb.csic.es/Gblocks/Gblocks.html) for automatically selecting conserved blocks in the alignment can be very useful.

7. It is recommended to use the option "correct for multiple substitutions" to provide better estimates of the distances. This correction takes into account that, when fairly distant sequences are compared, the observed differences are a minimal estimate of the real amount of mutations, since a single residue may have mutated several times.

8. Several models are available for performing maximum likelihood analysis, some differences in the length of the branches and, eventually, the topology of the tree can be expected when using different models. A general model, such as JTT, will work fairly well for most cases, but if we want to use the proper model, we can run PhyML with several models and use the one producing the best likelihood. Regarding the use of a gamma-distribution, choosing a four-classes gamma distribution with the alpha parameter and the proportion of invariable sites being estimated from the data usually gives good results.

9. The concept of orthology is central to comparative genomics. Orthologous genes are a special kind of homologous genes derived from an ancestral sequence through speciation only. Paralogues, in contrast, are homologous genes that derive from an ancestral sequence through events of gene duplication. Since orthology has a phylogenetic definition, the ideal way of determining orthology is by reconstructing the phylogeny of a given family and subsequently mapping on the tree the events of speciation and duplication to establish the orthology relationships.

10. Be aware of the so-called promiscuous domains, like those involved in signal transduction, that appear in very diverse functional and genomic contexts. Fusion events involving such domains usually are not informative.

11. This quantity may be altered, for example, when working with eukaryotic genomes where the distance between the genes is greater.

12. Yet another modification of the method allows for genes to be gained at certain lineages. The decision whether a single gene gain is preferred over multiple gene losses must be set in the form of a gain vs. loss threshold.

13. If the species tree is not fully resolved, the method still can be applied by representing the uncertainty with multiple furcations in the tree.

Acknowledgments Toni Gabalón is supported by a long-term fellowship from EMBO (LTF 402-2005). He thanks Martijn A. Huynen for introducing him to the field of computational protein function prediction.

References

1. Devo, D, Valencia A (2001) Intrinsic errors in genome annotation. Trends Genet 17:429–431

2. Iliopoulos I, Tsoka S, Andrade MA, Janssen P, Audit B, Tramontano A, Valencia A, Leroy C, Sander C, Ouzounis CA. (2001) Genome sequences and great expectations. Genome Biol 2: INTERACTIONS0001

3. Gabaldón T, Huynen MA (2004) Prediction of protein function and pathways in the genome era. Cell Mol Life Sci 61:930–944
4. Durbin, R., Eddy, S. R., Krogh, A., and Graeme, M. (1988) Biological sequence analysis: Probabilistic models of proteins and nucleic acids. Cambridge University Press, Cambridge
5. Thompson JD, Higgins DG, Gibson TJ (1994) CLUSTAL W: Improving the sensitivity of progressive multiple sequence alignment through sequence weighting, position-specific gap penalties and weight matrix choice. Nucleic Acids Res 22:4673–4680
6. Edgar RC (2004) MUSCLE: A multiple sequence alignment method with reduced time and space complexity. BMC Bioinformatics 5:113
7. Gabaldón T (2005) Evolution of proteins and proteomes, a phylogenetics approach. Evolutionary Bioinformatics Online 1:51–56
8. Guindon S, Gascuel O (2003) A simple, fast, and accurate algorithm to estimate large phylogenies by maximum likelihood. Syst Biol 52:696–704
9. Huynen MA, Bork P (1998) Measuring genome evolution. Proc Natl Acad Sci USA 95:5849–5856
10. Tatusov RL, Koonin EV, Lipman DJ (1997) A genomic perspective on protein families. Science 278:631–637
11. Zmasek CM, Eddy SR (2001) A simple algorithm to infer gene duplication and speciation events on a gene tree. Bioinformatics 17:821–828
12. Tatusov RL, Galperin MY, Natale DA, Koonin EV (2000) The COG database: A tool for genome-scale analysis of protein functions and evolution. Nucleic Acids Res 28:33–36
13. Tatusov RL, Fedorova ND, Jackson JJ, Jacobs AR, Kiryutin B, Koonin EV, Krylov DM, Mazumder R, Mekhedov SL, Nikolskaya AN, Rao BS, Smirnov S, Sverdlov AV, Vasudevan S., Wolf YI, Yin JJ, Natale DA (2003) The COG database: An updated version includes eukaryotes. BMC Bioinformatics 4:41
14. Birney E, Andrews D, Caccamo M et al (2006) Ensembl 2006. Nucleic Acids Res 34: D556–561
15. Alexeyenko A, Tamas I, Liu G, Sonnhammer EL (2006) Automatic clustering of orthologs and inparalogs shared by multiple proteomes. Bioinformatics 22:e9–e15
16. Dufayard JF, Duret L, Penel S, Gouy M, Rechenmann F, and Perriere G (2005) Tree pattern matching in phylogenetic trees: Automatic search for orthologs or paralogs in homologous gene sequence databases. Bioinformatics 21:2596–2603
17. Burns DM, Horn V, Paluh J, Yanofsky C (1990) Evolution of the tryptophan synthetase of fungi. Analysis of experimentally fused Escherichia coli tryptophan synthetase alpha and beta chains. J Biol Chem 265:2060–2069
18. Marcotte EM, Pellegrini M, Ng HL, Rice DW, Yeates TO., Eisenberg D (1999) Detecting protein function and protein-protein interactions from genome sequences. Science 285:751–753
19. Enright AJ, Iliopoulos I, Kyrpides NC, Ouzounis CA (1999) Protein interaction maps for complete genomes based on gene fusion events. Nature 402:86–90
20. Yanai I, Derti A, DeLisi C (2001) Genes linked by fusion events are generally of the same functional category: A systematic analysis of 30 microbial genomes. Proc Natl Acad Sci USA 98:7940–7945
21. Moreno-Hagelsieb G, Trevino V, Perez-Rueda E, Smith TF, Collado-Vides J (2001) Transcription unit conservation in the three domains of life: A perspective from Escherichia coli. Trends Genet 17:175–177
22. Dandekar T, Snel B, Huynen M, Bork P (1998) Conservation of gene order: A fingerprint of proteins that physically interact. Trends Biochem Sci 23:324–328
23. Overbeek RF, M D'Souza M, Pusch GD,. Maltsev N (1998) Use of contiguity on the chromosome to infer functional coupling. In Silico Biol 2:93–108
24. Blumenthal T (1998) Gene clusters and polycistronic transcription in eukaryotes. Bioessays 20:480–487
25. Spieth J, Brook, G, Kuersten S, Lea K, Blumenthal T (1993) Operons in C. elegans: Polycistronic mRNA precursors are processed by trans-splicing of SL2 to downstream coding regions. Cell 73:521–532

26. von Mering C, Huynen M, Jaeggi D, Schmidt S, Bork P, Snel B (2003) STRING: A database of predicted functional associations between proteins. Nucleic Acids Res 31:258–261

27. Pellegrini M, Marcotte EM, Thompson MJ, Eisenberg D, Yeates TO (1999) Assigning protein functions by comparative genome analysis: Protein phylogenetic profiles. Proc Natl Acad Sci USA 96:4285–4288

28. Galperin MY, Koonin EV (2000) Who's your neighbor? New computational approaches for functional genomics. Nat Biotechnol 18:609–613

29. Huynen M, Snel B, Lathe W, Bork P (2000) Exploitation of gene context. Curr Opin Struct Biol 10:366–370

30. Wu J, Kasif S, DeLisi C (2003) Identification of functional links between genes using phylogenetic profiles. Bioinformatics 19:1524–1530

31. Perna NT, Plunkett G III, Burland V, Mau B et al (2001) Genome sequence of enterohaemorrhagic *Escherichia coli* O157:H7. Nature 409:529–533

32. Blattner FR, Plunkett G III, Bloch CA, Perna NT et al (1997) The complete genome sequence of *Escherichia coli* K-12. Science 277:1453–1474

33. Gabaldón T, Huynen MA (2005) Lineage-specific gene loss following mitochondrial endosymbiosis and its potential for function prediction in eukaryotes. Bioinformatics 21, Suppl 2: ii144–ii50

34. Fryxell KJ (1996) The coevolution of gene family trees. Trends Genet 12:364–369

35. Pazos F, Valencia A (2001) Similarity of phylogenetic trees as indicator of protein-protein interaction. Protein Eng 14:609–614

Chapter 27

Design, Manufacture, and Assay of the Efficacy of siRNAs for Gene Silencing

Louise A. Dawson and Badar A. Usmani

Abstract Small interfering RNAs (siRNAs) have been widely exploited for nucleotide-sequence-specific posttranscriptional gene silencing, as a tool to investigate gene function in eukaryotes, and they hold promise as potential therapeutic agents. Conventionally designed siRNAs are 21-mers with symmetric 2-nt 3' overhangs that mimic intermediates (microRNAs or miRNAs) of the natural processing of longer dsRNA (double-stranded RNA). siRNAs are sequences with full complementarity to their target mRNA and can be generated by either chemical synthesis or processing of shRNAs (short hairpin RNAs) transcribed from DNA vectors. To minimize off-target effects, any homology to nontarget mRNA can be verified using the expressed sequence tag (EST) database for the relevant organism. Here, we provide a practical guide and an overview to the design and selection of effective and specifc siRNAs.

Keywords small interfering RNA, gene silencing, endothelin-converting enzyme, prostate cancer, invasion, stromal-epithelial interactions

1 Introduction

The silencing of specific genes by small interfering RNA (termed RNA interference or RNAi) has provided a revolutionary approach to studying gene function. While technologies such as homologous recombination and antisense targeting have been used successfully in reverse genetics, these are expensive and laborious. RNAi, in contrast, is comparatively rapid and, due to its remarkable potency, cost effective. The phenomenon of RNAi was first described in 1998 by Nobel laureates Andrew Fire and Craig Mello, who developed a method of gene silencing using specific double-stranded RNA in *Caenorhabditis elegans* [1]. Following on previous work indicating that both sense and antisense RNA were equally effective in suppressing the expression of specific genes in *C. elegans* [2], Fire and Mello discovered that sense and antisense RNAs had a synergistic effect on gene expression, with the dsRNA (double-stranded RNA) showing far greater potency than either RNA alone. Indeed, only a small amount of dsRNA was required to dramatically reduce the expression of the target gene. This elegant study shed mechanistic light on previous work in plants showing that exogenously added DNA could affect endogenous target gene expression [3] and plants combated RNA viruses by destroying their RNA [4]. The technique of RNAi was quickly adopted by investigators studying the genetics of *C. elegans*, with considerable success. The subsequent identification of short interfering RNAs (siRNAs) led to the adoption of RNAi to study gene function in mammals, revolutionizing this field of research [5]. The widespread adoption of the technique has led to an increasing commercialization of both dsRNA production and methods of delivery; indeed, prevalidated siRNAs are now widely available to study the function of an increasingly large array of genes.

Since the discovery of RNAi, remarkable progress has been made in understanding the underlying mechanisms. One of the first clues as to the mechanism of RNAi came from studies in *Drosophila* embryo extracts in which the ability of the lysates to synthesize luciferase from an exogenous mRNA was diminished following incubation with dsRNA [6]. This raised the possibility that the silencing effect was mediated by a nuclease targeted by the dsRNA to degrade homologous transcripts. Subsequently, this nuclease was identified in *Drosophila* as a complex known as RISC (RNA-induced silencing complex, see Fig. 27.1) [7]. It has since become clear that this complex is able to recognize target molecules following the conversion of the dsRNAs (including miRNAs and replicating viruses) to short interfering RNAs (siRNAs) produced by Dicer, a member of the RNase III family [8]. siRNAs are 21–23 nucleotides with 3' overhangs and 5' phosphate and 3' hydroxyl groups which are incorporated into the RISC complex. Following incorporation, the duplex siRNA is unwound and the antisense strand targets the complex to homologous sequences, mediating mRNA cleavage and gene silencing (see Fig. 27.1) [9]. The elucidation of this pathway, and in particular the identification of siRNAs, paved the way for the utilization of RNAi in mammalian cells. While the introduction of dsRNA into *C. elegans*, plants, and other species was extremely effective, this technique was not suitable for mammalian cells due to their antiviral response mechanisms to foreign dsRNA. The demonstration that siRNAs could be utilized to evoke RNAi without triggering nonspecific antiviral effects [5] opened the door for the use of RNAi-mediated gene silencing in mammalian cells. The ease of delivery of siRNAs into mammalian cells using standard transfection reagents and the potency and specificity of gene silencing has quickly led to RNAi becoming a routine technique for studying gene function in mammalian cells.

The ability to efficiently and selectively silence target gene expression has numerous applications. In a basic science setting, RNAi has been utilized successfully to study the function of single genes, both in vitro and in vivo. In addition, the identification of microRNAs (miRNAs), endogenously expressed small RNAs that evoke RNAi [10], led to the development of short hairpin RNAs (shRNAs), which mimic pre miRNA hairpins and can be transcribed from DNA vectors, allowing their stable expression and the use of retrovirus or

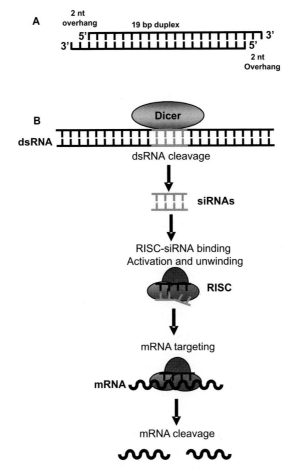

Fig. 27.1 The RNA interference pathway. (A) Short interfering RNA (siRNA) consisting of 21 nucleotides (nt), a 19 bp duplex with 3' dinucleotide unphosphorylated overhangs. (B) RNAi pathway. Double-stranded RNA is cleaved into short interfering (si)RNA by Dicer, a member of the RNase III family member. The siRNA is incorporated into the RNAi silencing complex (RISC). RISC is activated and the siRNA unwinds. RISC is targeted to the complementary sequence in the target mRNA and mediates mRNA cleavage and gene silencing

adenovirus as vectors [10]. This allows the construction of expression libraries of shRNAs targeting a range of genes, allowing RNAi to be used on a large, genomewide scale. Such large-scale studies are useful in studying biochemical pathways and in identifying and validating drug targets. In the clinic, efforts are underway to utilize RNAi as a therapeutic tool to target a number of diseases ranging from HIV to cancer. While promising results have been achieved using RNAi in mice [11], it remains to be seen whether it can be successfully utilized in humans, due to the difficulties in developing delivery strategies and the danger of off-target effects. Despite these limitations, however, it seems likely that RNAi will evolve from an extremely useful tool to study gene function in the laboratory into a therapeutic strategy in the future.

2   Materials

2.1   Oligonucleotides

1. siGLO™ RISC-free siRNA (Dharmacon, D-001600-01-20; Thermo Fischer Scientific UK). This is a nontargeting siRNA with a fluorescent label and impaired ability for RISC interaction. It is useful as both a negative control and a cotransfection marker.
2. siCONTROL™ Lamin A/C siRNA (Dharmacon, D-001050-01-20; Thermo Fischer Scientific UK). This is a positive silencing control for guaranteed silencing of Lamin A/C in human, mouse, and rat cells.
3. siGENOME™ *SMART*pool® reagent targeting endothelin-converting enzyme-1 (Dharmacon; Thermo Fischer Scientific UK). This is a mixture of four *SMART*selection™ designed siRNAs.
4. Custom synthesized individual siRNA duplex targeting ECE-1 using the siDESIGN™ Centre on the Dharmcon website; (www.dharmacon.com/sirna/sirna.aspx).

2.2   Lipid Mediated Oligonucleotide Transfection into Mammalian Cells

1. Mouse embryonic fibroblasts (STO; kindly provided by Prof. N. J. Maitland, Yorkshire Cancer Research Unit, University of York, UK). They are available from both the American Tissue Culture Collection (ATCC, CRL-1503) and the European Collection of Cell Cultures (ECACC, 86032003).
2. Dulbecco's Modified Eagles Medium (DMEM; BioWhittaker: BE12-604F; Wokingham, UK).
3. Trypsin/EDTA. (BioWhittaker, Wokingham, UK).
4. Phosphate buffered saline (PBS; BioWhittaker, Wokingham, UK).
5. Six-well plates (Nunc™, Denmark).
6. Oligofectamime™ (Invitrogen;, 12252-011; Paisley, UK).
7. Opti-MEM reduced serum medium (GIBCO, 31985; Paisley, UK).
8. Fetal bovine serum (FBS; BioWhittaker, Wokingham, UK).
9. Cell culture facility.

2.3   Real-Time PCR Analysis

1. RNeasy Kit (Qiagen, UK).
2. iScipt cDNA Synthesis Kit (Bio-rad 170-8890; UK, *see* **Note 1**).
3. iCycler thermal cycler (Bio-rad).
4. SYBR-Green Supermix (Bio-rad: 170-8880; U.K.): the kit contains KCl, Tris-HCl pH8.4, 0.4 mM of each dNTP, Taq DNA polymerase, $MgCl_2$, SYBR-Green, and fluoroscein.
5. 0.2-mL thin-wall PCR plates (AbGene, AB-0624; Surrey, UK).
6. Oligonucleotide primer pairs at 10 pmol (MWG Biotech, Ebersberg, Germany).

2.4   Western Analysis

2.4.1   Preparation of Cell Lysates

1. PBS (Oxoid, Basingstoke, UK).
2. Triple detergent lysis buffer: 50 mM Tris-HCl pH7.4, 150 mM NaCl, 0.5% (w/v) sodium deoxycholate, 0.1% (w/v) SDS, 0.02% (w/v) sodium azide, 1% (v/v) Nonidet P-40, 0.1 mM Na_3VO_4, EDTA-free protease inhibitor cocktail (1X; Roche) (pH 8.0) for the complete inhibition of serine and cysteine proteases.
3. Benzonase nuclease (Novagen).
4. 23G syringe needles (Tyco Healthcare, UK).
5. 1-mL syringes (Terumo, Belgium).
6. Bovine serum albumin (BSA; Sigma, Poole, UK).
7. Bicinchoninic acid (BCA; Sigma, Poole, UK).
8. 4% (w/v) $CuSO_4$ solution (Sigma, Poole, UK).
9. Microtiter plate reader (Anthos Labtec Instruments, Salzburg, Austria).

2.4.2   SDS-PAGE and Western Blotting

1. SDS gel equipment (Invitrogen, Paisley, UK).
2. 3-8% Tris-acetate precast gels (Invitrogen, EA0375BOX; Paisley, UK).
3. 1X LDS sample buffer (Invitrogen, NP0007; Paisley, UK).
4. 0.5 M DTT (Invitrogen, Paisley, UK).
5. SeeBlue Plus 2 prestained standards (Invitrogen, LC5625; Paisley, UK).
6. 1X NuPAGE Tris-acetate SDS running buffer (Invitrogen, LA0041; Paisley, UK).
7. NuPAGE antioxidant (Invitrogen, NP0005; Paisley, UK).
8. Blotting pads (Invitrogen, Paisley, UK).
9. Nitrocellulose membrane (Invitrogen, Paisley, UK).

10. Methanol (B.D.H. Chemicals Ltd, UK).
11. 1X NuPAGE transfer buffer (Invitrogen, NP0006-1, Paisley, UK).
12. Blotting module (Invitrogen, Paisley, UK).
13. Filter paper.
14. Ponceau red stain: 1% (w/v) in 30% (v/v) trichloroacetic acid and 30% (v/v) sulfosalicyclic acid (Sigma, Poole, UK).
15. TBST: 0.1% Tween-20 in 10 mM Tris-HCl, pH 7.4.
16. Blocking solution: TBST with 5% (w/v) milk powder and 1% (w/v) BSA.
17. Reference primary antibody (ECE-1 monoclonal antibody 1:100; kindly provided by Dr K. Tanzawa, Sankyo Research Laboratories, Tokyo).
18. Antimouse horseradish peroxidase-conjugated secondary antibody (Amersham Ltd, UK).
19. Enhanced chemiluminesence (ECL; Amersham Ltd., UK).
20. Saran wrap.
21. Cassette.
22. Autoradiography film (Kodak, UK).

2.5   Immunofluorescence Staining

1. Coverslips (B.D.H. Chemicals Ltd, UK).
2. PBS (Oxoid, Basingstoke, UK).
3. Methanol/acetone (1:1 ratio: B.D.H. Chemicals Ltd, UK).
4. Blocking buffer: TBS, 1 % (v/v) normal goat serum and 0.2 % (w/v) gelatin.
5. Reference primary antibody: ECE-1 monoclonal antibody (1:100; kindly provided by Dr. K. Tanzawa, Sankyo Research Laboratories, Tokyo).
6. Control IgG1 subclass antibody (Sigma, Poole, UK).
7. FITC conjugated antimouse IgG secondary antibody (Jackson ImmunoResearch Laboratories, USA).
8. 6-Diamidino-2-phenylindole (DAPI): 1:1000 in PBS from 0.5 mg/mL stock in DMSO (Sigma; Poole, UK).
9. Vectashield (Vector Laboratories, USA).
10. Nail varnish.
11. Olympus IX70 inverted wide-field fluorescence microscope.

2.6   Invasion Assay

1. Matrigel basement membrane (Becton Dickinson, 354234). BD Matrigel™ matrix is a solublized basement membrane preparation extracted from Engelbreth-Holm-Swarm mouse tumor mouse sarcoma (*see* **Note 2**).
2. BD Falcon™ cell culture inserts for a 24-well plate, 8.0-μm translucent PET membrane (Becton Dickinson, 353097).

3. BD Falcon™ cell culture insert companion plate, 24-well (Becton Dickinson, 353504).
4. DMEM containing 0.1 % bovine serum albumin (BSA).
5. PC-3 prostate epithelial cells were kindly provided by Prof. D. Nanus, Cornell University, New York. They also are available from both the ATCC, CRL-1435, and the ECACC, 90112714 (*see* **Note 3**).
6. 0.1 % crystal violet (Sigma, Poole, UK).

3   Methods

3.1   Design of oligonucleotides

Many websites are available for functional siRNA search and design for mammalian RNAi studies [12]. We designed siRNA duplexes to the ECE-1 sequence using the siDESIGN™ Centre on the Dharmacon website. The conventional design strategy (*see* **Note 4**) does the following:

1. Identifies a region that begins 75–100 bases downstream of the start codon to avoid nucleotide sequences occupied by regulatory or translational proteins and exon-exon junctions.
2. Identifies the first AA(N19)TT motif, where N19 represents the 19 nucleotide core duplex and AA and TT define the overhang sequences of the duplex (*see* **Note 5**).
3. Evaluates the GC content of the potential siRNA sequence. Ideally, the G:C ratio should be near to 50% or 30–70% according to Tuschl's guidelines [13].
4. Compares the candidate siRNA with known expressed sequence tags (ESTs) in publicly available databases to ensure specificity of targeting (e.g., www.ncbi.nlm.nih.gov/BLAST).

3.2   Synthesis of Oligonucleotides (Synthesis, Purification, Enhancing Modifications)

Chemical synthesis provides the best material of choice for introduction of siRNAs in vitro. Although the natural intermediates that siRNAs mimic carry a 5′ phosphate, the 5′ phosphate is omitted in chemical synthesis as a cellular kinase rapidly phosphorylates the siRNA once it is inside the cell. Dharmacon has developed siRNAs suitable for in vivo protocols, which include 2′ deprotection, desalting, and duplex annealing. These modifications enhance the persistence of the siRNA duplex inside the cell. Other modifications to consider are delivery, concentration, and dosage. For a complete review, see Nature Methods' "Focus on RNA Interference—A User's Guide" [14].

3.3   Experimental Design and Controls

To maximize the specificity and validity of any RNAi experiment, the researcher should

1. Include untreated cells to serve as a baseline standard.
2. Titrate the siRNA duplex being transfected to its lowest effective level; siRNA concentrations greater than 100 nM produce nonspecific off-target effects at a much higher frequency than concentrations of 20–100 nM.
3. Confirm that the target mRNA (northern analysis/real-time PCR) or protein (Western analysis) is specifically reduced by comparing to a nontargeting, non-functional siRNA control that will reveal any off-target effects.
4. Include a positive silencing control such as high-efficiency siRNAs targeted to known genes, such as lamin A/C, cyclophilin B.
5. Confirm that the observed experimental phenotype is not the result of off-target effects by using multiple siRNAs targeted to distinct regions of the same gene. Dharmacon's readily available siRNA SMARTpool may be an option.
6. If possible, rescue expression of the target gene by expressing it in a form resistant to the siRNA being used.

3.4   Methods of siRNA Delivery

siRNAs can be delivered using many protocols, including lipofection, electroporation, microinjection, and infection. For a lipid mediated oligonucleotide transfection into adherent mammalian cells the following protocol is used for one well of a six-well plate:

1. Culture STO cells routinely in 75-cm^2 flasks in 10 mL of DMEM medium supplemented with 10 % (v/v) FBS.
2. To seed the cells, remove the culture medium and wash twice with 10 mL PBS. Add 1 mL of 1X trpsin/EDTA and incubate at 37 °C until the cells detach from the flask. Stop the trypsin reaction by adding 8 mL of DMEM.
3. Count the cells and seed them into a six-well plate at 50% confluency 24 h prior to transfection (*see* **Note 6**).
4. For one well of a six-well plate, add 4 μL of Oligofectamime™ to a sterile tube containing 26 μL of Opti-MEM (*see* **Note 7**).
5. In a separate tube, dilute 4 μL of a 20 μM stock of the siRNA oligonucleotide duplex in 66 μL of Opti-MEM to give a 100 nM oligo concentration (based on 400 μL total volume).
6. Incubate the individual mixes for 10 min at room temperature.
7. Add the diluted Oligofectamime™ reagent to the diluted siRNA duplexes, mix gently, and incubate the pooled mixture at room temperature for 20 min.
8. For each six-well transfection add 300 μL of Opti-MEM to the mix, giving a final volume of 400 μL per well.
9. Wash the cells once with Opti-MEM.

10. Overlay 400 μL of the complex into each well of cells.
11. Incubate the cells for 4 h at 37 °C in a CO_2 incubator (5% CO_2).
12. Following incubation, add 2 mL of normal growth media containing three times the normal amount of FBS to the cells without removing the transfection mixture.
13. Assay cell extracts for gene activity 12–72 h post transfection.

3.5 Methods for Detecting the Efficacy of siRNA Gene Silencing

3.5.1 mRNA Detection

For a real-time (quantitative) PCR analysis (see Fig. 27.2),

1. Isolate total RNA using the RNeasy Kit following the manufacturer's instructions.
2. Quantify the RNA using a spectrophotometer at 260 nm. The purity of RNA can be estimated from the ratio of absorbance readings at 260 nm and 280 nm.
3. Perform first-strand synthesis using the iScript cDNA Synthesis Kit (Bio-rad). For a single reaction, pipette 1Xx iScript reaction mix, 1 μg of total RNA template, and 1 μL of iScript reverse transcriptase into a PCR tube and add ddH₂O to give a total volume of 20 μL for each reaction.
4. The reaction protocol is as follows: 25 °C for 5 min, 42 °C for 30 min, and 85 °C for 5 min.
5. Assay the resulting cDNA using real-time PCR. Real-time PCR was performed in 0.2-mL thin-wall PCR plates using the iCycler thermal cycler and carried out with SYBR-Green Supermix.
6. In a single reaction, include 1X SYBR-Green Supermix, 10 pmol of each primer and 10 ng of template in a total reaction volume of 20 μL (*see* **Note 8**). Prepare a master mix of SYBR Green Supermix containing the primers and probe. Add 18 μL to each well containing 2 μL of cDNA (*see* **Note 9**).
7. Heat the mixture initially to 95 °C for 3 min followed by 35 cycles with denaturation at 95 °C for 30 s, annealing at 57 °C for 30 s, and extension at 72 °C for 30 s. Perform each assay for unknown samples simultaneously with standard samples. Standard samples should consist of serial dilutions of candidate DNA (0.01 pg–100 pg). A reference gene such as GAPDH should be used (*see* **Note 10**).

3.5.2 Protein Detection

3.5.2.1   Western Analysis (Fig. 27.3)

For preparation of the cell lysates,

1. Wash cells three times in PBS and collect using a rubber policeman.

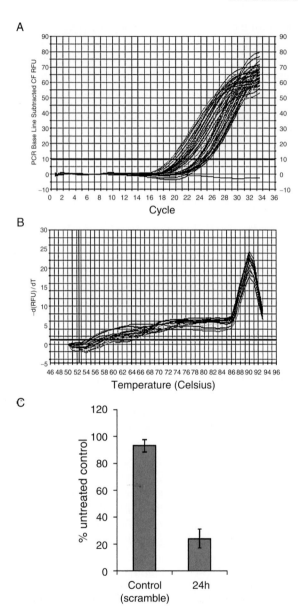

Fig. 27.2 Quantification of siRNA gene silencing for ECE-1 using real-time (quantitative) PCR analysis: (A) Amplification plot and (B) melt curve for ECE-1 mRNA in stromal cells. (C) Expression of ECE-1 in stromal cells normalized against GAPDH. Each bar represents the mean ±SEM of $n = 8$

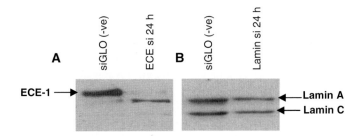

Fig. 27.3 Western blot analysis showing knock-down of ECE-1 expression in mouse stromal (STO) cells. (A) STO cells were transfected with 200 n*M* ECE-1 siRNA (B) and 200 n*M* Lamin siRNA (B) using Oligofectamine then incubated for 24 h. The siGLO siRNA was used as a negative control. Samples were resolved using SDS PAGE and transferred to a nitrocellulose membrane. Membranes were incubated in ECE-1C (1:200) and Lamin A/C (1:500) primary antibody. Protein expression was reduced following siRNA treatment

2. Centrifuge the cells at 190 *g* for 10 min.
3. Resuspend the resulting pellet in 200 μL of ice-cold triple detergent lysis buffer containing EDTA-free protease inhibitor cocktail and pass five times through a 23G needle.
4. To reduce sample viscosity, add 2 μL of benzonase nuclease and incubate the samples for 20 min on ice.
5. Centrifuge the samples at 8,500 *g* for 2 min and collect the supernatant.
6. To determine protein concentration using the the bicinchoninic acid (BCA) protein assay, first, generate a standard curve using bovine serum albumin (BSA). Add a CuSO$_4$/BCA solution to the sample, standard or blank, in a 96-well plate and incubate at 37 °C for 30 min. Measure the absorbance at 570 nm in a colorimetric plate reader.

For SDS-PAGE and Western blotting,

1. Dilute the protein isolated from whole cell lysates (10–40 μg) to 1.1X the desired concentration in 4X LDS sample buffer and ultrapure water.
2. Mix the resulting sample solution with reducing agent in a 9:1 ratio and incubate the samples at 70 °C for 10 min on a heating block.
3. Resolve the prepared samples by SDS-PAGE using Tris-acetate 3–8% (v/v) acrylamide gels. Remove all packaging from the gel and rinse the wells three times with 1X NuPAGE Tris-acetate SDS running buffer. Fill the wells with running buffer.
4. Place the gels in an electrophoresis unit with the well side facing inward then load 5 μL of prestained standard and 20 μL of each of the samples onto the gel.
5. Fill the inner chamber with 1X running buffer containing 0.5% (v/v) NuPAGE antioxidant. Fill the outer chamber with running buffer and resolve the proteins in the sample by electrophoresis at 150 V for 1 h.

6. Following electrophoresis soak the blotting pads, nitrocellulose membrane, and two pieces of filter paper in 1X NuPAGE transfer buffer containing 10% (v/v) methanol per gel and 0.5% (v/v) antioxidant. Position two blotting pads in the cathode core of the blotting module. Remove the Tris-acetate gels from their cassettes and excise the stacking gel. Assemble the transfer sandwich in the following order: presoaked filter paper, gel, nitrocellulose membrane, and filter paper. Place the sandwich on the blotting pads in the blot module and place a further two blotting pads on top of the sandwich. Position the blot module in the XCell Surelock Mini-Cell blotting unit and fill it with 1X transfer buffer. Fill the outer reservoir with water to act as a coolant. Transfer at 30 V for 1 h.

7. Following electrotransfer, assess protein-loading equality using Ponceau red dye. Destain by rinsing in distilled water (*see* **Note 11**).

8. Block nonspecific binding sites with TBST containing 5% (w/v) milk powder and 2% (w/v) BSA. Incubate the membranes for 1 h with agitation.

9. Add primary antibody for 2 h at room temperature with gentle rotation at an appropriate dilution in TBST containing 2% milk powder. ECE-1 (1:500).

10. Wash the membranes three times over 10 min in TBST.

11. Incubate with horseradish peroxidase-conjugated secondary antibody for 1 h at room temperature.

12. Wash the membranes three times over 10 min with TBST.

13. Visualize immunoreactive bands using enhanced chemiluminescence (ECL). Incubate the membrane in 6 mL of a 1:1 mix of detection solutions A and B provided for 1 min at room temperature.

14. Wrap the membrane in Saran wrap and expose it to autoradiography film.

3.5.2.2   Immunofluorescent Staining (see Fig. 27.4)

1. Sterilize the glass coverslips by dipping them in 70% ethanol and then flaming while loosely holding between forceps. Place one sterile coverslip in each well of a six-well plate.

2. Plate 2 mL of the cell suspension into the wells containing the coverslips at 60% confluency and allow the cells to adhere over 24 h.

3. Wash the cells twice with 2 mL PBS and fix and permeabilize for 10 min in 2 mL of freshly prepared methanol/acetone (1:1 ratio) at room temperature (*see* **Note 12**). Set up a humidified chamber consisting of a petri dish with a lid containing water-saturated filter paper on which a layer of parafilm is placed.

4. Place the coverslips cell face down onto 50 μL of blocking buffer, which is pipetted directly onto the parafilm in the humidified chamber to block nonspecific binding. Incubate the coverslips for 30 min at room temperature in the humidified chamber.

5. Incubate the coverslips in 50 μL of primary antibody at an appropriate dilution in blocking buffer for 2 h at room temperature in the humidified chamber. For the control antibody, *see* **Note 13**.

6. Wash the cells three times in PBS over 5 min on a shaker.

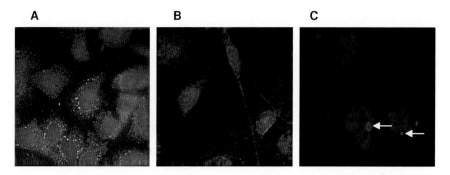

Fig. 27.4 Immunofluorescent staining showing knock-down of ECE-1 expression in mouse stromal cells. STO cells were transfected with (A) 200 n*M* siGLO siRNA (−ve control) or (B) 200 n*M* ECE-1 siRNA using Oligofectamine and incubated for 24 h. (C) siGLO is labeled with a fluorophore which can be used to assess transfection efficiency (white arrows). All cells were fixed and permeabilised in methanol/acetone and incubated with ECE-1 primary antibody followed by a FITC-conjugated secondary antibody. ECE-1 staining is punctate within the cytoplasm (A). The nuclei are stained with DAPI. Protein expression is reduced following siRNA treatment (B)

7. Incubate the coverslips for 30 min at room temperature with 50 µL of labeled secondary antibody (FITC conjugated antimouse IgG, 1:1000) diluted in blocking buffer (*see* **Note 14**).
8. Wash the cells once over 5 min in PBS and then add 1 mL of DAPI at 1.1000 dilution in PBS from 0.5 mg/mL stock in DMSO to the coverslips in the wells of a six-well plate and incubate for 1 min at room temperature (*see* **Note 15**).
9. Remove the DAPI immediately and wash the coverslips three times in PBS over 5 min on a shaker.
10. Invert and mount the coverslips onto glass slides using Vectashield and seal the edge of the coverslips using nail varnish.
11. Examine the cells using a fluorescent microscope. Images were captured using Delta Vision from Applied Precision.

3.5.2.3 Functional Assays (see Fig. 27.5)

1. Defrost the Matrigel overnight at 4 °C on ice. Matrigel will gel rapidly at 22–35 °C (Matrigel may even gel at slightly elevated temperatures in a refrigerator). Dilute the Matrigel to 250 µg/mL (v/v) with DMEM using precooled pipettes, plates, and tubes.
2. Place the plastic inserts into the wells of a 24-well companion plate using sterile forceps.
3. Layer 200 µL of Matrigel onto each cell culture insert housed in a 24-well companion plate and incubate overnight at 37 °C.
4. In parallel, seed 1 mL of STO cells at 100% confluency into the wells of a 24-well companion plate and incubate overnight at 37 °C.

A

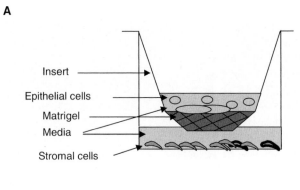

B

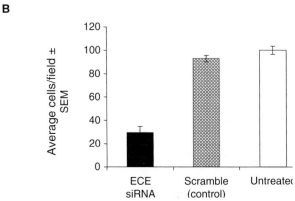

Fig. 27.5 Influence of STO cells treated with ECE and control (scramble) siRNA on PC-3 cell invasion. (A) Stromal cells are plated onto the bottom of the well and Matrigel is layered over the 8-μm pore insert. Both well and insert are incubated separately at 37 °C for 24 h. Next day, the insert is united with the well and epithelial cells are seeded over the Matrigel. Supplements are added to both well and insert and the assay is incubated overnight. (B) STO (stromal) cells transfected with ECE and scramble (control) siRNA were harvested 48 h post transfection. Treated STO cells and prostate cancer (PC-3) epithelial cells were plated as in (A). PC-3 cells invading through the Matrigel were counted after 24 h incubation. Each bar represents the mean value of eight fields counted

5. The following day, aspirate the spent growth medium from the STO cells, wash once with DMEM, and add 0.5 mL DMEM containing 0.1% (w/v) BSA.
6. Place the Matrigel-containing inserts into the wells of the companion plates containing the STO cells and add 250 μL of an 8×10^5 cell/mL prostate epithelial (PC-3) cell suspension in DMEM containing 0.1% (w/v) BSA to the insert containing Matrigel.
7. Set up control wells excluding STO cells. Replace the cells with 0.5 mL of DMEM containing 0.1% (w/v) BSA. Set up control wells excluding Matrigel. Replace the Matrigel with an equal volume of DMEM (250 μL).
8. Incubate the invasion assay overnight at 37 °C.
9. Remove the inserts from the wells and put them into a new 24-well companion plate.

10. Wash once in PBS both above and below the insert.
11. Fix the cells in 500 μL of 100% methanol for 10 min at room temperature. Add to both the top and bottom of the insert.
12. Remove the methanol using an aspirator and add 500 μL of 0.1% (w/v) crystal violet to each insert both top and bottom and incubate for 5 min at room temperature.
13. Remove the crystal violet and wash four times with distilled water or until the wash is completely colorless.
14. Remove noninvading cells and excess Matrigel from the surface of the Matrigel layer using a cotton bud (*see* **Note 16**).
15. Fill the wells with distilled water and count the cells that invaded to the underside of the inserts using a light microscope. The cells will be stained purple. Count four fields of view from each insert.

4   Notes

1. A cDNA synthesis kit should be purchased with respect to the thermal cycler being used for real-time PCR.
2. At room temperature, BD Matrigel matrix polymerizes to produce biologically active matrix material resembling the mammalian cellular basement membrane. It is stable for at least 3 months when kept frozen at –20 °C or 12 d at 37 °C. Avoid multiple freeze-thaws. Do not store in a frost-free freezer.
3. This protocol can be adapted for other invasive cell line models.
4. Dharmacon's conventional design method (www.dharmacon.com) selects 19–21 nucleotide targets followed by a nucleotide (BLAST) search. Due to the variable success rate of this method and to rule out off-target effects, investigators are advised to complement custom siRNA sequence studies with a Dharmacon commercial SMARTpool targeted to the same candidate gene.
5. If no suitable sequence can be found, the search can be relaxed to identify sequences with the motif NA(N21).
6. Optimal cell density differs between cell lines and should be determined empirically for each cell line. Transfection of oligonucleotides is very sensitive to cell density, and so it is important to maintain a standard seeding protocol to ensure reproducibility.
7. We compared a number of transfection reagents, such as Ribojuice, Dharmafect, and Fugene-6, but maximum transfection efficiency was observed with Oligofectamine.
8. The melt curve should be checked to ascertain specificity of primer by identifying a single peak. Multiple peaks imply poor primer specificity for target sequence.
9. Due to the sensitivity of quantitative PCR, pipetting errors can easily affect results. To avoid this prepare a master mix of SYBR Green Supermix containing both the primers and probe. For optimal reproducibility of replicate samples, the template DNA sample could be added to aliquots of the master mix.

10. Choose the reference gene carefully; for example, GAPDH expression fluctuates with hormone treatment.
11. Equivalent loading also can be checked by probing the membrane with an antibody against a housekeeping protein such as β-actin. Levels of housekeeping proteins remain constant.
12. Alternative fixing solutions can be employed, as some antibodies cannot tolerate acetone. Paraformaldehyde or formalin is less harsh on the membrane and often the method of choice for studying membrane proteins.
13. For monoclonal antibodies, the control antibody should be of the same IgG subclass, such as IgG1. For polyclonal antibodies, incubate with serum specific to the animal in which the polyclonal was raised.
14. A separate coverslip incubated with secondary antibody alone is a useful reference point for identifying background nonspecific antibody binding.
15. DAPI is a nuclear stain and useful for identifying the presence of cells when the reference protein is not visible.
16. Removing noninvading cells from the surface of the Matrigel layer prevents them being accidentally counted along with cells that invaded to the underside. After counting, wipe away invading cells to ensure no more cells are visible.

Acknowledgments   This work was supported by Yorkshire Cancer Research (YCR) and Prostate Cancer Research Foundation (PCRF).

References

1. Fire A et al (1998) Potent and specific genetic interference by double stranded RNA in *Caenorhabditis elegans*. Nature 391:806–811
2. Guo S, Kemphues KJ (1995) *par-1*, a gene required for establishing polarity in *C. elegans* embryos, encodes a putative Ser/Thr kinase that is asymmetrically distributed. Cell 81:611–620
3. Jorgensen R (1990) Altered gene expression in plants due to *trans* activations between homologous genes. Trends Biotechnol 8:340–344
4. Ruiz MT, Voinnet O, Baulcombe DC (1998) Initiation and maintenance of virus-induced gene silencing. Plant Cell 10:937–946
5. Elbashir SM et al (2001) Duplexes of 21- nucleotide RNAs mediate RNA interference in cultured mammalian cells. Nature 411:494–498
6. Tuschl T, Zamore PD, Lehmann R, Bartel DP, Sharp PA (1999) Targeted mRNA degradation by double-stranded RNA *in vitro*. Genes Dev 13:3191–3197
7. Hammond SM, Bernstein E, Beach D, Hannon GJ (2000) An RNA-directed nuclease mediates post-transcriptional gene silencing in *Drosophila* cells. Nature 404:293–296
8. Bernstein E, Caudy AA, Hammond SM, Hannon GJ (2001) Role for bidentate ribonuclease in the initiation step of RNA interference. Nature 409:363–366
9. Nykanen A, Haley B, Zamore PD (2001) ATP requirements and small interfering RNA structure in the RNA interference pathway. Cell 107:309–321
10. Paddison PJ, Caudy AA, Sachinandam R, Hannon GJ (2004) Short hairpin activated gene silencing in mammalian cells. Methods Mol Biol 265:85–100
11. Song E et al (2003) RNA interference targeting Fas protects mice from fulminant hepatitis. Nature Med 9:347–51

12. Ui-Tei K, Naito Y, Saigo K (2006) Essential notes regarding the design of functional siRNAs for efficient mammalian RNAi. J Biomed Biotechnol 65052: 1–8

13. Elbashir SM, Martinez J, Patkaniowska A, Lendeckel W, Tuschl T (2002) Analysis of gene function in somatic mammalian cells using small interfering RNAs. Methods 26:199–213

14. Kiermer V (ed). (2006) Focus on RNA interference—a user's guide. Nature Methods 3:669–719

Index

Printed in the United States of America